Heath Mathematics

Walter E. Rucker · Clyde A. Dilley · David Lowry

Contributing author: **Earl G. Ockenga**

D. C. Heath and Company

Lexington, Massachusetts Toronto

Illustrations Leo Abbett; Charlie Freeman/Sharon Kurlansky; George Hughes/ Gwen Goldstein; Mark Kelly/Sharon Kurlansky; Andrew Shiff

Design of front matter and Warmup: Verne Bowman

Photography Abacus/Slidemakers: 201; Sharon Beals: 224, 277, 299, 345; Bettman Archives, Inc.: 163; Fredrik Bodin: 82, 117, 178, 204, 284, 314; Andy Brilliant: 122, 135, 222, 236; Bullaty-Lomee/The Image Bank: 292, 293; Stuart Cohen: 194, 195, 256, 263, 286, 295, 316; Stuart Cohen/Stock, Boston: 139; M. DiGiacomo/The Image Bank: 42; Earth Images © 1980 Stephen Muir: 22; Burt Glinn/Magnum Photos, Inc.: 323; Hallinen/FPG: 252; Grant Heilman: 169, 349; Historical Picture Service: 199; Tim Holt/The Image Bank: 207; Bohdan Hrynewych: 238; N. Keller/Alpha: 369; Talbot D. Lovering: Cover; Tom Magno: 24, 34, 35, 120, 125, 237, 352; Mike Malyszko: 48, 278; Dan McCoy/Rainbow: 21; Ann McQueen: 251; NASA: 95; Julie O'Neil: 78, 79, 81, 91, 105, 113, 118, 133; Raft Foto: 177; Sack/Editorial Photocolor Archives: 38; A. Satterwhite/The Image Bank: 59; Cliff Sciappa/Picture Group, Inc.: 28, 29; Tom Stack/Tom Stack & Assoc.: 115; Deidra Delano Stead: 306, 307, 319, 347, 384; Gerald Smith/ Monkmeyer Press Photo Service: 235; Unimation Inc.: 181; © Zimmerman/Alpha: 65, 130

Published simultaneously in Canada.

Printed in the United States of America.

International Standard Book Number: 0-669-04161-0

Contents

Use the Warmup "magazine" to

- discover whether you need to review some whole number skills
- review and practice whole number skills
- apply your skills on puzzles and brain teasers

THE WHOLE NUMBER REVIEW
WARMUP
Test Your Skills
Addition Advice
Grab a Pencil
Multiplication Memo
Division Directions
Subtraction Suggestions
Broken Keys
THE Q ICK BR WN
F X J MPED VER
THE L ZY D GS.
OFF AND ON
You switched the calculator on. Then a friend switched it 23 times. Then you switched it 21 more times. Was the calculator on or off when you stopped?
NECK AND NECK
Did you know a sparrow has more neck bones than a giraffe?
It's true! I have 7 more neck bones than a giraffe. Together, we have 21 neck bones.
How many neck bones does a sparrow have?
WATCH OUT!
The watch shows the same time forward as backward. How many minutes before the watch will again show the same time forward as backward?

TEST YOUR SKILLS

Add.

1.	2.	3.	4.
25 +43	60 +34	46 +31	32 +72

5.	6.	7.	8.
72 23 +43	52 45 +41	52 25 +22	34 63 +71

ANY ERRORS? Read the ADDITION ADVICE. Then GRAB A PENCIL!

Addition Advice

	Add ones.	Add tens.
72 +51	72 +51 3	72 +51 123

GRAB A PENCIL!

Add.

1.	2.	3.	4.	5.
34 +83	53 +16	35 +72	34 +95	71 +56

6.	7.	8.	9.	10.
24 34 +60	43 23 +73	62 35 +81	32 45 +22	54 43 +71

CHECK YOURSELF

You should be able to find each answer in this list:

69 99 107 117 118 127 129 139 168 178

CAN YOU BEAT IT?

A player said these sums in 50 seconds.
Can you beat that? Write only your answers.

World's Worst Riddle

Crack the code to get the answer.

CODE:

a	b	c	d	...	z
1	5	9	13		101

RIDDLE:
What's black and white, white and black, and green?

ANSWER:
77-89-57 73-41-81-53-41-73
21-33-25-29-77-33-53-25
57-85-17-69 1
61-33-9-41-45-17

Who's Keeping Score?

The Mudville Dodgers played four games. The scores were:

game 1—9 to 8
game 2—7 to 4
game 3—3 to 2
game 4—10 to 6

The Dodgers scored a total of 22 runs. Which games did the Dodgers win?

SNEAKY COIN PROBLEM

Two coins are worth 35¢. One of the coins is not a quarter. What are the coins?

STUCK ON STAMPS

Cost 20¢

Cost 11¢

Cost 15¢

Cost ?

Broken Keys

It cost $5 to fix each broken key. How much would it cost to fix the broken letter keys?

BE A PRINTER

What capital letters can you print without lifting your pencil or retracing lines?

TEST YOUR SKILLS

Add.

1. 27 + 54	**2.** 48 + 48	**3.** 76 + 69	**4.** 58 + 43
5. 17 + 98 + 56	**6.** 24 + 73 + 67	**7.** 75 + 22 + 36	**8.** 38 + 39 + 46

ANY ERRORS? Read the ADDITION ADVICE. Then GRAB A PENCIL!

Addition Advice

GRAB A PENCIL!

Add.

1. 26 + 58	**2.** 17 + 37	**3.** 38 + 35	**4.** 69 + 72	**5.** 77 + 66
6. 28 + 49 + 36	**7.** 85 + 14 + 67	**8.** 46 + 83 + 76	**9.** 24 + 44 + 15	**10.** 65 + 12 + 56

CHECK YOURSELF

You should be able to find each answer in this list:

54 73 83 84 113 133 141 143 166 205

CAN YOU BEAT IT?

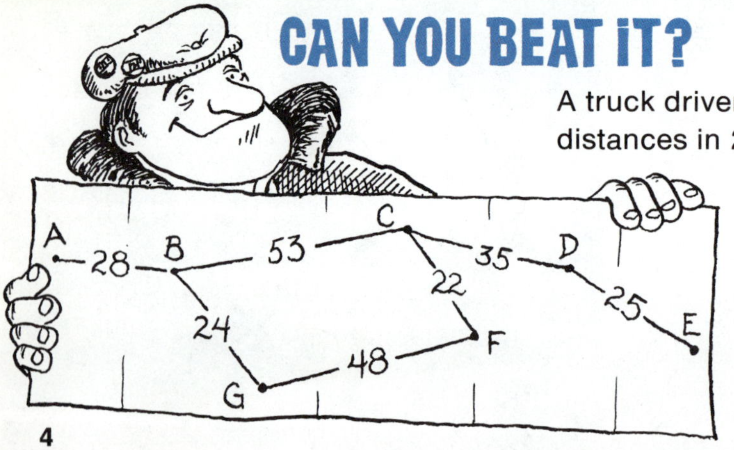

A truck driver found these route distances in 2 minutes. Can you beat that?

1. A to B to C
2. B to C to F
3. B to G to F
4. A to B to G
5. C to D to E
6. G to F to C

PICK A PAIR

MOVIE 14 MOVIE 15 adds to 29.

MOVIE 31 MOVIE 32 adds to 63.

MOVIE ? MOVIE ? adds to 177.

SCAVENGER HUNT

Find the second and third greatest numbers on these two pages.

Add your two numbers. Did you get a total of 382?

WALLPAPER SALE!

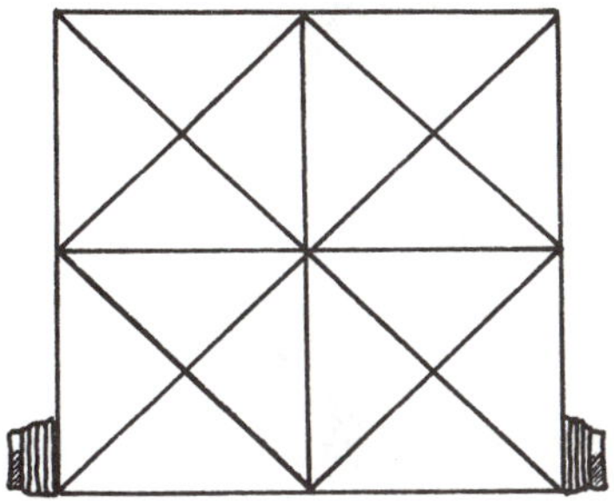

How many triangles can you count? (*Hint:* There are more than 40.)

WHO'S WRONG?

Sally, Tracy, and Rosie each threw 3 darts. Sally said she scored 30 points. Tracy said she scored 18 points. Rosie said she scored 26 points. Who miscalculated her score?

IT'S IN THE BAG!

How many black beads are inside the bag? (*Hint:* Look for the pattern on the string.)

WHO IS IT?

What is the name of the person sitting in front of the person sitting behind the person sitting in front of you?

TEST YOUR SKILLS

Subtract.

1. $78 - 25$	2. $87 - 20$	3. $96 - 34$	4. $88 - 52$
5. $77 - 31$	6. $69 - 25$	7. $68 - 47$	8. $64 - 30$

ANY ERRORS? Read the SUBTRACTION SUGGESTIONS. Then GRAB A PENCIL!

Subtraction Suggestions

Here's how to subtract in columns!

	Subtract ones.	Subtract tens.
86	86	86
−52	−52	−52
	4	34

GRAB A PENCIL!

Subtract.

1. $87 - 51$	2. $78 - 53$	3. $68 - 10$	4. $79 - 14$	5. $59 - 33$	6. $67 - 32$
7. $76 - 45$	8. $95 - 72$	9. $87 - 76$	10. $96 - 52$	11. $89 - 62$	12. $54 - 32$

CHECK YOURSELF: You should be able to find each answer in this list:

11 22 23 25 26 27 31 35 36 44 58 65

CAN YOU BEAT IT?

A flight attendant said these differences in 1 minute. Can you beat that? Write only your answers.

96 seats.
15 empty seats.
81 passengers on board.

1. $87 - 34$	2. $89 - 53$	3. $99 - 73$	4. $65 - 34$
5. $53 - 40$	6. $59 - 26$	7. $75 - 11$	8. $78 - 15$

BAD DEAL!

A woman with a $50 camera needed $75. She pawned the $50 camera for $40 and then sold the pawn ticket to a man for $35. Who lost money on this deal? How much money did he or she lose?

OFF AND ON

You switched the calculator on. Then a friend switched it 23 times. Then you switched it 21 more times. Was the calculator on or off when you stopped?

HOW MUCH MONEY?

CLUES:
Add the two prices and you get $40. Subtract the two prices and you get $10. What are the prices?

NECK AND NECK

Did you know a sparrow has more neck bones than a giraffe?

How many neck bones does a sparrow have?

Scavenger Hunt

a b c d … z
$1 $2 $3 $4 … $26

Find the most expensive 4-letter word on this page.

WATCH OUT!

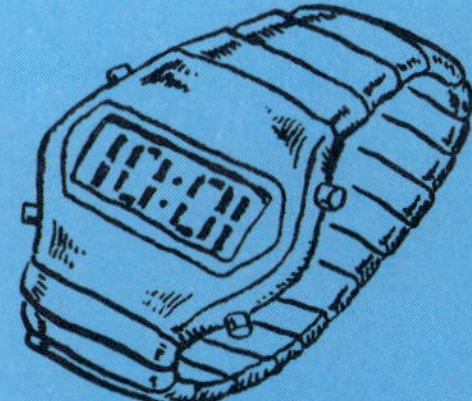

The watch shows the same time forward as backward. How many minutes before the watch will again show the same time forward as backward?

TEST YOUR SKILLS

Subtract.

1.	2.	3.	4.	5.
60	92	73	84	51
− 41	− 24	− 68	− 37	− 36

ANY ERRORS? Read the SUBTRACTION SUGGESTIONS. Then GRAB A PENCIL!

Subtraction Suggestions

	Not enough ones. So regroup a ten.	Subtract ones. Subtract tens.
83	7	7
	8̸ 13	8̸ 13
− 57	− 57	− 57
		26

GRAB A PENCIL!

Subtract.

1.	2.	3.	4.
61	47	71	66
− 23	− 19	− 27	− 27

5.	6.	7.	8.
80	45	22	83
− 49	− 39	− 18	− 49

9.	10.	11.	12.
70	73	84	41
− 23	− 15	− 27	− 14

CHECK YOURSELF

You should be able to find each answer in this list:

4	6	27	28
31	34	38	39
44	47	57	58

CAN YOU BEAT IT?

Find the errors.
It took a banker 2 minutes to correct the 3 errors. Can you beat that?

1.	2.	3.	4.	5.	6.
75	60	48	71	96	63
− 28	− 38	− 39	− 48	− 58	− 47
47	32	11	23	38	26

BEWARE OF SIGNS

Crack the code to get the message.

CODE:

1	2	3	4	5	6	7	8	9
a	b	c	d	e	f	g	h	i
j	k	l	m	n	o	p	q	r
s	t	u	v	w	x	y	z	

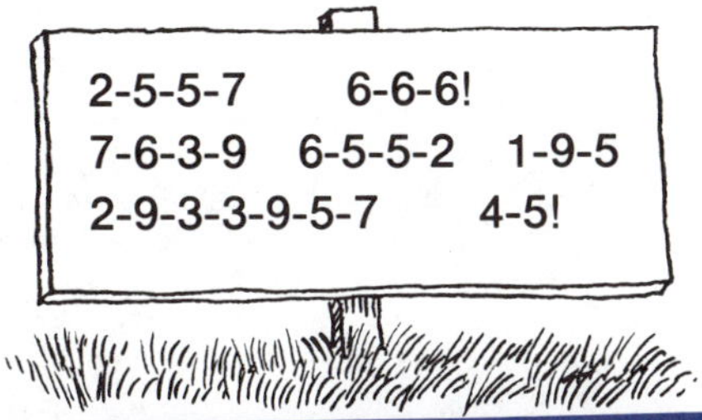

CAUGHT HOLDING THE BAG?

There are 13 bags of popcorn. If you take 5 of the bags, how many do you have?

SCAVENGER HUNT

Find the greatest number on these two pages. Find the second greatest number.

Subtract your two numbers. Did you get a difference of 1?

Mirror Sale!

What is the sale price?

FAMILY TIME

In the car there are 2 daughters, 2 mothers, 1 granddaughter, and 1 grandmother. What is the smallest number of people that can be in the car?

(*Hint:* there are fewer than 5 people in the car.)

TEST YOUR SKILLS

Multiply.

1. 32×3 2. 14×2 3. 71×6 4. 90×7

5. 61×4 6. 70×8 7. 84×2 8. 62×4

ANY ERRORS? Read the MULTIPLICATION MEMO. Then GRAB A PENCIL!

Multiplication Memo

Here's how to multiply in columns!

	Multiply ones.	Multiply tens.
74×2	$74 \times 2 = 8$	$74 \times 2 = 148$

GRAB A PENCIL!

Multiply.

1. 42×3 2. 20×3 3. 11×7 4. 21×4 5. 64×2

6. 73×3 7. 81×8 8. 72×3 9. 90×6 10. 82×2

11. 80×3 12. 50×5 13. 61×6 14. 52×4 15. 74×2

CHECK YOURSELF

You should be able to find each answer in this list:

60 77 84 126 128 148 164 208 216 219 240 250 366 540 648

CAN YOU BEAT IT?

A postal worker said these products in 55 seconds. Can you beat that? Write only your answers.

1. 4×9 2. 5×6 3. 8×8 4. 7×5 5. 9×7

6. 7×7 7. 6×6 8. 5×8 9. 3×9 10. 4×5

11. 2×7 12. 0×9 13. 9×9 14. 5×5 15. 6×7

WORLD'S 2nd WORST RIDDLE

Crack the code to get the answer.

CODE:

6 ×5	7 ×8	9 ×4	8 ×3	5 ×9	8 ×9	7 ×7	4 ×5
a	b	c	d	e	f	g	h
i	j	k	l	m	n	o	p
q	r	s	t	u	v	w	x

RIDDLE:
What's gray, then yellow, then gray, then yellow?

ANSWER:
30-72 45-24-45-20-20-30-72-24
56-49-24-24-30-72-49 24-49-49-72 30
20-30-24-24 49-30-24-20 30
24-30-72-24-45-24-30-49-72 30-72
20-30-36 45-49-45-24-20

BUYING GOLD

You have 56¢. How much more money do you need to buy 4 goldfish?

SQUARE ROUND-UP

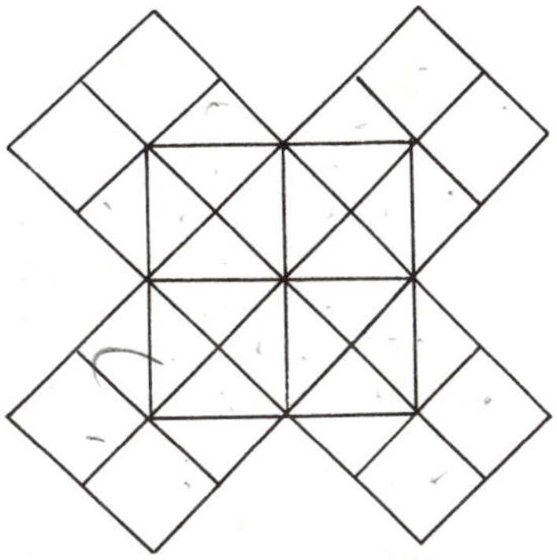

How many squares can you count? (*Hint:* There are more than 30.)

MONEY FACT

A quarter weighs 6 grams.

How much would a 5-dollar stack of quarters weigh?

AND THE WINNER IS...

Alice beat Beth by 10 seconds. Carla finished 8 seconds ahead of Daphne. Alice finished 6 seconds ahead of Daphne but 2 seconds behind Carla. Who won the race? Who finished last?

TEST YOUR SKILLS

Multiply.

1. 26×3	**2.** 18×4	**3.** 68×7	**4.** 26×8
5. 46×9	**6.** 88×3	**7.** 65×8	**8.** 35×6

ANY ERRORS? Read the MULTIPLICATION MEMO. Then GRAB A PENCIL!

Multiplication Memo

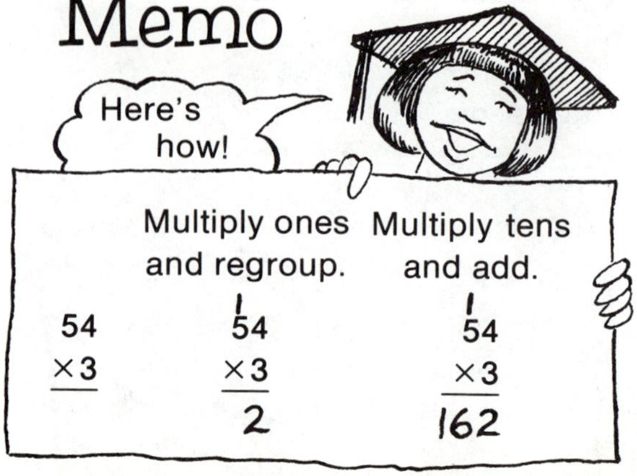

GRAB A PENCIL!

Multiply.

1. 29×3	**2.** 13×5	**3.** 83×7	**4.** 46×8	**5.** 89×8
6. 86×7	**7.** 35×6	**8.** 65×8	**9.** 58×4	**10.** 99×3

CHECK YOURSELF

You should be able to find each answer in this list:

65 87 210 232 297 368 520 581 602 712

CAN YOU BEAT IT?

A clerk counted boxes and multiplied to find the number of tacks. She took inventory in 1 minute. Can you beat that?

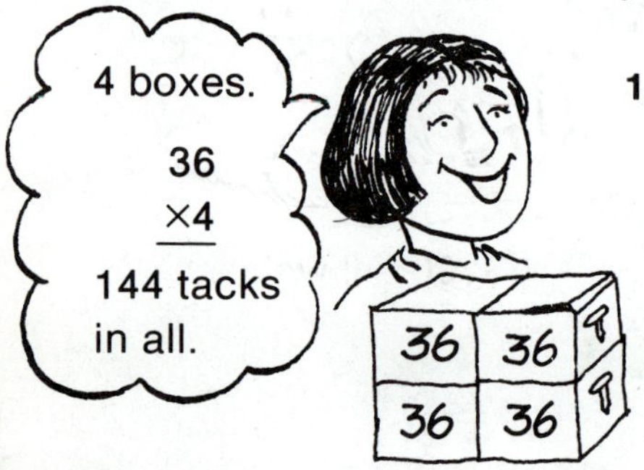

1.

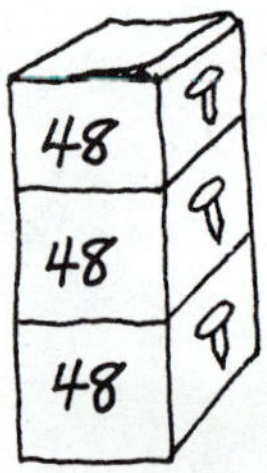

2.

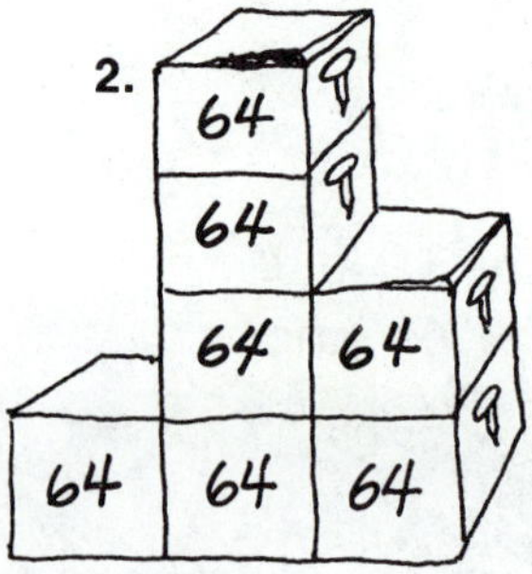

3.

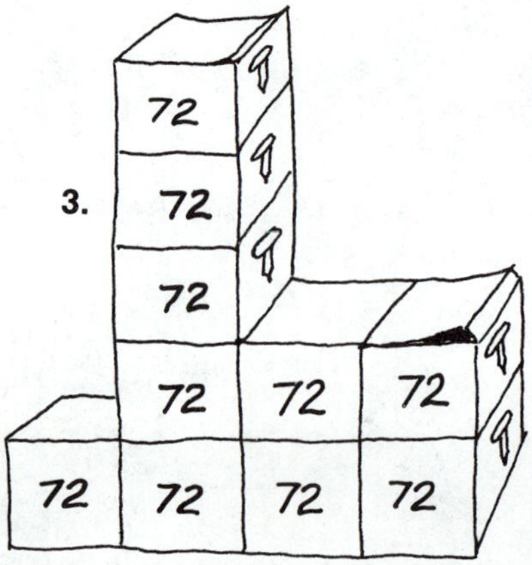

NO FAIR PEEKING!

How many marbles are in the jar?

CLUES:
There are more than 80 marbles.
There are fewer than 100 marbles.
The number is 7 times the sum of its digits.

HAPPY BiRTHDAY!

Tom's birthday is the day after tomorrow. Jean's birthday is two days before Tom's birthday. The day before yesterday was Friday. What day is Jean's birthday?

Sum Difference

What is the difference between four twenty-fives and four fives and a twenty?

GOING BROKE

SEPTEMBER

S	M	T	W	Th	F	S
		1	2	3	4	5
6	7	8	9	10	11	12
13	14	15	16	17	18	19
20	21	22	23	24	25	26
27	28	29	30			

Earn $10 on the 1st.
Spend $1 on the 2nd.
Earn $9 on the 3rd.
Spend $2 on the 4th.
Earn $8 on the 5th.
Spend $3 on the 6th.
Look for the pattern.
On what day will you have no money?

PERSONAL

Four times my age is 14 more than three times my age. How old am I?

Money Fact

A $1 bill weighs 1 gram.

What does a $100 bill weigh?

TEST YOUR SKILLS

Divide.

1. 5)33	2. 9)29	3. 7)46	4. 8)71	5. 6)49
6. 4)29	7. 6)50	8. 3)17	9. 9)65	10. 8)35

ANY ERRORS?

Read the DIVISION DIRECTIONS. Then GRAB A PENCIL!

Division Directions

	Recall facts.	Divide. Then multiply.	Subtract to get remainder.
8)34	8 ×3 = 24; 8 ×4 = 32; 8 ×5 = 40 ← Too big	4 over 8)34; 32	4 R2 over 8)34; −32; 2

GRAB A PENCIL!

Divide.

1. 7)32	2. 5)23	3. 6)25
4. 7)53	5. 8)45	6. 5)49
7. 8)27	8. 9)42	9. 7)48
10. 2)13	11. 5)29	12. 9)59
13. 6)53	14. 7)27	15. 4)35
16. 8)59	17. 9)35	18. 6)38

CHECK YOURSELF

You should be able to find each answer in this list:

3 R3	3 R6	3 R8	4 R1	4 R3	4 R4
4 R6	5 R4	5 R5	6 R1	6 R2	6 R5
6 R6	7 R3	7 R4	8 R3	8 R5	9 R4

CAN YOU BEAT IT?

A carpenter said these division facts in 1 minute. Can you beat that? Write only your answers.

1. 4)28	2. 8)16
3. 7)21	4. 8)48
5. 2)12	6. 6)24
7. 9)36	8. 4)24
9. 7)63	10. 3)12
11. 8)56	12. 8)64
13. 9)81	14. 3)27
15. 7)49	16. 8)40
17. 9)18	18. 5)0

FISHY CODE

Crack the code
to get the message.

CODE:

1	2	3	4	5	6	7	8	9
a	b	c	d	e	f	g	h	i
j	k	l	m	n	o	p	q	r
s	t	u	v	w	x	y	z	

6-9-1-8-8-6-6-2   3-1-4-5-9-1

2-1-2-5   7-9-3-2-3-9-5-1
6-6   2-8-5   6-5-5-1
2-8-1-2   7-5-2   1-5-1-7!

MONEY FACTS

2 grams   5 grams   6 grams

What coins are worth
70¢ and weigh 35 grams?

Hot Dog!

You want exactly 100 hot dogs for a picnic. What is the smallest number of packages you need to buy?

BRIGHT IDEA

Amy sold some light bulbs. Brian sold 16 more than Amy. Charles sold 11 fewer than Brian. Charles earned 96¢ selling light bulbs.

How many light bulbs did Amy sell?

NUMBER PLEASE

How many times was the digit 5 used in numbering the pages of a 128-page book?

TEST YOUR SKILLS

Divide.

1. 4)328	**2.** 3)189	**3.** 8)640
4. 7)217	**5.** 2)166	**6.** 5)155
7. 8)488	**8.** 3)249	**9.** 8)168

ANY ERRORS? Read the DIVISION DIRECTIONS. Then GRAB A PENCIL!

Division Directions

Here's how!

	Regroup. Then divide tens.	Divide ones.
6)246	4 6)246	41 6)246

GRAB A PENCIL!

Divide.

1. 4)128	**2.** 6)366	**3.** 9)270	**4.** 4)360	**5.** 5)400
6. 8)568	**7.** 3)249	**8.** 2)108	**9.** 3)156	**10.** 6)306
11. 7)350	**12.** 2)146	**13.** 5)205	**14.** 8)160	**15.** 4)324
16. 3)159	**17.** 4)124	**18.** 9)819	**19.** 2)184	**20.** 5)300

CHECK YOURSELF

You should be able to find each answer in this list:

20 30 31 32 41 50 51 52 53 54
60 61 71 73 80 81 83 90 91 92

CAN YOU BEAT IT?

A scorekeeper computed these scoring averages in 1 minute. Can you beat that?

6 games played. 126 total points.

That's a 21-point scoring average.

Games played	Total points
1. 9	279
2. 8	240
3. 5	255
4. 4	128
5. 3	123
6. 2	126
7. 3	246
8. 8	328

DIVE IN!

WANTED: Lifeguards. Earn $4 an hour. Set your own hours. Call 555-6226.

Mike, Jodi, and Cindy took the job. At the end of the first week, Mike had earned $52 more than Jodi, and Cindy had earned $28 less than Mike. How many more hours did Cindy work than Jodi?

DOWN ON THE FARM

You bought some pigs and chickens. You spent a total of $24. The animals you bought had a total of 22 legs. How many pigs did you buy?

PENNY PINCHER

How much do you save per package by buying the special?

STAMP FACT

There are twelve 1¢ stamps in a dozen.

How many 4¢ stamps are there in a dozen?

DON'T GET STUCK!

How much money for all the boxes of tacks?

TEST YOUR SKILLS

Divide.

1. 6)78	**2.** 4)96	**3.** 7)91
4. 3)84	**5.** 5)80	**6.** 4)52
7. 3)72	**8.** 8)96	**9.** 6)84

ANY ERRORS?

Read the DIVISION DIRECTIONS. Then GRAB A PENCIL.

Division Directions

	Divide tens.	Regroup. Then divide ones.
2)76	3 2)76 -6 1	38 2)76 -6 16 -16 0

GRAB A PENCIL!

Divide.

1. 3)78	**2.** 7)98	**3.** 7)84	**4.** 4)72	**5.** 6)66
6. 2)96	**7.** 7)91	**8.** 3)87	**9.** 5)80	**10.** 4)68
11. 2)54	**12.** 4)92	**13.** 5)95	**14.** 3)84	**15.** 2)74
16. 3)45	**17.** 4)84	**18.** 4)100	**19.** 2)60	**20.** 2)76

CHECK YOURSELF

You should be able to find each answer in this list: 11 12 13 14 15 16 17 18 19 21 23 25 26 27 28 29 30 37 38 48

CAN YOU BEAT IT?

A recreation director did these divisions in 2 minutes. Can you beat that?

1. 3)72	**2.** 7)98	**3.** 3)57
4. 4)64	**5.** 6)90	**6.** 4)56
7. 8)96	**8.** 5)75	**9.** 4)76
10. 6)96	**11.** 5)70	**12.** 4)96

FRUIT SALAD

Cost 63¢

Cost 62¢

Cost 37¢

Cost ?

TIME'S UP!

A clock is 10 minutes slow and losing 5 minutes an hour. How many hours until the clock is 1 hour slow?

Healthy Hares

If 10 rabbits eat 60 carrots in 3 days, 5 rabbits will eat 20 carrots in how many days?

MAP FACTS

Kansas City is east of Denver.
Lufkin is south of Kansas City.
Lufkin is in what direction from Denver?

SOUP'S ON!

Jan bought 3 cans of soup with exactly 5 coins. Did she buy 3 large cans or 3 small cans? List the 5 coins Jan used.

FAIR SHARE

Adrian bought 7 slices of pizza and Todd bought 2 slices of pizza. When Curt joined them, they shared the 9 slices equally. Curt paid Adrian 96¢ for his share. How much money should Todd pay Adrian?

A good score on this test means you are ready to start Chapter 1.

POSTTEST

	1.	2.	3.	4.	5.
A.	$46 + 12$	$36 + 40$	$54 + 74$	$82 + 63$	$61 + 52 + 34$
B.	$36 + 17$	$35 + 35$	$74 + 89$	$78 + 28$	$32 + 94 + 97$
C.	$78 - 37$	$47 - 12$	$56 - 24$	$83 - 43$	$59 - 57$
D.	$84 - 36$	$70 - 24$	$71 - 67$	$73 - 36$	$92 - 58$
E.	21×4	23×3	41×6	80×4	61×7
F.	28×3	16×6	74×8	39×6	87×7
G.	$3\overline{)26}$	$8\overline{)43}$	$4\overline{)37}$	$7\overline{)61}$	$5\overline{)48}$
H.	$6\overline{)126}$	$4\overline{)288}$	$5\overline{)350}$	$2\overline{)146}$	$3\overline{)249}$
I.	$8\overline{)96}$	$3\overline{)81}$	$5\overline{)85}$	$6\overline{)84}$	$7\overline{)98}$

Score with Honors

0 wrong

1 wrong

2 wrong

Whole Numbers

1

Large numbers

MOUNT ST. HELENS
BLOWS ITS TOP

On May 18, 1980, in the state of Washington, Mount St. Helens exploded. The explosion blew 400,000,000 tons of dust into the earth's atmosphere and killed 2,000,000 animals, including many birds and fish. Blown down or shredded were 3,200,000,000 board feet of timber. This amount of timber could build enough four-bedroom homes to provide living space for half the 4,130,163 people living in the state of Washington.

A place-value table can help you read and write large numbers.

Trillions			Billions			Millions			Thousands			Ones		
Hundreds	Tens	Ones	Hundreds	Tens	Ones	Hundreds	Tens	Ones	Hundreds	Tens	Ones	Hundreds	Tens	Ones
								4	1	3	0	1	6	3

Standard numeral: **4,130,163**

Read as "Four million, one hundred thirty thousand, one hundred sixty-three"

We started at the right and marked off groups of three digits, called **periods,** with commas.

To read a large number, think about the value of each period.

trillions	billions	millions	thousands	ones
4,	**372,**	**053,**	**007,**	**203**

four trillion, three hundred seventy-two billion, fifty-three million, seven thousand, two hundred three

EXERCISES

A calculator display doesn't show commas. Write the number using commas. Then write the number in words.

1. 43856 **2.** 703821 **3.** 3506789 **4.** 62135089

Read. Then write the number in words.

5. 195,396 **6.** 528,030 **7.** 401,006 **8.** 7,821,534

9. 9,060,358 **10.** 38,297,438 **11.** 51,703,004,000 **12.** 7,020,350,413,000

Write each state population as a standard numeral.

13. Alaska: four hundred thousand, four hundred eighty-one

14. Oklahoma: three million, twenty-five thousand, two hundred sixty-six

15. Alabama: three million, eight hundred ninety thousand, sixty-one

Tens, hundreds, and thousands

In order to count money, a bank teller puts it in stacks of the same kind of bill.

Here are some **multiples** of 10:

10, 20, 30, 40, 50, 60, 70

Here are some multiples of 100:

100, 200, 300, 400, 500, 600

Multiples of 10 (or 100) that follow one after the other are called **consecutive** multiples of 10 (or 100).

90 and 100 are consecutive multiples of 10.

430 and 440 are consecutive multiples of 10.

3000 and 3100 are consecutive multiples of 100.

Can you give two consecutive multiples of 1000?

EXERCISES

Give the next three multiples of 10.

1. 50 2. 70 3. 90 4. 160 5. 380 6. 990 7. 1610 8. 2580

Give the next three multiples of 100.

9. 200 10. 500 11. 900 12. 2300 13. 3800 14. 18,700 15. 26,000 16. 39,800

Give the next three multiples of 1000.

17. 3000 18. 7000 19. 14,000 20. 19,000 21. 28,000 22. 98,000

Write the two consecutive multiples of 10 that the given number is between.

23. 53	**24.** 82	**25.** 95	**26.** 236	**27.** 758
50, 60	**28.** 996	**29.** 137	**30.** 297	**31.** 599
32. 1235	**33.** 1364	**34.** 2375	**35.** 4991	**36.** 6998

Write the two consecutive multiples of 100 that the given number is between.

37. 358	**38.** 526	**39.** 908	**40.** 1542	**41.** 1973
42. 2713	**43.** 2974	**44.** 7820	**45.** 1529	**46.** 9280
47. 11,248	**48.** 15,943	**49.** 23,993	**50.** 96,421	**51.** 84,601

Write the two consecutive multiples of 1000 that the given number is between.

52. 2816	**53.** 5601	**54.** 9309	**55.** 12,386	**56.** 51,742
57. 52,174	**58.** 68,031	**59.** 74,395	**60.** 80,861	**61.** 99,376
62. 123,452	**63.** 215,395	**64.** 109,465	**65.** 596,645	**66.** 899,904

Here is how to write a standard numeral as an **expanded numeral.** Notice the multiples that are used.

526,378 =	500,000 +	20,000 +	6000 +	300 +	70 +	8
A multiple of:	100,000	10,000	1000	100	10	1

★ **Give the expanded numeral.**

67. 758	**68.** 2356	**69.** 4829	**70.** 6238
71. 39,274	**72.** 83,261	**73.** 295,378	**74.** 563,174

Rounding

A teenager built a pyramid with 71,825 pennies. The pyramid was 19 inches high and had a 19-inch-by-19-inch-square base. He stacked the coins without using any glue.

To round the number of pennies to the nearest 100,

consider the consecutive multiples of 100 that the number is between:

71,800 **71,825** **71,900**

↑ multiples of 100 ↑

71,825 is nearer 71,800. So, round down to 71,800.

To round the number of pennies to the nearest 1000,

consider the consecutive multiples of 1000 that the number is between:

71,000 **71,825** **72,000**

↑ multiples of 1000 ↑

71,825 is nearer 72,000. So, round up to 72,000.

When a number is halfway between two multiples, round up to the greater multiple.

EXERCISES

Round to the nearest 10.

1. 38	**2.** 55	**3.** 84	**4.** 93	**5.** 96
6. 132	**7.** 189	**8.** 265	**9.** 411	**10.** 629

Round to the nearest 100.

11. 429	**12.** 568	**13.** 350	**14.** 736	**15.** 958
16. 1108	**17.** 1564	**18.** 2319	**19.** 3650	**20.** 4849

Round to the nearest 1000.

21. 5361	**22.** 7832	**23.** 4099	**24.** 6500	**25.** 9734
26. 16,235	**27.** 24,658	**28.** 31,500	**29.** 42,515	**30.** 63,496

Step 1. Find the thousands place.

67,523

Step 2. Look 1 place to the right. If the digit is 5 or greater, round "up."

67,523 → 68,000

Use the shortcut to round the following numbers.
Round to the nearest 100.

31. 780 **32.** 923 **33.** 5280 **34.** 650 **35.** 14,350

Round to the nearest 1000.

36. 8500 **37.** 27,499 **38.** 186,501 **39.** 399,846

Round to the nearest 100,000.

40. 386,543 **41.** 2,867,321 **42.** 14,977,000 **43.** 16,450,000

The table shows the number of each kind of coin minted in the United States in 1980.

44. Round each number of coins to the nearest million.

45. Round each number of coins to the nearest ten million.

Kind of coin	*Number minted in 1980*
penny	12,553,610,000
nickel	1,095,380,000
dime	1,454,090,000
quarter	1,153,740,000
half-dollar	76,820,000
dollar	89,460,000

Comparing numbers

The first hot-air balloon was built in 1783 by two Frenchmen. Today hot-air ballooning is a sport enjoyed by many.

Hot-air balloonists compete in several events. One event is spot-landing. The balloons must take off and land during a specified time and land near a specified spot. The balloon that lands closest to the spot wins.

Suppose one balloon missed the spot by 1526 m and another missed by 1578 m. To determine which balloon wins, the numbers are compared.

Since both have the same number of digits, we start at the left and compare the digits that are in the same place.

Since 2 < 7

1526 < 1578

is less than

Since 7 > 2

1578 > 1526

is greater than

EXERCISES

< or >?

1. 864 ● 865	**2.** 700 ● 699	**3.** 999 ● 1000
4. 907 ● 970	**5.** 6493 ● 6385	**6.** 4863 ● 987
7. 88,235 ● 91,740	**8.** 64,291 ● 64,289	**9.** 71,892 ● 53,461
10. 80,007 ● 78,999	**11.** 354,621 ● 703,945	**12.** 406,735 ● 406,673
13. 488,183 ● 487,997	**14.** 800,000 ● 799,000	**15.** 99,999 ● 100,000

16. Here are the results of a spot-landing event. Rank the balloons from 1 through 6. First place (least number) gets a rank of 1.

Name of balloon	*Meters landed from spot*
America	786
Big Apple	770
Easy Floater	909
Explorer	859
Free Spirit	913
High Rise	768

17. The winner of a cross-country event is the balloon that floats the farthest during a specified time. Rank the balloons from 1 through 6. First place (greatest number) gets a rank of 1.

Name of balloon	*Kilometers traveled*
America	1380
Big Apple	1275
Easy Floater	1363
Explorer	1196
Free Spirit	1394
High Rise	1282

Adding numbers

Estimate a sum before you add.

EXAMPLE.

589 + 323

To find the sum, we add in each column.

Step 1. Add ones and regroup.

```
   1
  589
 +323
    2
```

Step 2. Add tens and regroup.

```
  1 1
  589
 +323
   12
```

Step 3. Add hundreds.

```
  11
  589
 +323
  912
```

Is the sum near the estimate?

EXERCISES

First estimate the sum. Then add.

1. 373 + 499	**2.** 287 + 566	**3.** 675 + 184	**4.** \$8.26 + 1.37	**5.** \$6.85 + 2.37
6. 6482 + 7358	**7.** 5761 + 3897	**8.** 9473 + 6745	**9.** \$42.35 + 38.49	**10.** \$20.58 + 39.79

Add.

11. 56,248 + 39,746	**12.** 51,742 + 39,165	**13.** 80,375 + 29,654	**14.** 503,498 + 128,745	**15.** 736,429 + 167,893

16. 467 + 465 **17.** 897 + 399 **18.** 858 + 79

19. 4235 + 3849 **20.** 2058 + 3979 **21.** 637 + 2847

22. \$9.76 + \$3.85 **23.** \$6.74 + \$.32 **24.** \$87.29 + \$5.62

25. 64,328 + 7463 **26.** 80,406 + 594 **27.** 95,216 + 3798

Add.

28. 623 + 507 + 31

29. 847 + 521 + 477

30. 46 + 873 + 29

31. 703 + 259 + 384 + 801

32. 591 + 538 + 32 + 960

33. 48 + 272 + 65 + 716

34. 5258 + 3940 + 7132 + 8679

35. 5366 + 170 + 8422 + 7359

36. 524 + 7183 + 37 + 7438

Match. Find the sums by estimating.

37. 1536 + 294	**a.** 1185
38. 399 + 582	**b.** 1422
39. 908 + 695	**c.** 1297
40. 911 + 386	**d.** 1830
41. 620 + 314 + 488	**e.** 2080
42. 375 + 529 + 281	**f.** 981
43. 692 + 575 + 813	**g.** 1603

44. **EIGHT IS ENOUGH**

Here is how to arrange 8 eights to get a sum of 280. Show how to arrange 8 eights to get a sum of 1000.

$$\begin{array}{r} 88 \\ 88 \\ 88 \\ 8 \\ +\ 8 \\ \hline 280 \end{array}$$

KEEPING SKILLS SHARP

Round to the nearest 100.

1. 496

2. 506

3. 739

4. 850

5. 967

6. 1421

7. 1530

8. 2674

9. 3961

10. 5382

Round to the nearest 1000.

11. 6250

12. 8109

13. 9950

14. 10,072

15. 26,439

16. 37,825

17. 48,954

18. 59,500

19. 60,972

20. 79,036

Subtracting numbers

You can estimate a difference by rounding each number to the nearest 100.

823 − 294

To find the difference, we subtract in each column.

Step 1. Not enough ones. Regroup 1 ten for 10 ones.

$$\begin{array}{r} 8\overset{1}{\not{2}}\overset{1}{3} \\ -294 \\ \hline \end{array}$$

Step 2. Subtract ones.

$$\begin{array}{r} 8\overset{1}{\not{2}}\overset{1}{3} \\ -294 \\ \hline 9 \end{array}$$

Step 3. Not enough tens. Regroup 1 hundred for 10 tens.

$$\begin{array}{r} \overset{7}{\not{8}}\overset{11}{\not{2}}\overset{1}{3} \\ -294 \\ \hline 9 \end{array}$$

Step 4. Subtract tens.

$$\begin{array}{r} \overset{7}{\not{8}}\overset{11}{\not{2}}\overset{1}{3} \\ -294 \\ \hline 29 \end{array}$$

Step 5. Subtract hundreds.

$$\begin{array}{r} \overset{7}{\not{8}}\overset{11}{\not{2}}\overset{1}{3} \\ -294 \\ \hline 529 \end{array}$$

EXERCISES

First estimate the difference. Then subtract.

1. $\begin{array}{r} 385 \\ -162 \\ \hline \end{array}$

2. $\begin{array}{r} 531 \\ -283 \\ \hline \end{array}$

3. $\begin{array}{r} 8263 \\ -3288 \\ \hline \end{array}$

4. $\begin{array}{r} 5082 \\ -2659 \\ \hline \end{array}$

5. $\begin{array}{r} \$4.42 \\ -3.65 \\ \hline \end{array}$

6. $\begin{array}{r} \$12.86 \\ -7.53 \\ \hline \end{array}$

7. $\begin{array}{r} \$46.32 \\ -22.78 \\ \hline \end{array}$

8. $\begin{array}{r} \$58.10 \\ -47.64 \\ \hline \end{array}$

9. 617 − 356
10. 420 − 274
11. 913 − 187
12. 561 − 73
13. 2686 − 1347
14. 5934 − 359
15. 8842 − 7966
16. 7938 − 67
17. $42.22 − $16.79
18. $34.23 − $7.78

Sometimes you have to regroup more than once before you can subtract.

Step 1. Not enough ones. No tens. Regroup 1 hundred for 10 tens.	**Step 2.** Regroup 1 ten for 10 ones.	**Step 3.** Subtract.
602 −357	602 −357	602 −357

Subtract.

19.	20.	21.	22.	23.	24.
503 −265	200 −35	400 −138	3107 −1288	8000 −2134	2800 −766

25. 702 − 384 **26.** 804 − 796 **27.** 800 − 463

28. 8003 − 5236 **29.** 7021 − 1673 **30.** 4000 − 3906

31. 406 − 58 **32.** 612 − 174 **33.** 500 − 84

34. 8013 − 278 **35.** 5963 − 4877 **36.** 3051 − 79

9000 −4763	→ →	8999 −4762

Use the shortcut to find the differences.

37.	38.	39.	40.	41.
800 −274	7000 −2975	10,000 −7,358	23,000 −18,654	36,000 −17,368

42. 1 + 3 = ? **43.** 1 + 3 + 5 = ? **44.** 1 + 3 + 5 + 7 = ?

45. *Guess:* What is the sum of the first ten odd numbers? Check your guess.

Problem solving—too much information

In his coin collection, Jean's grandfather has 39 Indian Head pennies, 12 Flying Eagle pennies, and 9 Liberty pennies. How many more Indian Head pennies than Flying Eagle pennies does he have?

These steps can help you solve the problem:

1. Study the problem until you understand it.

   Find the question.

   How many more Indian Head pennies than Flying Eagle pennies does he have?

   Find the information you need to answer the question.

   39 Indian Head pennies,
   12 Flying Eagle pennies

2. Plan what to do and do it.

   Subtract 12 from 39.

$$\begin{array}{r} 39 \\ -12 \\ \hline 27 \end{array}$$

3. Answer the question and check your answer.

   He had 27 more Indian Head pennies than Flying Eagle pennies.

   The answer seems right because 39 is about 40, 12 is about 10, and $40 - 10 = 30$.

Quantities Minted and Prices of Some Carson City Silver Dollars					
Year	*Quantity minted*	*Fine (F)*	*Very fine (VF)*	*Extra fine (EF)*	*Almost uncirculated (AU)*
1878	2,212,000	$35	$40	$45	$55
1879	756,000	$50	$85	$250	$550
1880	591,000	$55	$70	$80	$125
1881	296,000	$85	$95	$120	$140

A chart usually has more information than you need. Use this chart to solve problems 1 through 4.

1. How many more Carson City silver dollars were minted in 1880 than in 1881?

2. How much more is an 1879 EF dollar worth than an 1878 AU dollar?

3. What is the total value of an 1878 VF, an 1879 VF, an 1880 VF, and an 1881 VF dollar?

4. How many Carson City silver dollars were minted from 1878 through 1881?

Solve.

5. In 1880, 12,601,355 silver dollars were minted in Philadelphia, 5,305,000 in New Orleans, 8,900,000 in San Francisco, and 591,000 in Carson City.

   a. How many more silver dollars were minted in Philadelphia than in New Orleans?

   b. How many of these silver dollars were *not* minted in Philadelphia?

6. In 1882, 1,133,000 Carson City silver dollars were minted, and in 1883, 1,204,000 Carson City silver dollars were minted. An uncirculated 1882 dollar is worth $180, and an uncirculated 1883 dollar is worth $135.

   a. How many more Carson City silver dollars were minted in 1883 than in 1882?

   b. How much are 3 uncirculated 1882 dollars worth?

Practice exercises

Add.

1. $\begin{array}{r}529\\+788\\\hline\end{array}$ 2. $\begin{array}{r}3564\\+4296\\\hline\end{array}$ 3. $\begin{array}{r}75,287\\+38,615\\\hline\end{array}$ 4. $\begin{array}{r}71,492\\+29,970\\\hline\end{array}$ 5. $\begin{array}{r}637,483\\+540,518\\\hline\end{array}$

6. $\begin{array}{r}38\\9\\72\\+41\\\hline\end{array}$ 7. $\begin{array}{r}96\\44\\18\\+53\\\hline\end{array}$ 8. $\begin{array}{r}328\\252\\767\\+833\\\hline\end{array}$ 9. $\begin{array}{r}5937\\2615\\7875\\+1483\\\hline\end{array}$

Subtract.

10. $\begin{array}{r}529\\-286\\\hline\end{array}$ 11. $\begin{array}{r}415\\-379\\\hline\end{array}$ 12. $\begin{array}{r}603\\-258\\\hline\end{array}$ 13. $\begin{array}{r}700\\-675\\\hline\end{array}$ 14. $\begin{array}{r}3002\\-1758\\\hline\end{array}$

15. $\begin{array}{r}9000\\-2654\\\hline\end{array}$ 16. $\begin{array}{r}18,132\\-7,964\\\hline\end{array}$ 17. $\begin{array}{r}32,109\\-14,836\\\hline\end{array}$ 18. $\begin{array}{r}81,034\\-7,526\\\hline\end{array}$ 19. $\begin{array}{r}638,002\\-59,708\\\hline\end{array}$

(15 − 7) + 2	**15 − (7 + 2)**
8 + 2	15 − 9
10	6

Compute.

20. (78 + 52) − 29
21. 78 + (52 − 29)
22. (358 − 156) − 78
23. 358 − (156 − 78)
24. (900 − 532) + 53
25. 900 − (532 + 53)
26. (98 + 36) − (70 + 19)
27. (163 − 129) + (800 − 593)
28. (6381 − 592) + (753 + 96)
29. (3974 + 2831) − (3000 − 714)
30. (4983 + 627) − (40 − 8)
31. (295 + 6041) − (985 − 26)
32. (2539 − 109) + (863 − 92)
33. (4409 − 1283) + (62 − 19)
34. (7513 − 4998) − (253 + 1097)
35. (610 + 5347) − (1983 − 625)

Problem solving

1. Study and understand.
2. Plan and do.
3. Answer and check.

Use the map to solve.

1. How far is
   a. Stanton from Fairview?
   b. the shortest route from Bethany to Ashland?

2. How much farther is Durant
   a. from Ashland than from Fairview?
   b. from Fairview than from Millbury?

3. What city are you in if you are
   a. 104 miles from Millbury?
   b. 124 miles from Deerfield?

4. How far from Millbury are you if you are 35 miles
   a. southwest of Deerfield?
   b. west of Durant?

★ 5. Mr. Harris took the shortest route from Deerfield to Durant. Ms. Shaw took the shortest route from Fairview to Millbury. They started at the same time and drove at the same speed. How many miles from Durant were they when they passed each other?

★ 6. Between which 2 cities is there a red house if the nearest city to the red house is 40 miles away?

Multiplying by a 1-digit number

A chair lift can carry 960 people each hour. How many people can it carry in 8 hours?

To find the answer, we can multiply.

Step 1. Multiply ones.

$$\begin{array}{r} 960 \\ \times 8 \\ \hline 0 \end{array}$$

Step 2. Multiply tens and regroup.

$$\begin{array}{r} {}^{4} \\ 960 \\ \times 8 \\ \hline 80 \end{array}$$

Step 3. Multiply hundreds and add.

$$\begin{array}{r} {}^{4} \\ 960 \\ \times 8 \\ \hline 7680 \end{array}$$

The chair lift can carry 7680 people in 8 hours.

The answer is reasonable because 960×8 is about 1000×8, or 8000.

EXERCISES

First estimate the product. Then multiply.

1. 78 × 4	**2.** 59 × 6	**3.** 73 × 8	**4.** 85 × 5	**5.** 94 × 7
6. 292 × 5	**7.** 311 × 7	**8.** 529 × 9	**9.** 614 × 8	**10.** 673 × 6

Multiply.

11. $5.96 × 4

12. $7.84 × 3

13. $9.21 × 8

14. $7.56 × 6

15. $7.92 × 7

16. 3821 × 5

17. 7456 × 3

18. 9378 × 6

19. 2916 × 4

20. 5304 × 2

21. 82,771 × 8

22. 65,382 × 7

23. 59,307 × 6

24. 68,259 × 9

25. 73,006 × 5

26. 563 × 3

27. 958 × 6

28. 1814 × 7

29. 6048 × 5

30. 3254 × 5

31. 61,078 × 4

32. 93,865 × 9

33. 12,875 × 8

Solve.

34. Jeffrey bought a half-day lift ticket for $7.50 and rented skis for $11.65. What was the total cost?

35. Sonja had $20. She paid $8.50 for a 30-minute ski lesson. How much money did she have then?

36. A one-day chair-lift ticket costs $12.75. How much would it cost for 3 days?

37. There are two hills. One is 783 meters long and the other is 1002 meters long. How much longer is the second hill?

38. Garrett rented skis for $11.65, bought a chair-lift ticket for $12.75, and paid $8.50 for a 30-minute ski lesson. How much did he spend?

39. TWO-HANDED CALCULATOR

You can use your fingers to multiply by 9. Here is how to find 8 × 9:

Hold your hands in front of you with your fingers extended. Since you are multiplying by 8, fold down the 8th finger from the left. The picture shows how you can see the product, 72.

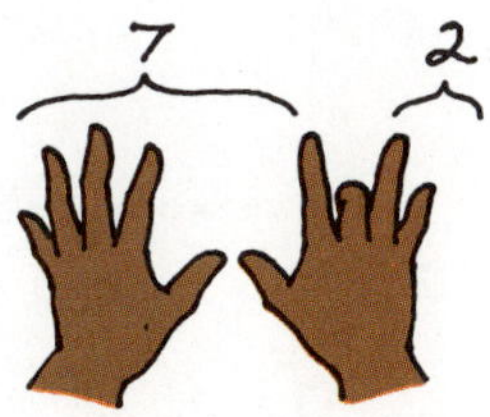

Try this with 3 × 9, 7 × 9, etc.

Multiplying by multiples of 10, 100, and 1000

$$42 \times 10 = 420$$

$$58 \times 100 = 5800$$

$$136 \times 1000 = 136{,}000$$

What are the shortcuts for multiplying by 10, 100, and 1000?

To multiply a number by 30, you can first multiply the number by 3 and then multiply the answer by 10.

$$\begin{array}{r} 46 \\ \times 30 \\ \hline 1380 \end{array} \qquad \begin{array}{r} 46 \\ \times 3 \\ \hline 138 \\ \times 10 \\ \hline 1380 \end{array} \qquad \begin{array}{r} 46 \\ \times 30 \\ \hline 1380 \end{array}$$

This is an example of the **associative (grouping) property of multiplication.**

$$46 \times 30 = 46 \times (3 \times 10)$$
$$= (46 \times 3) \times 10$$

EXERCISES

Give each product.

1. 37×10
2. 37×100
3. 37×1000
4. 562×10
5. 562×100
6. 562×1000
7. 7563×10
8. 7563×100
9. 7563×1000
10. 80×10
11. 80×100
12. 80×1000
13. 700×10
14. 700×100
15. 700×1000
16. 1000×10
17. 1000×100
18. 1000×1000

Multiply.

19. 57×20
20. 57×200
21. 57×2000
22. 570×200
23. 638×30
24. 638×300
25. 638×3000
26. $63{,}800 \times 30$
27. 598×80
28. 473×50
29. 675×60
30. 341×90
31. 462×300
32. 296×500
33. 800×400
34. 300×700
35. 5673×3000
36. 9846×700
37. 5034×8000
38. 7735×9000

Solve.

39. A 46-kilometer race was run in 4 hours 37 minutes. How many minutes is that?

40. A 6925-meter line of dominoes was set up in 130 hours. The dominoes fell at a speed of 65 meters per minute. How many meters of dominoes fell in 30 minutes?

KEEPING SKILLS SHARP

Estimate to find the two incorrect answers.

1. a. $87 + 25 = 112$
   b. $674 + 288 = 862$
   c. $84 + 78 + 69 = 331$
   d. $839 + 416 + 727 = 1982$

Add.

2. $78 + 29$
3. $356 + 75$
4. $298 + 574$
5. $893 + 277$
6. $3654 + 2985$
7. $4736 + 989$
8. $98 + 26 + 59$
9. $375 + 192 + 468$
10. $26 + 35 + 44 + 29$
11. $529 + 638 + 466 + 583$

Multiplying by a 2-digit number

In a 15-mile race, runners are required to run 60 laps of a track that is 440 yards long. On a practice run, Steven ran 38 laps. How many yards did he run?

To find the answer, you multiply 440×38.

To multiply by 38, you multiply by 8, multiply by 30, and add.

Step 1. Multiply 440 by 8 ones.

$$\begin{array}{r} 440 \\ \times 38 \\ \hline 3520 \end{array}$$

yards run in 8 laps

Step 2. Multiply 440 by 3 tens.

$$\begin{array}{r} 440 \\ \times 38 \\ \hline 3520 \\ 1320 \end{array}$$

yards run in 30 laps

Step 3. Add.

$$\begin{array}{r} 440 \\ \times 38 \\ \hline 3520 \\ 1320 \\ \hline 16,720 \end{array}$$

yards run in 38 laps

Steven ran 16,720 yards.

This is an example of the **distributive property of multiplication over addition.**

$$\begin{aligned} 440 \times 38 &= 440 \times (8 + 30) \\ &= (440 \times 8) + (440 \times 30) \end{aligned}$$

EXERCISES

Multiply.

1. 58 ×12	**2.** 65 ×46	**3.** 74 ×33	**4.** 92 ×69	**5.** 89 ×52
6. 156 ×54	**7.** 474 ×53	**8.** 238 ×75	**9.** 506 ×24	**10.** 821 ×86
11. 734 ×49	**12.** 593 ×77	**13.** 658 ×92	**14.** 291 ×37	**15.** 756 ×54
16. 3214 ×68	**17.** 5978 ×86	**18.** 6359 ×75	**19.** 4618 ×34	**20.** 3921 ×93
★ **21.** 53,826 ×38	★ **22.** 75,346 ×46	★ **23.** 93,580 ×68	★ **24.** 296,374 ×89	★ **25.** 729,108 ×27

Solve.

26. Laurie ran 47 laps of a track that was 440 yards long. How many yards did she run?

27. On a 440-yard track, Alan ran 26 laps and Mark ran 25 laps. How many more yards did Alan run?

28. On a 440-yard track, Maureen ran 13 more laps than Sharon. How much farther did Maureen run?

29. A mile is 1760 yards. How many yards are in 15 miles?

30. Which is the greater distance: 8 miles or 30 laps of a 440-yard track? How much greater?

31. A marathon is a race that is usually 26 miles 385 yards. How many yards long is a marathon?

★ 32. SECOND HAND-CALCULATOR

You can use your fingers to find the product of 9 and a 2-digit number *if* the tens digit of the 2-digit number is less than the ones digit.

Example: 9 × 27

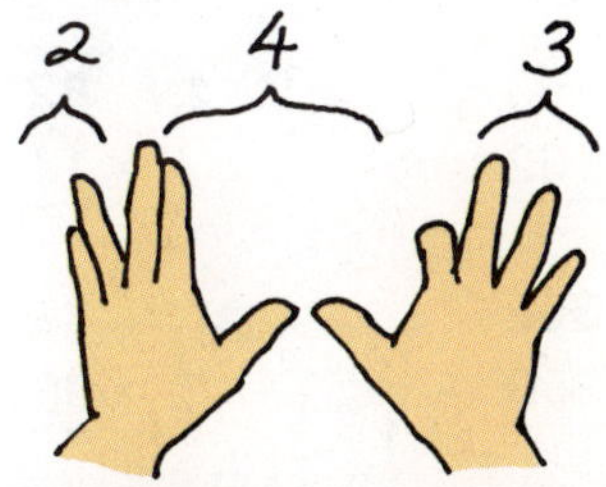

Make a space after the second finger from the left. Put down the seventh finger from the left. Read the answer: 243. Try this method with other problems. Check.

Multiplying by a 3-digit number

Study these examples for multiplying by a 3-digit number.

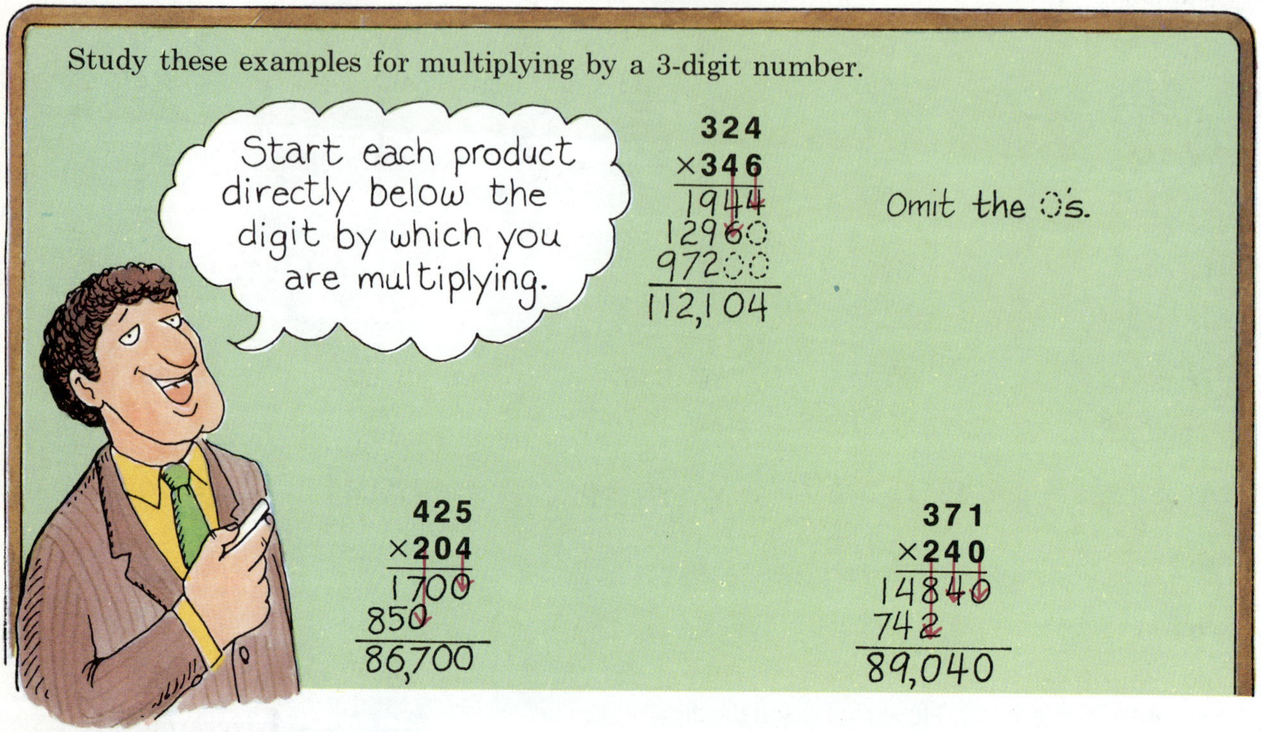

If you always write each product directly below the digit by which you are multiplying, you will make few mistakes.

EXERCISES
Multiply.

1. 753 × 162
2. 821 × 143
3. 742 × 234
4. 658 × 255
5. 291 × 368
6. 362 × 248
7. 455 × 937
8. 258 × 342
9. 615 × 839
10. 727 × 189
11. 1538 × 429
12. 1965 × 364
13. 2742 × 295
14. 3953 × 338
15. 5829 × 426

Multiply.

16. 523 ×306 3138 1569	17. 421 ×603	18. 592 ×402	19. 603 ×501	20. 304 ×606
	21. 571 ×630	22. 473 ×220	23. 217 ×300	24. 546 ×207
25. 3614 ×403	26. 2983 ×258	27. 7521 ×560	28. 6438 ×352	29. 2059 ×803
30. 5246 ×293	31. 3958 ×405	32. 7042 ×200	33. 5063 ×540	34. 9384 ×398
35. 5231 ×203	36. 4205 ×346	37. 3617 ×450	38. 4283 ×700	39. 7961 ×209

Solve.

40. There were 512 tickets sold for a city baseball game. Each ticket sold for $2.25. What was the total in ticket sales?

41. In Campbell Junior High there are 27 more girls than boys. There are 342 girls. What is the total enrollment?

42. Marcia bought 12 golf balls for 95¢ each. She gave the clerk $20. How much change did she receive?

43. Kirk wanted to buy 4 pairs of wool socks that cost $4.69 each. He had $18.25. How much more money did he need?

44. Find the missing numbers.

(1 × 8) + 1 = 9

(12 × 8) + 2 = 98

(123 × 8) + 3 = 987

(? × ?) + ? = 9876

(? × ?) + ? = 98,765

Exponents

There are $2 \times 2 \times 2$, or 2^3, baseballs.

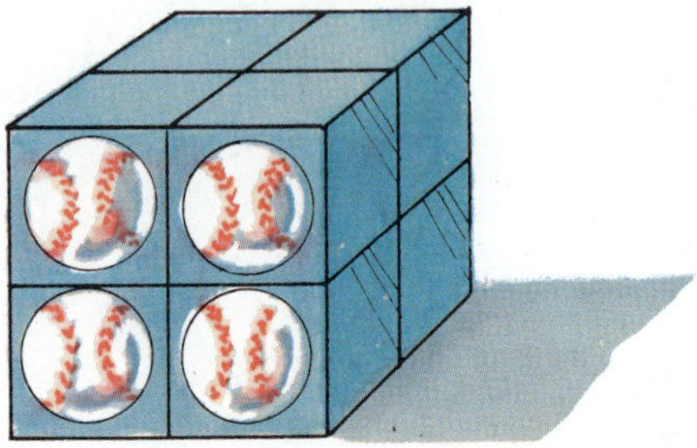

The 3 is called an **exponent.** The exponent tells the number of times the **base,** 2, is used as a factor.

Here are some other examples:

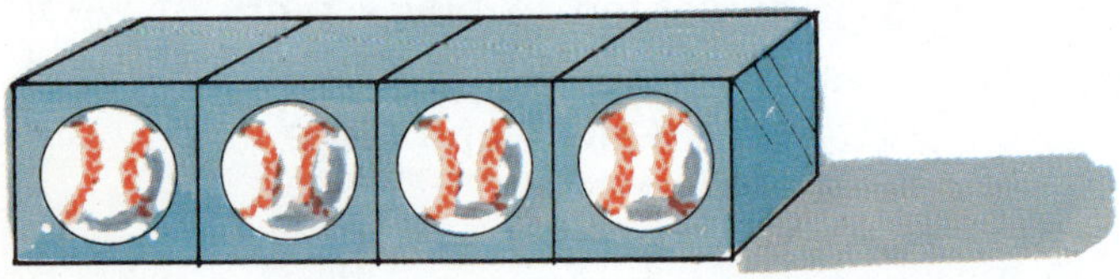

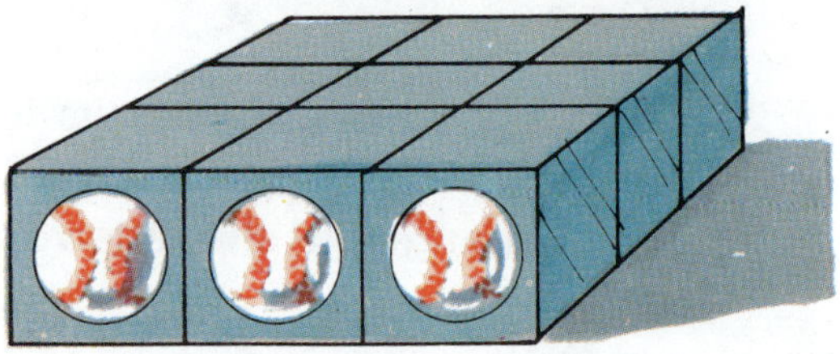

$\mathbf{4 = 4^1}$

"four to the first power"

$\mathbf{3 \times 3 = 3^2}$

"three to the second power" or "three squared"

$\mathbf{10 \times 10 \times 10 \times 10 = 10^4}$

"ten to the fourth power"

$\mathbf{2 \times 2 \times 2 \times 2 \times 2 = 2^5}$

"two to the fifth power"

EXERCISES

Write, using an exponent.

1. $2 \times 2 \times 2$ **2.** 3×3 **3.** $4 \times 4 \times 4 \times 4 \times 4$ **4.** $5 \times 5 \times 5$

5. 6×6 **6.** $7 \times 7 \times 7$ **7.** 10 **8.** 10×10

9. $10 \times 10 \times 10$ **10.** $10 \times 10 \times 10 \times 10$ **11.** $10 \times 10 \times 10 \times 10 \times 10$

Give the standard numeral.

12. 2^4 ∘∘ ($2 \times 2 \times 2 \times 2$) **13.** 4^2 **14.** 9^2 **15.** 3^3 **16.** 5^4 **17.** 2^6

18. 6^3 **19.** 20^2 **20.** 10^3 **21.** 8^1 **22.** 5^3 **23.** 4^5 **24.** 1^6

Copy and complete.

	Power of 10	*Factors*	*Standard numeral*	*Number of 0's in standard numeral*
	10^1	10	10	1
25.	10^2	10×10	100	?
26.	10^3	$10 \times 10 \times 10$	?	?
27.	10^4	?	?	?
28.	10^5	?	?	?
29.	10^6	?	?	?
30.	10^7	?	?	?
31.	10^8	?	?	?

32. In your complete table, compare the exponent of 10 with the number of 0's in the standard numeral. Are they the same?

Give the missing base. ***Hint:*** **Guess. Then check with a calculator.**

33. $\blacksquare^2 = 676$ **34.** $\blacksquare^2 = 5476$ **35.** $\blacksquare^2 = 3025$

36. $\blacksquare^3 = 4096$ **37.** $\blacksquare^3 = 13{,}824$ **38.** $\blacksquare^3 = 35{,}937$

Dividing by a 1-digit number

Carol, Joan, and Susan were given a box containing 113 records for working at a neighbor's garage sale. They decided to share the records equally.

To find how many records each should get, we can divide:

$$113 \div 3$$

Step 1. Not enough hundreds. Think about regrouping 1 hundred for 10 tens.

```
3)113
```

Step 2. Divide tens. Subtract.

```
   3
3)113
  -9
   2
```

Step 3. Regroup and divide ones.

```
   37 R2
3)113
  -9
   23
  -21
    2
```

The remainder is 2.

Each girl would get 37 records. They would have to decide who would get the remaining 2 records.

EXERCISES

Divide.

1. $2\overline{)246}$	**2.** $2\overline{)137}$	**3.** $3\overline{)58}$	**4.** $4\overline{)64}$	**5.** $5\overline{)95}$
6. $8\overline{)345}$	**7.** $4\overline{)826}$	**8.** $7\overline{)367}$	**9.** $9\overline{)875}$	**10.** $9\overline{)146}$
11. $6\overline{)935}$	**12.** $2\overline{)134}$	**13.** $7\overline{)534}$	**14.** $5\overline{)947}$	**15.** $3\overline{)816}$
16. $8\overline{)\$5.68}$	**17.** $7\overline{)\$3.85}$	**18.** $6\overline{)\$9.54}$	**19.** $9\overline{)\$2.07}$	**20.** $4\overline{)\$5.84}$
21. $2\overline{)3946}$	**22.** $5\overline{)5108}$	**23.** $3\overline{)3742}$	**24.** $2\overline{)5965}$	**25.** $7\overline{)3782}$
26. $3\overline{)3914}$	**27.** $4\overline{)5678}$	**28.** $6\overline{)4071}$	**29.** $4\overline{)3049}$	**30.** $8\overline{)9246}$
31. $7\overline{)7003}$	**32.** $4\overline{)5208}$	**33.** $3\overline{)4631}$	**34.** $9\overline{)2403}$	**35.** $6\overline{)1149}$
36. $5\overline{)31,604}$	**37.** $2\overline{)78,251}$	**38.** $5\overline{)39,283}$	**39.** $8\overline{)75,691}$	**40.** $8\overline{)64,238}$
41. $6\overline{)90,177}$	**42.** $7\overline{)72,104}$	**43.** $7\overline{)63,029}$	**44.** $4\overline{)98,106}$	**45.** $3\overline{)75,236}$
★ **46.** $2\overline{)129,634}$	★ **47.** $5\overline{)792,381}$	★ **48.** $9\overline{)691,085}$	★ **49.** $6\overline{)397,615}$	★ **50.** $9\overline{)263,842}$

Solve.

51. At a 3-day garage sale, the Romeros sold their used merchandise for \$382.11. What was the average amount they collected per day?

52. Janet bought 8 record albums that sold for 35¢ each and 6 paperback books that sold for 15¢ each. How much money did she spend?

53. At the garage sale there were 25 comic books in each of 13 boxes. How many comic books were there?

54. A box of 25 comic books sold for \$2.75 and one comic book sold for 15¢. How much cheaper was it to buy a box than to buy 25 single comic books?

55. Luis Romero sold a radio to Miguel for \$7.50. Miguel gave Luis \$10. How much change did Luis give Miguel?

56. The Romeros sold 2 bicycles for \$12 each and 3 bicycles for \$19 each. What was the average price of a bicycle?

Dividing by a 2-digit number

Study this example of dividing by a 2-digit number.

EXAMPLE. 57)2056

Step 1. Decide where the first digit of the quotient will be.

```
      □
57)2056
```

20 is less than 57, so the first digit of the quotient goes over the 5.

Step 2. Round the divisor and divide.

(60)

```
      3
57)2056
```

The first guess is 3, because 3×60 is a little less than 205.

Step 3. Multiply by 3.

```
      3
57)2056
  -171
    34
```

3 works!

Step 4. Regroup and divide.

(60)

```
     35
57)2056
  -171
    346
```

The next guess is 5, because 5×60 is a little less than 346.

Step 5. Multiply by 5.

```
     35
57)2056
  -171
    346
   -285
     61
```

61 is greater than 57. 5 is too small!

Step 6. Try 6 because 5 is too small.

```
     36 R4
57)2056
  -171
    346
   -342
      4
```

6 works!

EXERCISES

Divide.

1. 18)593	**2.** 24)842	**3.** 31)900
4. 25)675	**5.** 32)942	**6.** 48)876
7. 38)965	**8.** 11)527	**9.** 41)993
10. 52)1404	**11.** 68)2584	**12.** 72)2088
13. 41)3157	**14.** 77)2464	**15.** 92)4324
16. 79)6080	**17.** 28)4044	**18.** 81)4536
19. 49)17,736	**20.** 52)38,608	**21.** 51)18,342
22. 74)21,000	**23.** 89)41,625	**24.** 35)11,138
25. 22)94,870	**26.** 45)12,816	**27.** 36)14,928
★ **28.** 71)365,986	★ **29.** 82)274,164	★ **30.** 39)500,813

Solve.

31. Four backpackers agreed to divide 68 kilograms of supplies evenly. How much should each backpacker carry?

32. If a large pizza costs $4.75 and a pitcher of root beer costs $1.25, how much do 3 pizzas and 2 pitchers of root beer cost?

33. Karl earned $4.95, $5.74, and $3.88 on the three days he worked. He needs $20 to buy a football. How much more does he need to earn?

KEEPING SKILLS SHARP

Estimate to find the two incorrect answers.

1. a. $\begin{array}{r} 602 \\ -298 \\ \hline 404 \end{array}$

b. $\begin{array}{r} 716 \\ -399 \\ \hline 317 \end{array}$

c. $\begin{array}{r} 4002 \\ -1977 \\ \hline 2925 \end{array}$

d. $\begin{array}{r} 5024 \\ -3155 \\ \hline 1869 \end{array}$

Subtract.

2. $\begin{array}{r} 275 \\ -188 \\ \hline \end{array}$

3. $\begin{array}{r} 352 \\ -167 \\ \hline \end{array}$

4. $\begin{array}{r} 4203 \\ -1735 \\ \hline \end{array}$

5. $\begin{array}{r} 6000 \\ -2044 \\ \hline \end{array}$

6. $\begin{array}{r} 6523 \\ -1741 \\ \hline \end{array}$

7. $\begin{array}{r} 3998 \\ -1689 \\ \hline \end{array}$

Estimating in division

Many birds migrate long distances. The most well-traveled bird is the Arctic tern. It migrates from the North Pole to the South Pole and back. One Arctic tern flew the one-way distance of 20,000 kilometers in 65 days.

Mike did this division to find about how many kilometers the Arctic tern averaged per day.

```
      37
65)20,000
  -195
     500
    -455
      45
```

about 38 kilometers

Notice that Mike made a common mistake. He forgot a zero in the quotient. Check his work to see where the zero goes.

He should have estimated the quotient first. Then he would have noticed the mistake.

Here is one way to estimate the quotient:

Step 1. Write the first digit of the quotient in the correct place.

```
     3
65)20,000
```

Step 2. Write zeros in the remaining places.

```
     300
65)20,000
```

The quotient is between 300 and 400.

EXERCISES

Estimate the quotient, then divide.

1. 29)8903	2. 62)3999	3. 83)8062	4. 18)4327
5. 91)62,384	6. 57)11,138	7. 39)80,035	8. 42)93,152
9. 62)25,361	10. 65)57,938	11. 48)43,205	12. 74)96,792
13. 37)24,312	14. 24)76,128	★ 15. 17)480,395	★ 16. 48)908,172

Solve. If necessary, round your answer to the nearest whole number.

17. A blue-wing teal that was banded (had a ring put around its leg) in northern Alberta, Canada, was seen in Venezuela exactly 30 days later. If this distance is about 6100 kilometers, how many kilometers per day were averaged?

18. A lesser yellowlegs flew from Cape Cod, Massachusetts, to the French West Indies in 6 days. It averaged 509 kilometers per day. How far did it fly?

19. The gray-cheeked thrush flies from northern Florida to northwest Alaska, a distance of 6500 kilometers, in 31 days. How many kilometers does it average per day?

20. Robins have been known to fly from Iowa to Alaska, a distance of about 4800 kilometers, in 78 days. How many kilometers do they average per day?

21. The blackpolled warbler migrates from Florida to western Alaska in the spring. It averages 56 kilometers per day for the first 29 days and then 320 kilometers per day for the next 15 days. How far is that?

22. Golden plovers hold the nonstop record. They fly from northeast Canada to South America nonstop, a distance of 3800 kilometers. If they fly 190 kilometers per day, how many days do they fly without stopping?

Dividing by a 3-digit number

EXAMPLE. $328\overline{)89{,}356}$

Step 1. Divide and subtract.

```
        2
328)89,356
   -656
    237
```

Step 2. Regroup.

```
        2
328)89,356
   -656
    2375
```

Step 3. Divide and subtract.

```
        27
328)89,356
   -656
    2375
   -2296
      79
```

Step 4. Regroup.

```
        27
328)89,356
   -656
    2375
   -2296
      796
```

Step 5. Divide and subtract.

```
        272 R140
328)89,356
   -656
    2375
   -2296
      796
     -656
      140
```

EXERCISES

Divide.

1. $112\overline{)35,274}$
2. $208\overline{)69,782}$
3. $115\overline{)59,386}$
4. $234\overline{)74,261}$
5. $326\overline{)37,821}$
6. $402\overline{)59,784}$
7. $453\overline{)60,715}$
8. $526\overline{)52,830}$
9. $704\overline{)42,065}$
10. $628\overline{)32,978}$
11. $750\overline{)74,216}$
12. $378\overline{)93,852}$
13. $288\overline{)756,031}$
14. $813\overline{)258,921}$
15. $529\overline{)653,184}$
16. $478\overline{)953,165}$
17. $937\overline{)742,167}$
18. $653\overline{)406,314}$
19. $892\overline{)556,395}$
20. $960\overline{)814,638}$

Solve.

21. 74 weeks
How many days?
22. 595 days
How many weeks?
23. 38 days
How many hours?
24. 888 eggs
How many dozen?
25. 768 months
How many years?
26. 2700 minutes
How many hours?
27. 11,680 days
How many years?
[Use 365 days
for 1 year.]
28. 15 years
How many days?
29. 5424 inches
How many feet?
30. 2238 feet
How many yards?
31. 492 feet
How many inches?
32. 2844 inches
How many yards?
33. 591 pounds
How many ounces?
34. 52 yards
How many inches?
35. 864 ounces
How many pounds?
36. 123 tons
How many pounds?

Properties of addition and multiplication

In this lesson we will review some properties of addition and multiplication. Remembering the properties can sometimes help with computing.

ADDITION

The Adding 0 Property

The sum of any number and 0 is the number.

$358 + 0 = 358$

The Commutative Property of Addition

Changing the order of the addends does not change the sum.

$267 + 139 = 139 + 267$

The Associative Property of Addition

Changing the grouping of the addends does not change the sum.

$(18 + 37) + 13 = 18 + (37 + 13)$

MULTIPLICATION

The Multiplying by 1 Property

The product of any number and 1 is the number.

$594 \times 1 = 594$

The Commutative Property of Multiplication

Changing the order of the factors does not change the product.

$68 \times 57 = 57 \times 68$

The Associative Property of Multiplication

Changing the grouping of the factors does not change the product.

$(28 \times 4) \times 25 = 28 \times (4 \times 25)$

The Distributive Property

$19 \times (8 + 2) = (19 \times 8) + (19 \times 2)$

EXERCISES

See if you can compute each answer in your head by using one of the properties above.

$128 + (75 + 25)$

$23 \times (4 \times 25)$

1. $(128 + 75) + 25$
2. $(225 + 185) + 15$
3. $(23 \times 4) \times 25$
4. $(627 + 45) + 55$
5. $(47 \times 5) \times 20$
6. $(245 \times 250) \times 4$
7. $(13 \times 6) + (13 \times 4)$
8. $(23 \times 3) + (23 \times 7)$
9. $(52 \times 60) + (52 \times 40)$

Order of operations

Barry and Doreen each simplified the expression

$$5 + 2 \times 6$$

Notice that they got different answers. The answers are different because Barry added first and Doreen multiplied first.

To be sure that an expression has only one value, mathematicians have agreed on some rules for the order of operations:

1. First, do the operations within the grouping symbols (parentheses).
2. Next, work from left to right doing the multiplications and divisions.
3. Finally, work from left to right doing the additions and subtractions.

Study these examples. The order of operations is shown by the numbered arrows.

$6 + 8 \div 2 = 10$ (1: $8 \div 2 = 4$; 2: $6 + 4$)

$3 \times 2 + 12 \div 4 = 9$ (1: $3 \times 2 = 6$; 2: $12 \div 4 = 3$; 3: $6 + 3$)

$24 - (6 - 4) = 22$ (1: $6 - 4 = 2$; 2: $24 - 2$)

EXERCISES

Simplify.

1. $8 \times 7 - 3$
2. $12 + 8 \div 2$
3. $15 \div 5 - 2$
4. $24 \div 2 \times 3$
5. $20 - 6 - 4$
6. $8 + 4 \div 4$
7. $25 \times (8 - 3)$
8. $(15 + 50) \div 5$
9. $48 \div (6 \times 2)$
10. $15 + 6 \times 2 - 1$
11. $3 + 6 \times 8 \div 2$
12. $8 \times 2 + 6 \times 3$
13. $28 + 6 \times 12 \div 3$
14. $36 - 6 + 21 \div 3$
15. $2 + 8 \times 13 - 6$
16. $(2 + 8) \times 13 - 6$
17. $2 + 8 \times (13 - 6)$
18. $(2 + 8) \times (13 - 6)$
19. ★ $42 \div (3 + 3) \times 8 - 5$
20. ★ $(59 + 61) \times 21 + 40$
21. ★ $65 \times 18 + 26 \div 2$

Problem solving—more than one way to solve

1. Study and understand.
2. Plan and do.
3. Answer and check.

Benjamin had $50. He bought a pair of roller skates for $37 and a pair of socks for $5. How much money did he have left?

Kenneth's way to solve.

$$\begin{array}{r} \$37 \\ +\ \ 5 \\ \hline \$42 \end{array}$$ *total spent*

$$\begin{array}{r} \$50 \\ -42 \\ \hline \$\ \ 8 \end{array}$$ *amount left*

Benjamin had $8 left.

Barbara's way to solve.

$$\begin{array}{r} \$50 \\ -37 \\ \hline \$13 \end{array}$$ amount left after buying skates

$$\begin{array}{r} \$13 \\ -\ \ 5 \\ \hline \$\ \ 8 \end{array}$$ amount left after buying socks

Benjamin had $8 left.

EXERCISES

One way to solve each problem is given. Find another way to solve it.

1. John had $22.45. He bought a helmet for $19.75. Then he earned $10 for raking leaves. How much money did he have then?

$$\begin{array}{r} \$22.45 \\ -19.75 \\ \hline \$\ \ 2.70 \end{array} \qquad \begin{array}{r} \$2.70 \\ +10.00 \\ \hline \$12.70 \end{array}$$

2. Martha bought 2 jerseys for $7.98 each and 2 elbow pads for $1.65 each. How much did she spend?

$$\begin{array}{r} \$\ 7.98 \\ \times 2 \\ \hline \$15.96 \end{array} \qquad \begin{array}{r} \$1.65 \\ \times 2 \\ \hline \$3.30 \end{array} \qquad \begin{array}{r} \$15.96 \\ +\ 3.30 \\ \hline \$19.26 \end{array}$$

Solve.

3. The Fast Wheels roller-skating track charges $2.25 for tickets on weekdays. Three friends went skating on Monday and on Thursday of one week. How much did they spend that week for tickets?

4. The Fast Wheels roller-skating track charges $2.25 for tickets on weekdays and $2.75 for tickets on weekends. Four girls went skating on Tuesday and on Sunday of one week. How much did they spend that week for tickets?

5. The Fast Wheels roller-skating track charges $2.25 for tickets on weekdays and $2.75 for tickets on weekends. Five boys went skating on Wednesday. How much did they spend for tickets?

6. Deborah skated for 1 hour a day on Monday, Wednesday, and Friday and for 3 hours a day on Saturday and Sunday. How many minutes of skating did she average per day?

7. Yobo skated 1 hour 15 minutes for 5 days and 1 hour 30 minutes for 3 days. How long would he have to skate on another day to average 85 minutes per day?

8. The blue course at Fast Wheels roller-skating track is 345 yards long. The green course is 317 yards long. Derek skated the blue course 12 times and Teresa skated the green course 12 times. How much farther did Derek skate?

9. Mark skated the 345-yard course at Fast Wheels 15 times in 20 minutes. How many seconds did he average per time?

10. Julian skated the 345-yard course at Fast Wheels 5 times and the 317-yard course 5 times. How far did he skate?

CHAPTER CHECKUP

Write the standard numeral. [pages 22–23]

1. seventeen million
2. two hundred billion
3. eight trillion, six thousand

Round [pages 24–27]

4. 7583 to the nearest hundred.
5. 23,488 to the nearest thousand.
6. 8,754,321 to the nearest hundred thousand.

Add. [pages 30–31, 36]

7. $578 + 466$
8. $9835 + 2768$
9. $293 + 476 + 351$
10. $36{,}748 + 867 + 9{,}988$

Subtract. [pages 32–33, 36]

11. $8321 - 5673$
12. $702 - 427$
13. $5000 - 2354$
14. $603{,}508 - 168{,}739$

Multiply. [pages 38–45]

15. 2783×6
16. 53×26
17. 482×321
18. 568×306

Give the standard numeral. [pages 46–47]

19. 3^4 20. 5^2 21. 6^3 22. 1^4 23. 2^5 24. 10^3

Divide. [pages 48–52, 54–55]

25. $9\overline{)4763}$
26. $8\overline{)6449}$
27. $38\overline{)58{,}361}$
28. $405\overline{)19{,}360}$

Simplify. [page 57]

29. $6 + 10 \div 2$
30. $(35 - 6) \times 4$
31. $24 - (7 + 1) \times 2$

Solve. [pages 34–35, 37, 53, 58–59]

32. Carl had 12 flowers in each of 35 bunches. He sold 16 bunches. How many bunches were not sold?

33. Joan bought 3 shirts for $7 each and a sweater for $9.95. How much change did she get from $40?

CHAPTER PROJECT

1. Find out how many telephone numbers each of your classmates has memorized. (The data could be listed on the chalkboard.)
2. Compute the average number of telephone numbers known. Give your answer to the nearest whole number.
3. How many classmates know
   a. more telephone numbers than the average?
   b. fewer telephone numbers than the average?
   c. twice as many telephone numbers as the average?
   d. half as many telephone numbers as the average?
4. Make a bar graph like the one shown here.

How Many Telephone Numbers We Know

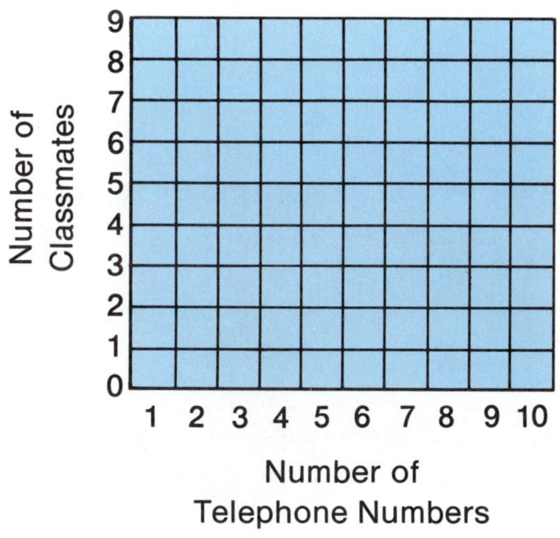

CHAPTER REVIEW

Round to the nearest thousand.

thousands place

63,672 ⇨ 64,000

This digit is 5 or greater.

Round to the nearest hundred.

1. 673
2. 650
3. 89,341

Round to the nearest thousand.

4. 7388
5. 2500
6. 67,832

```
  1 1
  358
+ 267
  625
```

Add.

7. 563 + 288
8. 463 + 297
9. 5864 + 2995
10. 87,351 + 25,489
11. 29 + 37 + 56
12. 463 + 297 + 341
13. 589 + 678 + 838
14. 2975 + 467 + 5986 + 321

```
  2 9
  3 0 2
- 1 7 7
  1 2 5
```

Subtract.

15. 526 − 238
16. 732 − 581
17. 6731 − 2951
18. 56,342 − 8,756
19. 904 − 656
20. 5000 − 276
21. 65,002 − 14,975
22. 30,204 − 16,578

```
    321
    ×34
   1284   ← 4 × 321
   9630   ← 30 × 321
  10914
```

Multiply.

23. 475 × 8
24. 36 × 43
25. 57 × 65
26. 394 × 62
27. 592 × 326
28. 548 × 602

```
(40)      73 R29
   39)2876
     -273
       146
      -117
        29
```

Divide.

29. 8)5287
30. 7)4921
31. 23)5261
32. 53)16,165
33. 148)16,673
34. 321)26,543

CHAPTER CHALLENGE

CLUES

1. How many grams does a sphere weigh?
2. How many grams does a cube weigh?
3. How many grams does a cone weigh?
4. How many grams does a cylinder weigh?
5. How many cubes weigh as much as 3 spheres?
6. How many cylinders weigh as much as 5 cones?

1. a b c d 2. a b c d 5. a b c d 6. a b
7. a b c d 8. a b c d 9. a b c d 10. a b c d 11. a b c d 12. a b

MAJOR CHECKUP
STANDARDIZED FORMAT

Choose the correct letter.

1. The standard numeral for four billion, twenty-five thousand is

a. 4,025,000
b. 4,000,025,000
c. 4,000,000,025,000
d. none of these

2. Write 7,000,000,000,000 in words.

a. seven million
b. seven quintillion
c. seven billion
d. none of these

3. Round 48,377 to the nearest thousand.

a. 48,000
b. 50,000
c. 49,000
d. none of these

4. Add.

$$\begin{array}{r} 632 \\ 577 \\ +298 \\ \hline \end{array}$$

a. 1507
b. 1497
c. 1407
d. none of these

5. Subtract.

$$\begin{array}{r} 5297 \\ -3898 \\ \hline \end{array}$$

a. 2399
b. 1399
c. 1409
d. none of these

6. Subtract.

$$\begin{array}{r} 6000 \\ -3956 \\ \hline \end{array}$$

a. 3044
b. 3043
c. 2044
d. none of these

7. Multiply.

$$\begin{array}{r} 37 \\ \times 26 \\ \hline \end{array}$$

a. 962
b. 296
c. 926
d. none of these

8. Multiply.

$$\begin{array}{r} 372 \\ \times 305 \\ \hline \end{array}$$

a. 13,020
b. 2976
c. 113,460
d. none of these

9. Divide.

$25\overline{)4036}$

a. 161 R11
b. 160 R36
c. 275
d. none of these

10. Divide.

$37\overline{)7548}$

a. 240
b. 24
c. 204
d. none of these

11. Records cost $5 each. Tapes cost $9 each. How much do 4 records cost?

a. $56 **b.** $20
c. $14 **d.** none of these

12. Paul ran 400 meters 5 times in 4 minutes. How many meters did he average per minute?

a. 500 **b.** 100
c. 320 **d.** none of these

Decimals—Addition and Subtraction

2

Tenths and hundredths

Decimal fractions (or just **decimals**) are written using our place-value system. In this system the value of each place is one tenth the value of the place to its left. To write decimals, we extend the place-value system to the right of the ones place. The digits in this table show how many unit squares are shaded.

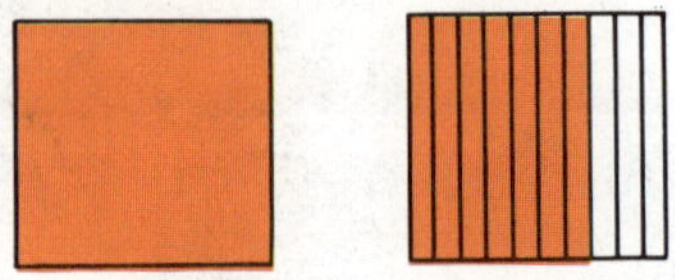

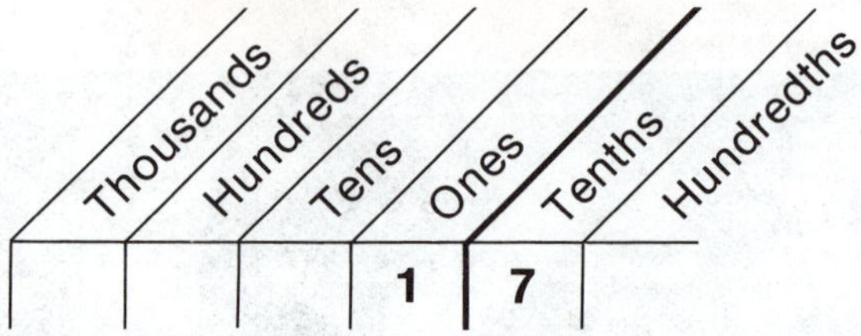

The standard numeral is written like this:

1.7

A decimal point separates the ones place from the tenths place.

Study this example.

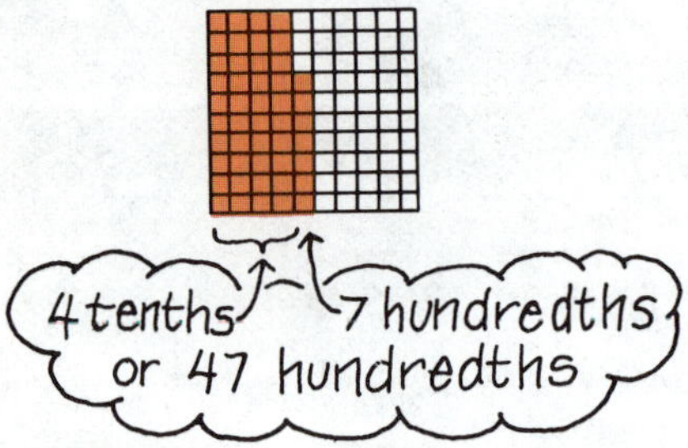

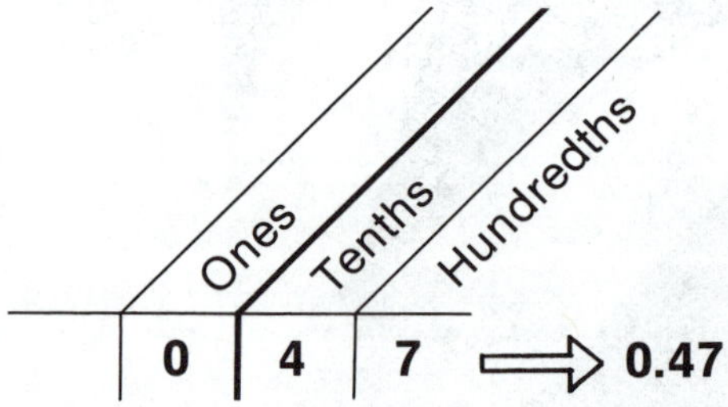

Read as "forty-seven hundredths."

EXERCISES

Read aloud.

1. 0.3 **2.** 0.03 **3.** 0.9 **4.** 0.09 **5.** 0.27 **6.** 0.58 **7.** 0.30

8. 2.6 **9.** 3.06 **10.** 5.36 **11.** 4.67 **12.** 24.35 **13.** 67.08 **14.** 235.43

How many unit squares are shaded?

15. 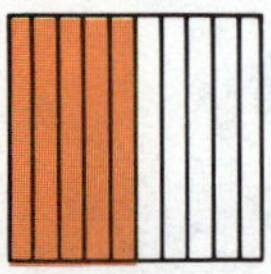

16. 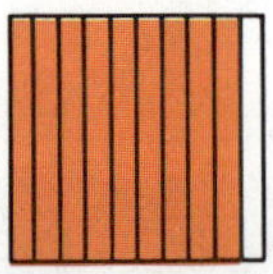

17. 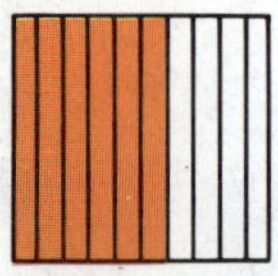

18. 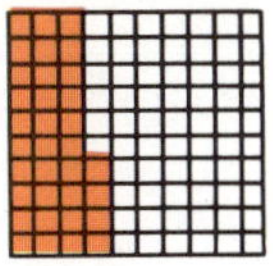

19. 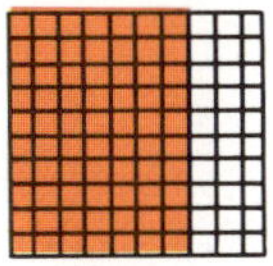

20. 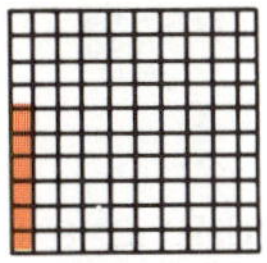

21. 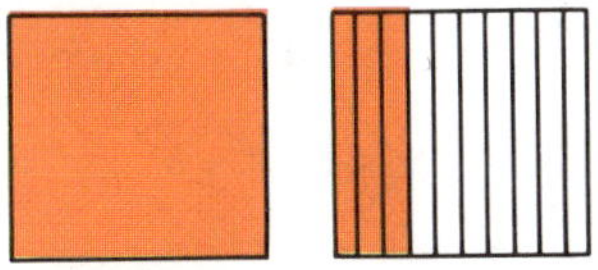

22. 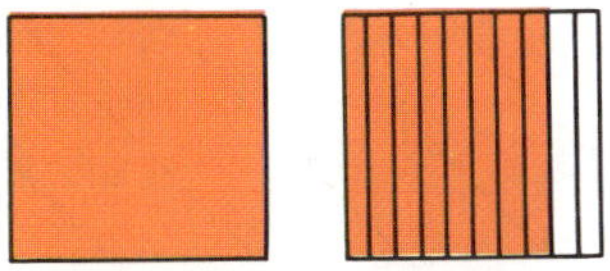

23. 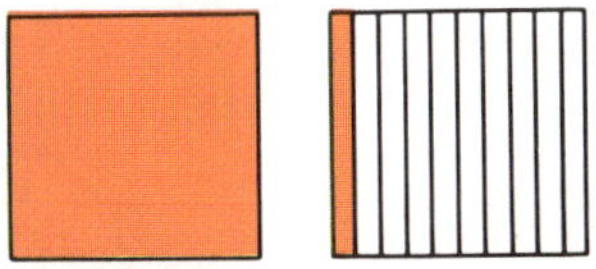

24. 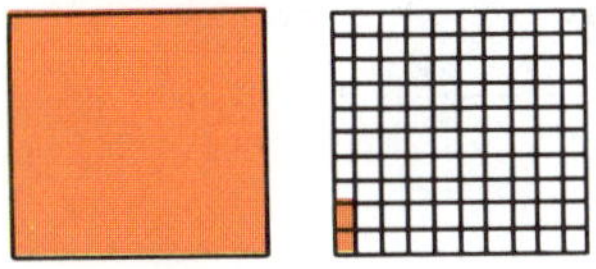

25. 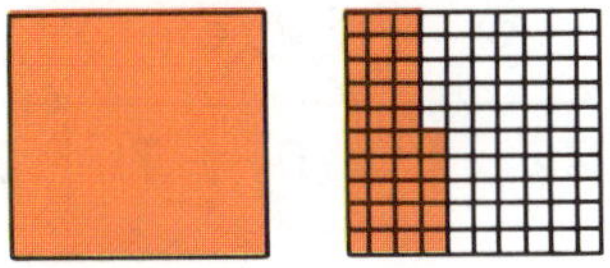

26. 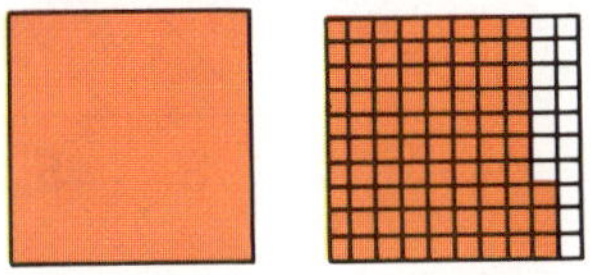

Give the standard numeral.

27. eight tenths
28. eight hundredths
29. eight hundred
30. two and three tenths
31. six and five hundredths
32. fourteen and fourteen hundredths
33. two hundred three
34. two hundred and three tenths
35. five and six hundredths
36. five hundred six
37. ten and three tenths
38. one hundred and three hundredths

Copy the numeral, and place the decimal point so that the statement makes sense.

39. "In a bicycle race I averaged 1575 miles per hour."
40. "It was 934° Fahrenheit on the day I went swimming."
41. "At the beginning of the seventh grade I weighed 9854 pounds."
42. "For lunch I ate 175 hamburgers."

More about decimals

The place-value system can be extended as far as needed in both directions.

Here is how to read a decimal:

1. Read the whole-number part. ⟶ thirty-five thousand, six hundred twenty-one
2. Say "and" for the decimal point. ⟶ and
3. Read the fraction part as if it were a whole number. Then say the name of the place of the last digit. ⟶ seven hundred thirty-five ten-thousandths

EXERCISES

Read aloud.

1. 0.008	**2.** 0.0008	**3.** 0.00008	**4.** 0.000008	**5.** 0.035
6. 0.0035	**7.** 0.0352	**8.** 0.165	**9.** 0.2734	**10.** 0.00231
11. 6.007	**12.** 0.603	**13.** 600.003	**14.** 0.827	**15.** 800.027
16. 627.0627	**17.** 43.1675	**18.** 5000.0005	**19.** 67.359	**20.** 45.0036

Write the standard numerals.

21. eight thousand	**22.** eight thousandths	**23.** nine hundredths
24. nine hundred	**25.** twenty-four thousandths	**26.** twenty-four hundredths

27. eighty-seven ten-thousandths

28. nine hundred twenty-six millionths

29. six hundred three thousandths

30. six hundred and three thousandths

31. five hundred forty thousandths

32. five hundred and forty thousandths

33. two hundred thousandths

34. two hundred and two hundredths

Look for a pattern.
Give the next three numbers.

35. 0.4, 0.5, 0.6, . . .

36. 1.2, 1.3, 1.4, . . .

37. 7.8, 7.9, 8.0, . . .

38. 5.6, 5.7, 5.8, . . .

39. 0.02, 0.03, 0.04, . . .

40. 0.27, 0.28, 0.29, . . .

41. 0.003, 0.004, 0.005, . . .

42. 0.057, 0.058, 0.059, . . .

43. 4.143, 4.144, . . .

44. 6.056, 6.057, . . .

Give a number that is between the two numbers.

45. 59.8, 60.0

46. 0.32, 0.34

47. 5.6, 5.7

48. 7.53, 7.54

49. 9.00, 9.01

50. 0.305, 0.306

★ 51. 2, 2.1

★ 52. 3.99, 4

53. DETECTIVE HUNT

On one of these two pages there is a number whose tens digit is 1 more than its hundredths digit. What is the number?

KEEPING SKILLS SHARP

Use estimation to find the incorrect answer.

1. a. $\begin{array}{r} 58 \\ \times 17 \\ \hline 986 \end{array}$

b. $\begin{array}{r} 413 \\ \times 203 \\ \hline 83{,}839 \end{array}$

c. $\begin{array}{r} 502 \\ \times 59 \\ \hline 37{,}618 \end{array}$

d. $\begin{array}{r} 714 \\ \times 386 \\ \hline 275{,}604 \end{array}$

Multiply.

2. $\begin{array}{r} 705 \\ \times 8 \\ \hline \end{array}$

3. $\begin{array}{r} 126 \\ \times 9 \\ \hline \end{array}$

4. $\begin{array}{r} 521 \\ \times 28 \\ \hline \end{array}$

5. $\begin{array}{r} 906 \\ \times 49 \\ \hline \end{array}$

6. $\begin{array}{r} 2837 \\ \times 242 \\ \hline \end{array}$

7. $\begin{array}{r} 5673 \\ \times 804 \\ \hline \end{array}$

Practice exercises

Match these numbers with the points on the number lines.

1. 6.3 **2.** 7.1

3. 5.9 **4.** 6.5

5. 5.4 **6.** 6.8

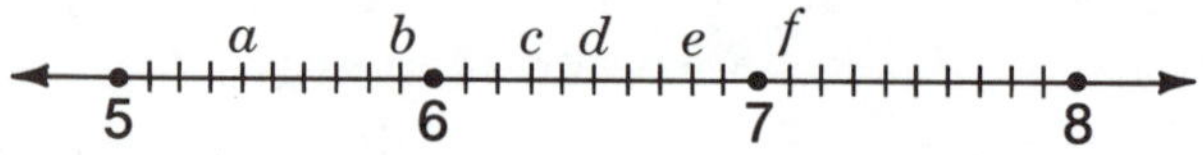

7. 0.5 **8.** 1.2

9. 0.7 **10.** 2.1

11. 0.8 **12.** 2.6

13. 0.26 **14.** 0.33

15. 0.09 **16.** 0.16

17. 0.19 **18.** 0.21

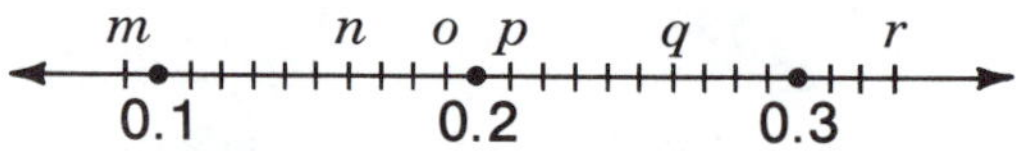

millions
↓
6,800,000

6.8 million

How many millions?

19. 8,200,000

20. 5,700,000

21. 12,800,000

22. 45,600,000

23. 68,750,000

Change to standard numerals.

★ **24.**

★ **25.**

Computers in stores

The packages for most products sold in stores are marked with a Universal Product Code (UPC). This combination of light and dark bars can be "read" by an optical scanner which sends the code to a computer. The computer translates the code into the number of the product, then uses the number to search its memory for the price of the product. This price is printed on the cash-register receipt.

The following code is similar to the UPC. It will help you understand how the UPC works.

SIMPLIFIED BAR CODE

Rule 1. There are three bar widths:

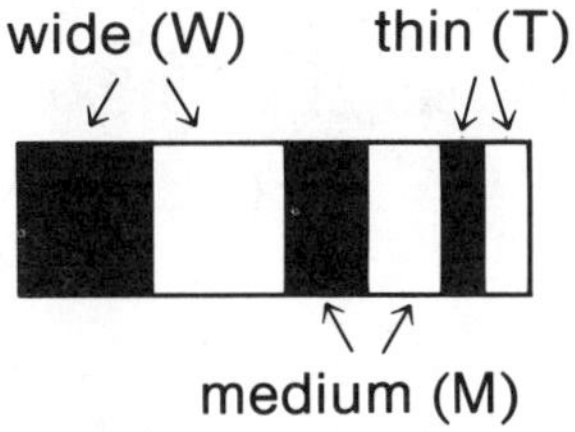

Rule 2. Bars alternate dark and light and are read from left to right.

Rule 3. Each four consecutive bars (dark-light-dark-light, starting from the left) represent a digit according to the width-sequence code given in the table.

CODE	
Width sequence	*Digit*
WMTT	0
MMMT	1
MTMM	2
TWTT	3
TTWM	4
TMWT	5
TTTW	6
TWTM	7
TMTW	8
WTTM	9

In this coded 6-digit number, the first digit has the width sequence TTWM and therefore represents 4.

1. What is the width sequence for the second digit?
2. What is the second digit?
3. What is the width sequence for each of the other digits?
4. What is the 6-digit number?

Comparing and ordering decimals

National League Batting Champs 1975–1980		
Year	*Player*	*Average*
1975	Bill Madlock	.354
1976	Bill Madlock	.339
1977	Dave Parker	.338
1978	Dave Parker	.334
1979	Keith Hernandez	.344
1980	Bill Buckner	.324

To compare the batting averages of Bill Madlock in 1976 and Dave Parker in 1977, we compare decimals.

We can compare these decimals by comparing the digits that are in the same places.

Bill Madlock (1976) — same — Dave Parker (1977)

.339 > **.338**

Bill Madlock had the greater batting average.

Here is another example of comparing decimals:

52.08 > **6.74**

EXERCISES

< or >?

1. 0.2 ● 0.3
2. 0.06 ● 0.05
3. 0.009 ● 0.008
4. 0.2 ● 0.02
5. 0.004 ● 0.03
6. 0.1 ● 0.08
7. 0.006 ● 0.2
8. 0.3 ● 0.004
9. 0.82 ● 0.85
10. 0.74 ● 0.71
11. 5.38 ● 5.83
12. 6.75 ● 6.57
13. 38.21 ● 38.54
14. 71.69 ● 71.7
15. 5.936 ● 59.36
16. 782.76 ● 780.99
17. 83.5 ● 8.76
18. 0.759 ● 1
19. 2 ● 1.899
20. 7.564 ● 7.099
21. 8.63 ● 16.58
22. 47.02 ● 9.385
23. 7.216 ● 72.16
24. 53.06 ● 5.899

Team Pitching	*Earned-Run Average*
Boston	3.99
New York	3.29
Kansas City	3.49
Chicago	3.94
California	3.89
Minnesota	4.05
Detroit	4.29
Oakland	3.29
Cleveland	3.84
Baltimore	3.17
Texas	3.90
Milwaukee	4.34

25. Which team had the lowest earned-run average?
26. Which team had the highest earned-run average?
27. Which had the lower earned-run average, Detroit or Milwaukee?
28. List the teams in order from lowest to highest earned-run average.

29. WHO'S ON FIRST?

The names of the first baseman, shortstop, catcher, and pitcher are Sid, Cindy, Dave, and Rita, but not necessarily in that order.

The pitcher's batting average is less than Cindy's average. Dave's batting average is highest, and he is a good friend of the shortstop. Rita wears a mask.

What is the name of the first baseman?

Rounding

Round decimals just as you would round whole numbers.

To round 12.73 to the nearest whole number, think about where 12.73 is on the number line.

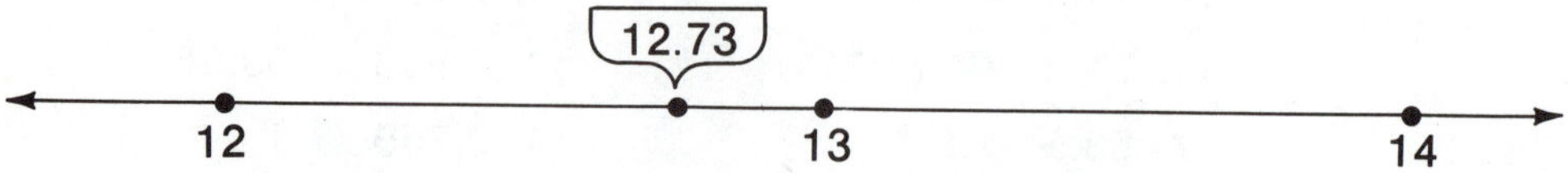

It is nearer 13. So, 12.73 rounded to the nearest whole number (or nearest one) is 13.

Here is a rounding shortcut.

Round to the nearest one.

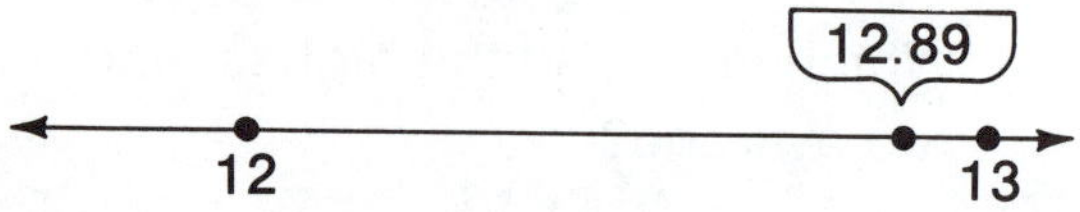

Round to this place.

12.89 **12.89**

Round to the nearest hundredth.

Round to this place.

12.902 **12.902**

EXERCISES

Round to the nearest one.

1. 3.4 **2.** 3.7 **3.** 15.1 **4.** 15.9 **5.** 2.2 **6.** 2.9

7. 10.9 **8.** 11.3 **9.** 12.6 **10.** 22.15 **11.** 9.09 **12.** 18.61

13. 15.50 **14.** 26.08 **15.** 30.75 **16.** 43.62 **17.** 39.74 **18.** 41.49

Round to the nearest tenth.

19. 0.73 **20.** 0.77 **21.** 0.84 **22.** 0.53 **23.** 0.69 **24.** 0.91

25. 0.95 **26.** 2.38 **27.** 4.09 **28.** 6.55 **29.** 8.34 **30.** 9.05

31. 18.58 **32.** 13.62 **33.** 25.74 **34.** 38.75 **35.** 52.53 **36.** 49.96

Round to the nearest hundredth.

37. 4.673 **38.** 21.987 **39.** 5.635 **40.** 4.8133 **41.** 2.7758

42. 0.078 **43.** 0.0977 **44.** 19.989 **45.** 30.205 **46.** 24.652

47. 32.096 **48.** 27.357 **49.** 18.444 **50.** 42.721 **51.** 34.398

Round to the nearest thousandth.

52. 2.8359 **53.** 5.7461 **54.** 8.4965 **55.** 0.8295 **56.** 3.6182

57. 6.0996 **58.** 7.7598 **59.** 4.2996 **60.** 5.2651 **61.** 3.3842

62. 15.8065 **63.** 12.3142 **64.** 24.7896 **65.** 33.4006 **66.** 47.3799

The speed record for a bicycle is 204.734 kilometers per hour. (A lead car had a mounted windshield to break the wind.) Round the number to the nearest

67. hundredth.

68. whole number.

69. tenth.

A speed record for a skateboard was 86.886 kilometers per hour. Round the number to the nearest

70. tenth.

71. hundredth.

72. whole number.

PROJECT

Look up the winning speed records for the Indianapolis 500. In what ways have the speeds changed? Make a line graph to show what you have found.

Adding decimals

To add decimals, we line up the decimal points and add in the same way that we add whole numbers.

5.6 + 4.97 = ?

Step 1. Add.

$$\begin{array}{r} 5.60 \\ +4.97 \\ \hline 7 \end{array}$$

Step 2. Add and regroup.

$$\begin{array}{r} \overset{1}{5}.60 \\ +4.97 \\ \hline .57 \end{array}$$

Step 3. Add.

$$\begin{array}{r} \overset{1}{5}.60 \\ +4.97 \\ \hline 10.57 \end{array}$$

EXERCISES

Add.

1. 4.73 +2.13	**2.** 15.9 +23.5	**3.** 0.73 +0.187	**4.** 27.56 +38.2	**5.** 16.05 +38.57
6. 5.37 +2.9	**7.** 90.6 +53.4	**8.** 0.425 +0.295	**9.** 6.711 +0.32	**10.** 34.47 +96.36
11. 7.84 +2.67	**12.** 7.56 +8.17	**13.** 18.90 +3.5	**14.** 29.76 +31.90	**15.** 762.8 +375.74

16. 3.6586 + 9.825
Hint: Line up decimal points.

17. 76.7 + 38.57

18. 79.46 + 8.98

19. 56.37 + 6.57

20. 70.69 + 27.39

21. 6543.2 + 38.2

22. 781.06 + 74.68

23. 9.2305 + 9.1753

24. 0.72961 + 0.29683

25. 834.75 + 42.8

26. 0.02 + 31.4

27. 3.6 + 0.94

28. 48 + 5.06 29. 25.8 + 3.79 30. 226.4 + 15.84

31. 119.6 + 289 32. 39.54 + 7.621 33. 4.328 + 0.7493

34. 0.593 + 16.48 35. 0.0295 + 0.0088 36. 3.526 + 48.79

Solve.

37. To practice for the cross-country team, Carol ran 4.5 kilometers each morning and 6.8 kilometers each evening. How far did she run each day?

38. John ran the Deer Trail Course in 12.84 minutes and the Pine Valley Course in 11.96 minutes. How many minutes did he run?

39. Kim ran an 11.81-kilometer course. Sarah ran a 5.78-kilometer course twice. Who ran farther?

40. During the first three days of training, each member of the cross-country team was supposed to run 35 kilometers. John ran a 17.28-kilometer course twice. Did he run far enough?

41. Carol ran an 11.81-kilometer course and a 9.65-kilometer course. Kim ran a 10.68-kilometer course twice. Who ran farther?

42. OUT OF BREATH

Anita ran 5 kilometers more than Carla. Beth ran 2 kilometers less than Carla. The 3 girls ran a total of 21 kilometers. How many kilometers did Beth run?

KEEPING SKILLS SHARP

Estimate to find the incorrect answer.

1. a. $\begin{array}{r} 37 \\ 42\overline{)1554} \end{array}$

b. $\begin{array}{r} 201 \\ 56\overline{)1176} \end{array}$

c. $\begin{array}{r} 302 \\ 17\overline{)5134} \end{array}$

d. $\begin{array}{r} 401 \\ 68\overline{)27{,}268} \end{array}$

Divide.

2. $8\overline{)5928}$
3. $6\overline{)6832}$
4. $4\overline{)7906}$
5. $9\overline{)8396}$
6. $35\overline{)2874}$
7. $52\overline{)1930}$
8. $47\overline{)3555}$
9. $66\overline{)25{,}028}$
10. $157\overline{)63{,}825}$
11. $243\overline{)43{,}016}$

Subtracting decimals

The average yearly precipitation in Birmingham, Alabama, is 135 centimeters and in Honolulu it is 55.7 centimeters.

We subtract to find the difference.

Step 1. Line up the decimal points.

Since 135 = 135.0, write a zero in the tenths place.

```
 135.0
 -55.7
```

Step 2. Regroup and subtract.

```
    4 1
 135.0
 -55.7
    .3
```

Step 3. Regroup and subtract.

```
  2 14 1
 135.0
 -55.7
   9.3
```

Step 4. Subtract.

```
  2 14 1
 135.0
 -55.7
  79.3
```

The average yearly precipitation in Birmingham is 79.3 centimeters more than in Honolulu.

EXERCISES

Subtract.

1. 59.6 − 13.5
2. 8.42 − 0.26
3. 72.6 − 7.8
4. 9.68 − 3.14
5. 85.3 − 28.9
6. 0.314 − 0.109
7. 60.2 − 28.9
8. 5.19 − 3.74
9. 0.302 − 0.195
10. 9.64 − 3.98

11. 5.918 − 0.265

12. 72.95 − 3.98

13. 0.7426 − 0.4285

14. 86.52 − 32.97

15. 646.4 − 289.6

16. 61.83 − 59.34

17. 6.882 − 2.817

18. 6.752 − 0.395

19. 93.71 − 0.78

20. 87.25 − 3.95

21. 70.2 − 26.14

22. 581.68 − 59.38

23. 74.729 − 27.351

24. 723.16 − 56.87

25. 95.305 − 4.26

26. 16.3 − 5.491

27. 28.9 − 16.03

28. 90.06 − 5.178

29. 263.3 − 59.648

30. 520.4 − 9.871

Average Monthly Precipitation
(in centimeters)

City	*Jan.*	*Feb.*	*Mar.*	*Apr.*	*May*	*June*	*July*	*Aug.*	*Sept.*	*Oct.*	*Nov.*	*Dec.*
Birmingham, Ala.	12.7	13.5	15.3	11.4	8.6	10.2	13.2	12.4	8.5	7.6	8.9	12.7
Chicago, Ill.	4.8	4.1	6.9	7.6	9.4	10.4	8.6	8.1	6.9	7.1	5.6	4.8
Honolulu, Hawaii	9.7	8.4	7.4	3.3	2.5	0.8	1.0	2.3	2.5	4.6	5.6	7.6
Little Rock, Ark.	13.2	10.9	12.2	12.4	13.5	9.1	8.4	7.1	8.1	7.4	10.4	10.4
Miami, Fla.	5.1	4.8	5.8	9.9	16.3	18.8	17.3	17.8	24.1	20.8	7.1	4.3
New York, N.Y.	8.4	7.1	10.2	8.6	9.4	8.4	9.4	11.2	9.9	7.8	8.6	8.4

Use the chart to solve.

31. What is the yearly precipitation in Chicago?

32. Which city gets more precipitation in a year, Chicago or Honolulu?

33. How much precipitation does New York get during the first half of the year? During the second half? How much more during the second half?

34. Which city gets more precipitation in a year, Little Rock or Miami? How much more?

Estimating sums and differences

If you estimate first, you will know whether your answer is reasonable.

$$\begin{array}{r} 6 \\ +5 \\ \hline 11 \end{array} \quad \begin{array}{r} 6.1 \\ +4.96 \\ \hline 11.06 \end{array} \qquad\qquad \begin{array}{r} 15 \\ -9 \\ \hline 6 \end{array} \quad \begin{array}{r} 15.40 \\ -9.27 \\ \hline 6.13 \end{array}$$

EXERCISES

First estimate each sum. Then add.

1. $3.55 + 2.5$
2. $72.6 + 39.4$
3. $0.23 + 4.583$
4. $5.41 + 23.8$
5. $18.917 + 9.6$
6. $9.66 + 3.15$
7. $81.4 + 0.56$
8. $4.307 + 0.853$
9. $9.85 + 15$
10. $6.73 + 2.496$

First estimate each difference. Then subtract.

11. $5.37 - 3.85$
12. $29.4 - 0.6$
13. $2.18 - 1.4$
14. $3.92 - 2.18$
15. $62.5 - 36.974$
16. $7.21 - 2.943$
17. $74 - 6.951$
18. $5.84 - 0.39$
19. $3.43 - 2.583$
20. $6.89 - 2.948$

First estimate. Then compute.

21. $5.9 + 38.74$
22. $63.5 - 29$
23. $85 - 3.645$
24. $(2.8 + 3) - 1.64$
25. $(5.84 - 2.6) + 0.59$
26. $(7.4 - 2.34) - 2$
27. $(9.6 + 7.8) - 6.4$
28. $9.6 + (7.8 - 6.4)$
29. $(15.2 - 9.48) - 4.6$
30. $15.2 - (9.48 - 4.6)$
31. $(47.6 - 3.82) + 15.79$
32. $47.6 - (3.82 + 15.79)$

Estimate to find the two incorrect answers. Then compute the correct answers.

★ 33. $(48.3 - 16.7) + 8.3 = 39.9$

★ 34. $64.9 - (8.5 + 23.7) = 42.7$

★ 35. $(9.6 - 3.25) + (8.7 - 3.96) = 11.09$

★ 36. $(18.2 + 4.6) - (9.21 + 8.5) = 15.43$

SKILL GAME

Tell which is closer to the target number, 50.

1. a. [5][2]. + [7].[6] − [8].[4] → 50
b. [7].[8] + [4][5]. − [2].[6] → 50

2. a. [9].[3] + [4][7]. − .[6][8] → 50
b. [7][8]. + [6].[9] − [4][3]. → 50

Play the game.

1. Make a table like this.

[ ][ ] + [ ][ ] − [ ][ ] → 50

2. A game leader will mix up these digit cards. The leader draws a card, replaces it, and continues until 6 cards have been drawn.

[0] [1] [2] [3] [4] [5] [6] [7] [8] [9]

3. As each card is drawn, write the digit somewhere in your table.

4. Place a decimal point anywhere you wish in each pair of digits.
For example:

John's table:

[6][0]. + .[4][7] − [6].[3] → 50

Karen's table:

[4][6]. + [3].[6] − .[0][7] → 50

5. The player closest to the target number, 50, wins the game. Who came closer, John or Karen?

Mathematics in careers

Payroll clerks use computers to prepare employee paychecks. After computing the employee's salary, the payroll clerk deducts some money for federal income tax and social security. (Some employees may choose to have more money deducted from their salary for such things as savings bonds, group medical insurance, group life insurance, etc.)

The following tables show the amounts to be deducted for federal income tax and social security.

Federal Income Tax Withholding (Weekly Payroll)

Not Married				Married						
Weekly salary		Number of exemptions		Weekly salary		Number of exemptions				
At least	But less than	0	1	At least	But less than	0	1	2	3	4
$200	$210	$33.60	$29.10	$200	$210	$26.20	$22.70	$19.20	$15.80	$12.30
210	220	36.20	31.20	210	220	28.10	24.50	21.00	17.60	14.10
220	230	38.80	33.80	220	230	30.20	26.30	22.80	19.40	15.90
230	240	41.40	36.40	230	240	32.30	28.30	24.60	21.20	17.70
240	250	44.00	39.00	240	250	34.40	30.40	26.40	23.00	19.50
250	260	46.60	41.60	250	260	36.50	32.50	28.50	24.80	21.30
260	270	49.20	44.20	260	270	38.60	34.60	30.60	26.60	23.10
270	280	51.80	46.80	270	280	40.70	36.70	32.70	28.60	24.90
280	290	54.80	49.40	280	290	42.80	38.80	34.80	30.70	26.70
290	300	57.80	52.10	290	300	45.10	40.90	36.90	32.80	28.80

Social Security Tax Withholding (Weekly Payroll)

At least	But less than	Tax withheld	At least	But less than	Tax withheld	At least	But less than	Tax withheld
$199.50	$201.50	$13.43	$233.50	$235.50	$15.71	$267.50	$269.50	$17.99
201.50	203.50	13.57	235.50	237.50	15.85	269.50	271.50	18.12
203.50	205.50	13.70	237.50	239.50	15.98	271.50	273.50	18.26
205.50	207.50	13.84	239.50	241.50	16.11	273.50	275.50	18.39
207.50	209.50	13.97	241.50	243.50	16.25	275.50	277.50	18.53
209.50	211.50	14.10	243.50	245.50	16.38	277.50	279.50	18.66
211.50	213.50	14.24	245.50	247.50	16.52	279.50	281.50	18.79
213.50	215.50	14.37	247.50	249.50	16.65	281.50	283.50	18.93
215.50	217.50	14.51	249.50	251.50	16.78	283.50	285.50	19.06
217.50	219.50	14.64	251.50	253.50	16.92	285.50	287.50	19.20
219.50	221.50	14.77	253.50	255.50	17.05	287.50	289.50	19.33
221.50	223.50	14.91	255.50	257.50	17.19	289.50	291.50	19.47
223.50	225.50	15.04	257.50	259.50	17.32	291.50	293.50	19.60
225.50	227.50	15.18	259.50	261.50	17.45	293.50	295.50	19.73
227.50	229.50	15.31	261.50	263.50	17.59	295.50	297.50	19.87
229.50	231.50	15.44	263.50	265.50	17.72	297.50	299.50	20.00
231.50	233.50	15.58	265.50	267.50	17.86	299.50	301.50	20.10

EXERCISES

Suppose you are the payroll clerk. Compute the take-home pay for the week for each employee.

(*Hint:* Deduct the income tax and social security tax from the weekly salary.)

1. Name: Dean Illig (married)
Exemptions: 3
Weekly salary: $268

2. Name: James Guerin (single)
Exemptions: 1
Weekly salary: $285.75

3. Name: Vincent Vezza (married)
Exemptions: 2
Weekly salary: $285.75

4. Name: Pamela Dickey (married)
Exemptions: 4
Weekly salary: $279.65

5. Name: Sandra Kaiser (single)
Exemptions: 0
Weekly salary: $256.80

★ **6.** Name: John Atkocaitis (married)
Exemptions: 3
Weekly salary: 40 hours @ $6.92 an hour

Problem solving

In drag racing, the car that travels 440 yards in the shortest time wins the race.

Two decimals are used to show a car's performance. They are written like this:

8.72/162.54

The first decimal tells the time in seconds that the car took to travel the distance. This time is called the *elapsed time.* The second decimal tells the speed (in miles per hour) of the car when it reached the finish line. This speed is called the *terminal speed.*

Here are the results of one race:

Car	*Driver*	*Time/Speed*
Comet	Lee	8.71/164.38
Early Finish	Cantrell	8.68/165.93
Good Bye	Walker	8.64/170.08
Lightning II	Staton	8.67/167.43
Power Plus	Barry	8.73/160.42
Smokie	Burbank	8.72/161.57

1. How many cars finished the race in less than 8.70 seconds?
2. How many cars had a terminal speed that was more than 164.5 miles per hour?
3. Which car won the race?
4. What was the terminal speed of the winning car?
5. What was the elapsed time of the car that placed last?
6. Find the difference of the elapsed times of the first- and last-place cars.
7. What was the difference between the terminal speeds of the first- and the second-place cars?
8. What was the difference between the terminal speeds of the second- and third-place cars?

Here are the results of another race:

Car	*Driver*	*Time/Speed*
So Long	Austin	8.95/174.05
Flash By	Summers	9.04/173.65
Silver Streak	Hicks	8.73/169.33
Quick Shot	Devine	8.61/170.25
Lightning I	Mahoney	8.83/168.71

9. Who won in this class?
10. Who came in second? Who came in last?
11. By how many seconds did Austin beat Summers?
12. Who had the fastest terminal speed?
13. How much faster was Devine's terminal speed than Mahoney's?
14. What was the difference between the fastest and second-slowest terminal speeds?
15. Did the driver with the fastest terminal speed win the race?
16. ★ Explain why the driver with the fastest terminal speed might not win the race.

CHAPTER CHECKUP

Give the standard numeral. [pages 66–69]

1. two and six tenths
2. six and three hundredths
3. forty-two thousandths
4. two hundred six thousandths
5. two hundred and six thousandths
6. five hundred four ten-thousandths

< **or** >**?** [pages 72–73]

7. 0.03 ● 0.004
8. 73.2 ● 73.4
9. 2.999 ● 3
10. 5.7346 ● 5.3382

Round 15.6599 to the nearest [pages 74–75]

11. tenth.
12. one.
13. hundredth.
14. thousandth.

Add. [pages 76–77, 80–81]

15. $\begin{array}{r} 5.7 \\ +3.8 \\ \hline \end{array}$

16. $\begin{array}{r} 3.98 \\ +0.47 \\ \hline \end{array}$

17. $\begin{array}{r} 2.08 \\ +5.26 \\ \hline \end{array}$

18. 3.48 + 1.962

19. 25.37 + 0.4876

Subtract. [pages 78–81]

20. $\begin{array}{r} 7.4 \\ -3.8 \\ \hline \end{array}$

21. $\begin{array}{r} 5.93 \\ -2.86 \\ \hline \end{array}$

22. $\begin{array}{r} 30.2 \\ -19.5 \\ \hline \end{array}$

23. 4.32 − 1.837

24. 25 − 2.68

Solve. [pages 82–85]

25.

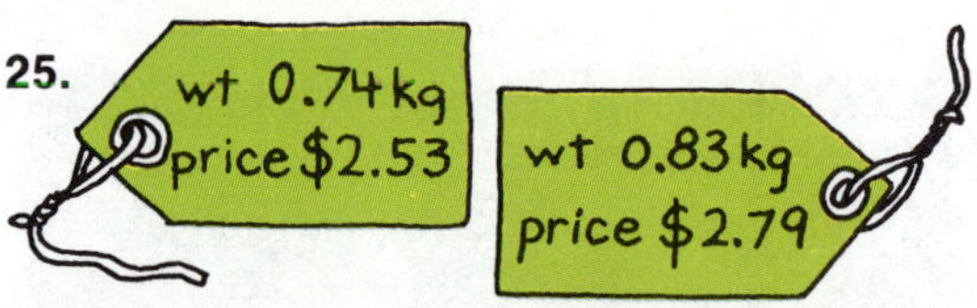

a. What is the total cost?
b. What is the total weight?

26.

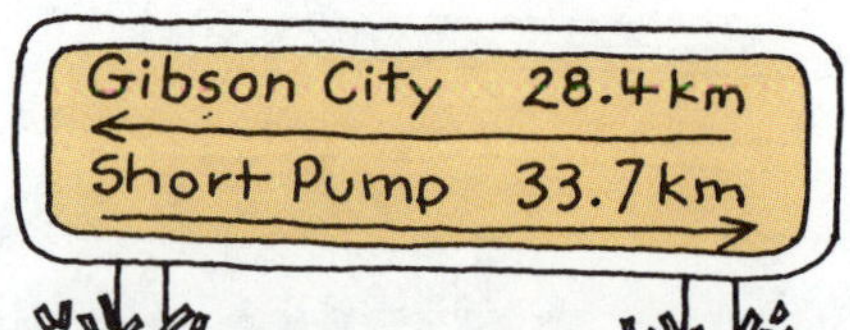

How far apart are the two towns?

CHAPTER PROJECT

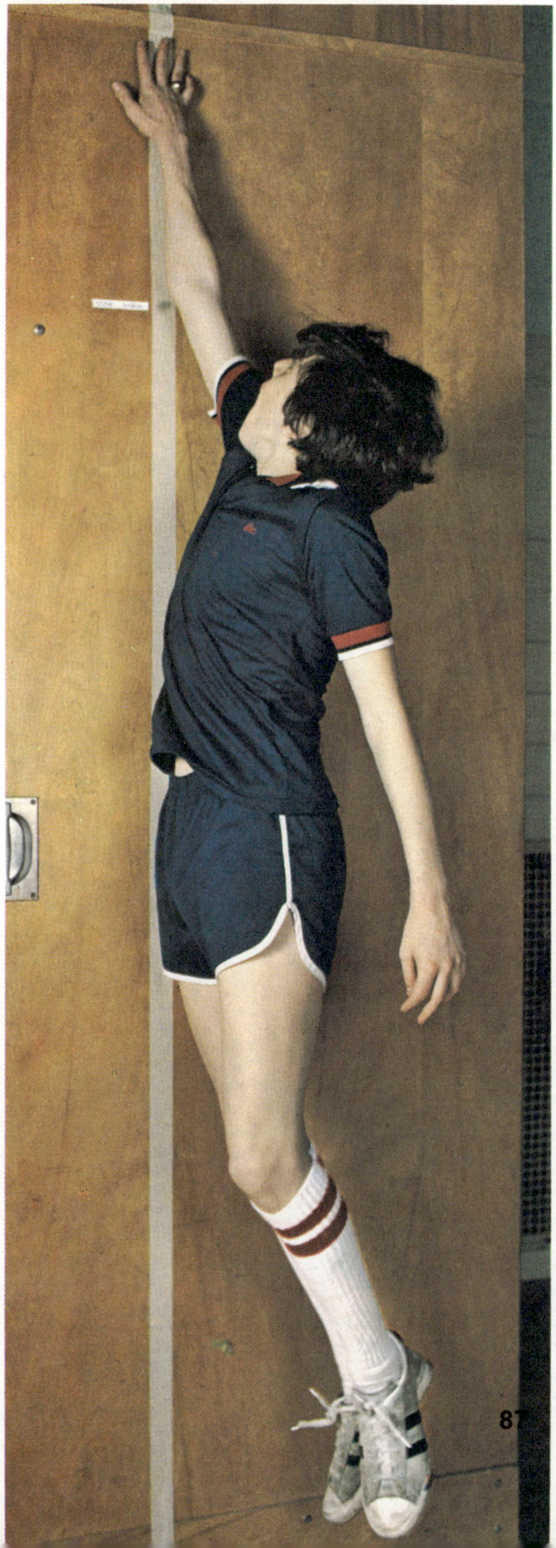

Think about standing under a basketball net and jumping straight up as high as you can. The distance that you could jump beyond your normal reach is your vertical jump. First guess what it would be. Then follow these steps to measure it.

1. Blacken the end of a middle finger with a pencil. While standing *flat-footed,* mark as high as you can reach on a paper strip that is taped to the wall. Measure the height of your mark to the nearest centimeter. Convert the height to meters.

2. From a standing position, jump to see how high you can mark on the same paper. Measure the distance to the nearest centimeter and convert it to meters.

3. Subtract to find your vertical jump in meters.

4. **a.** When all your classmates have measured their vertical jumps, list them on the chalkboard. Show the results on a graph like this:

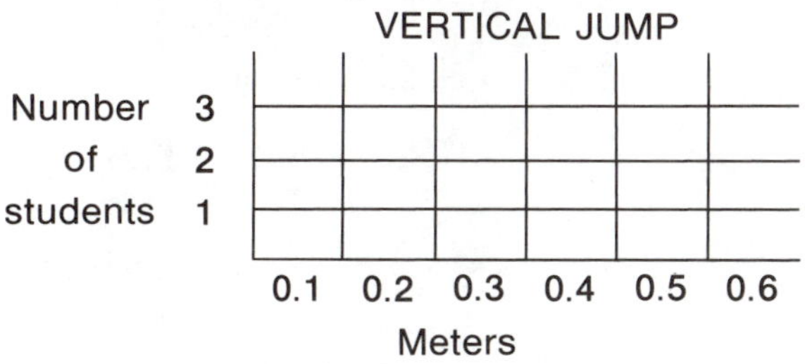

   b. List some facts shown by your graph.

CHAPTER REVIEW

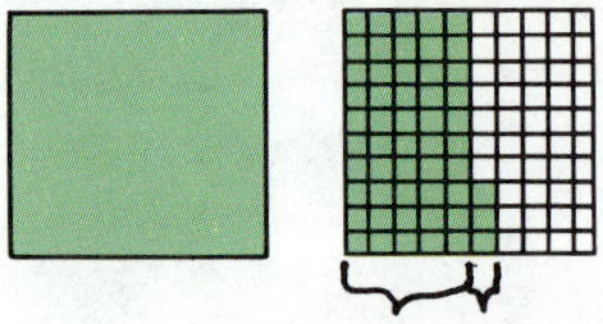

1 one — 5 tenths — 3 hundredths

1.53

How many unit squares are shaded?

1. 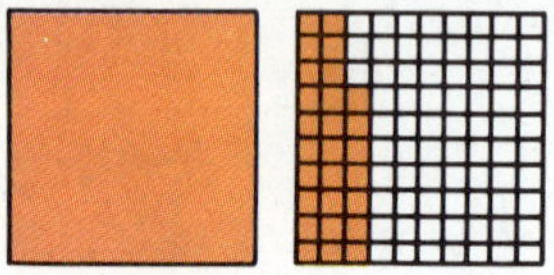

2. 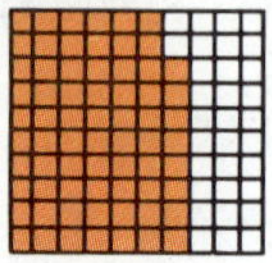

3. 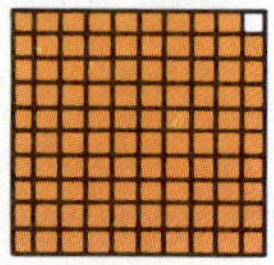

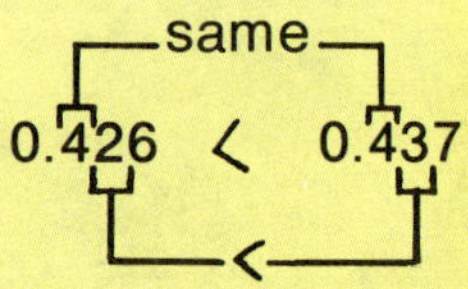

< or >?

4. 0.3 ● 0.7
5. 0.85 ● 0.76
6. 0.37 ● 0.21
7. 0.662 ● 0.667
8. 1.42 ● 1.52
9. 12.76 ● 12.73

Round to the nearest 1.

ones place ↓
43.6
↑ 5 or greater

44

Round to the nearest 1.

10. 26.3
11. 38.9
12. 47.5
13. 6.09
14. 12.499
15. 117.511

Round to the nearest 0.1.

16. 0.67
17. 0.61
18. 0.65
19. 13.341
20. 28.784
21. 31.976

$$\begin{array}{r} \overset{1}{4}.6 \\ +3.7 \\ \hline 8.3 \end{array}$$

Add.

22. $\begin{array}{r} 5.9 \\ +8.7 \\ \hline \end{array}$

23. $\begin{array}{r} 6.57 \\ +3.88 \\ \hline \end{array}$

24. 5.25 + 0.776

25. 0.843 + 27.9

$$\begin{array}{r} \overset{5}{\cancel{6}}.\overset{9\,1}{\cancel{0}}3 \\ -2.87 \\ \hline 3.16 \end{array}$$

Subtract.

26. $\begin{array}{r} 8.77 \\ -3.98 \\ \hline \end{array}$

27. $\begin{array}{r} 0.608 \\ -0.419 \\ \hline \end{array}$

28. 36.05 − 2.166

29. 14.035 − 6.27

CHAPTER CHALLENGE

•CLUES•

1. Which weighs more, the football or the soccer ball? How much more?
2. Which weighs more, the baseball or the football? How much more?
3. Which weighs more, the baseball or the soccer ball? How much more?
4. What is the weight of the baseball?
5. What is the weight of the football?
6. What is the weight of the soccer ball?

1. a b c d 2. a b c d 5. a b c d 6. a b
7. a b c d 8. a b c d 9. a b c d 10. a b c d 11. a b c d 12. a b

MAJOR CHECKUP
STANDARDIZED FORMAT

Choose the correct letter.

1. The standard numeral for three hundred million, forty-two is

a. 342,000,000
b. 300,000,042
c. 3,000,042
d. none of these

2. Round 783,499 to the nearest thousand.

a. 783,000
b. 784,000
c. 780,000
d. none of these

3. Add.

$$\begin{array}{r} 2965 \\ +4763 \\ \hline \end{array}$$

a. 6628
b. 7628
c. 7728
d. none of these

4. Subtract.

$$\begin{array}{r} 6003 \\ -4297 \\ \hline \end{array}$$

a. 1706
b. 2706
c. 2816
d. none of these

5. Multiply.

$$\begin{array}{r} 47 \\ \times 28 \\ \hline \end{array}$$

a. 1216
b. 470
c. 1316
d. none of these

6. Divide.

$6\overline{)3716}$

a. 619 R2
b. 620 R6
c. 618
d. none of these

7. The standard numeral for two hundred and nine thousandths is

a. 200.009
b. 0.209
c. 0.0209
d. none of these

8. Round 67.751 to the nearest one.

a. 67
b. 67.8
c. 68
d. none of these

9. Add.

3.98 + 42.7

a. 46.68
b. 82.5
c. 8.25
d. none of these

10. Subtract.

$$\begin{array}{r} 5.007 \\ -0.389 \\ \hline \end{array}$$

a. 5.718
b. 5.382
c. 4.618
d. none of these

11. Records cost $5 each. Tapes cost $9 each. How much more do 2 records cost than 1 tape?

a. $1 **b.** $4
c. $19 **d.** none of these

12. Alan bought 5 12¢ stamps and 20 18¢ stamps. What was the total cost of the stamps?

a. $3 **b.** $3.60
c. $7.50 **d.** none of these

Decimals—Multiplication and Division

Multiplying decimals

Janet drove 19.5 kilometers per hour for 3.5 hours. How far did she drive?

We can multiply to find out how far she drove.

To estimate the product, we can round each number to the nearest whole number and multiply.

$$\begin{array}{r} 19.5 \\ \times 3.5 \\ \hline \end{array} \qquad \begin{array}{r} 20 \\ \times 4 \\ \hline 80 \end{array}$$

The product should be about 80.

Here is how to find the product of two decimals:

Step 1. Multiply the numbers as you would multiply whole numbers.

$$\begin{array}{r} 19.5 \\ \times 3.5 \\ \hline 975 \\ 585 \\ \hline 6825 \end{array}$$

Step 2. Count the digits to the right of the decimal points.

$$\begin{array}{rl} 19.5 & \text{1 digit} \\ \times 3.5 & \text{+1 digit} \\ \hline 975 & \\ 585 & \\ \hline 6825 & \text{2 digits} \end{array}$$

Step 3. Count off the same number of digits to place the decimal point in the product.

$$\begin{array}{r} 19.5 \\ \times 3.5 \\ \hline 975 \\ 585 \\ \hline 68.25 \end{array}$$

Is the product "near" the estimate?

Janet drove 68.25 kilometers.

EXERCISES

Choose the product by estimating.

1. 3.9×6.8

a. 0.2652
b. 2.652
c. 26.52

2. 90.6×0.75

a. 0.67950
b. 67.950
c. 679.50

3. 49.6×11

a. 0.5456
b. 54.56
c. 545.6

Multiply.

4. 3.9×0.8	5. 5.7×0.1	6. 6.3×0.9	7. 7.8×0.8	8. 7.2×0.2	9. 9.1×0.6
10. 5.26×0.2	11. 3.28×0.5	12. 7.42×0.1	13. 9.06×0.4	14. 7.52×0.9	15. 3.91×0.3
16. 6.7×0.42	17. 0.57×3.5	18. 0.29×0.64	19. 79×6.3	20. 8.2×0.51	21. 65×0.39
22. 368×2.7	23. 5.04×32	24. 67.3×4.1	25. 5.82×0.73	26. 6.42×8.1	27. 75.9×0.36
28. 7.52×1.24	29. 9.38×1.03	30. 26.1×2.41	31. 5.18×72.5	32. 93.6×29.3	33. 8.53×6.78

Solve.

34. One lap around the Little Indy track is 1.2 kilometers. How far is 2.5 laps around the track?

35. The winning cart averaged 37.5 kilometers per hour. It took 1.6 hours to complete the race. How many kilometers long was the race?

36. It costs $3.60 an hour to drive on the Little Indy practice track. How much would it cost to practice for 1.75 hours?

37. In a 48.5-kilometer race, the winning cart averaged 32.4 kilometers per hour for the first 1.25 hours. Then it made a pit stop. How many kilometers did it have to go after the pit stop?

38. **SWITCH IT**

Switch two numbers so that the product along each side of the triangle is 6.

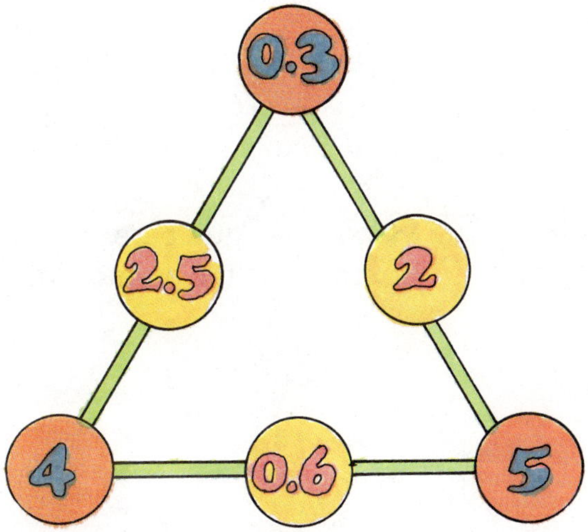

Zeros in multiplication

Sometimes when you multiply decimals you must write one or more zeros in the product in order to put the decimal point in the correct place. Study these examples.

$$\begin{array}{rl} \mathbf{0.06} \leftarrow & 2 \text{ digits} \\ \times \mathbf{0.5} \leftarrow & +1 \text{ digit} \\ \hline 0.030 \leftarrow & 3 \text{ digits} \end{array} \qquad \begin{array}{rl} \mathbf{0.32} \leftarrow & 2 \text{ digits} \\ \times \mathbf{0.03} \leftarrow & +2 \text{ digits} \\ \hline 0.0096 \leftarrow & 4 \text{ digits} \end{array}$$

EXERCISES

Multiply.

1. 0.02×0.3	**2.** 0.11×0.5	**3.** 2.1×0.04	**4.** 3.1×0.12	**5.** 0.67×0.03	**6.** 0.05×0.4
7. 0.35×3.6	**8.** 0.35×0.6	**9.** 0.79×0.08	**10.** 0.35×0.006	**11.** 55.7×0.38	**12.** 5.76×5.4
13. 55.7×2.31	**14.** 5.7×0.008	**15.** 0.002×0.13	**16.** 0.006×0.05	**17.** 0.538×12.7	**18.** 58.9×3.56
19. 38.9×2.84	**20.** 7.56×3.06	**21.** 9.21×71.4	**22.** 0.704×55.6	**23.** 36.5×1.83	**24.** 0.825×26.5
25. 3.78×1.52	**26.** 0.942×3.75	**27.** 68.3×2.91	**28.** 0.754×0.169	**29.** 3.82×61.5	**30.** 3.82×19.6

31. Place the numbers in the boxes to get the greatest possible product.

Problem solving

1. An object on the moon weighs about 0.17 of what it would weigh on earth. How much would a 185-pound astronaut weigh on the moon?

2. The astronauts of Apollo 11 collected moon rocks that weighed 49 pounds back on earth. How much did these rocks weigh on the moon? (See exercise 1.)

3. A lunar lander weighs 15,235 pounds on earth and 2590 pounds on the moon. How much less does it weigh on the moon?

4. The diameter of the moon is 2160 miles and the diameter of the earth is 7930 miles. How many miles greater is the diameter of the earth?

5. At its closest point (perigee) the moon is 221,588 miles from earth. At its farthest point (apogee) it is 252,862 miles from earth. How many miles closer is the moon at perigee than at apogee?

6. The average distance from the earth to the moon (center to center) is 14,005 miles less than the distance at apogee. What is the average distance between earth and moon? (See exercise 5.)

7. The moon travels around the earth at an average speed of 2299 miles per hour. How far does the moon travel around the earth in a day?

★ 8. The moon takes about 29 days 12 hours to complete one orbit of the earth. How long is the orbital path? (See exercise 7.)

★ 9. Apollo 11 stayed on the moon for 21 hours 36 minutes 21 seconds. Apollo 17 stayed a record 75 hours. How much longer was Apollo 17 on the moon?

★ 10. In Kansas City the moon rose at about 8:00 P.M. on July 1 and at about 8:50 P.M. on July 2. When would the moon next rise between 8:00 P.M. and 9:00 P.M. in Kansas City?

Some multiplication shortcuts

When you multiply by 10, 100, or 1000, you get a product that is greater than the number you multiplied.

Multiplying by 10

$$\begin{array}{r} 2.783 \\ \times 10 \\ \hline 27.830 \end{array}$$

$2.783 \times 10 = 27.83$

Multiplying by 100

$2.783 \times 100 = 278.3$

Multiplying by 1000

$2.783 \times 1000 = 2783.$

When you multiply by 0.1, 0.01, or 0.001, you get a product that is less than the number you multiplied.

Multiplying by 0.1

$$\begin{array}{r} 13.7 \\ \times 0.1 \\ \hline 1.37 \end{array}$$

$13.7 \times 0.1 = 1.37$

Multiplying by 0.01

$13.7 \times 0.01 = 0.137$

Multiplying by 0.001

$13.7 \times 0.001 = 0.0137$

EXERCISES
Give each product.

1. 4.3×1
2. 4.3×10
3. 4.3×100
4. 4.3×1000
5. 4.3×0.1
6. 4.3×0.01
7. 4.3×0.001
8. 4.3×0.0001

9. 28×1
10. 28×0.1
11. 28×0.01
12. 28×0.001
13. 28×10
14. 28×100
15. 28×1000
16. $28 \times 10{,}000$

17. 23.751×1
18. 23.751×10
19. 23.751×100
20. 23.751×1000
21. 23.751×0.1
22. 23.751×0.01
23. 23.751×0.001
24. 23.751×0.0001

Give each product.

25. 475.3×0.01 **26.** 621.5×10 **27.** 53.87×100

28. 6.28×0.01 **29.** 0.5836×1000 **30.** 4725.6×0.001

31. 6.7×0.001 **32.** 6.7×1000 **33.** 0.906×1000

Complete.

34. 6.3 × __?__ = 63

35. 6.3 × __?__ = 0.63

36. 7.25 × __?__ = 72.5

37. 7.25 × __?__ = 0.725

38. 526.3 × __?__ = 52.63

39. 526.3 × __?__ = 5.263

40. 0.84 × __?__ = 0.0084

41. 0.673 × __?__ = 67.3

42. 5683 × __?__ = 5.683

43. __?__ × 100 = 93.4

44. __?__ × 10 = 86.72

45. __?__ × 0.01 = 87.53

Solve.

46. 1 meter = 100 centimeters

The longest mustache is **2.59** meters long. How many centimeters is that?

47. 1 centimeter = 0.01 meter

The longest beard is **533.4** centimeters long. How many meters is that?

KEEPING SKILLS SHARP

Multiply.

1. 817×28

2. 923×74

3. 654×39

4. 548×803

5. 209×650

6. 457×362

Divide.

7. $18\overline{)894}$

8. $72\overline{)5364}$

9. $34\overline{)7091}$

10. $675\overline{)8502}$

11. $281\overline{)2168}$

12. $863\overline{)47,317}$

13. $196\overline{)60,829}$

14. $417\overline{)50,000}$

Problem solving

Scientists believe that there are about 100,000 different kinds of butterflies. Since there are so many kinds, some different types are hard to tell apart. For example, the Monarch and the Viceroy look so much alike that birds can't tell them apart. As soon as a young bird learns that Monarchs taste bad, it refuses to dine on either Monarchs or Viceroys.

Notice that the pictures do not show actual size. You will need a centimeter ruler for exercises 1 through 6.

Monarch

For actual size, multiply measurement by 1.2.

Viceroy

For actual size, multiply measurement by 1.2.

Measure to the nearest 0.1 centimeter and compute the wingspan of each butterfly. Round each wingspan to the nearest 0.1 cm.

1. Monarch

2. Viceroy

3. Tiger Swallowtail

4. Buckeye

5. Mourning Cloak

6. American Copper

7. How much greater is the wingspan of the Tiger Swallowtail than the wingspan of the Mourning Cloak? *Hint:* Use your answers to exercises 3 and 5.

8. The smallest butterfly known is the Western Pygmy Blue. It has a wingspan of 1.4 cm. How much greater is the wingspan of the American Copper?

9. The largest butterfly found in North America is the Giant Swallowtail. Its wingspan is 9.3 times as great as the wingspan of the Western Pygmy Blue. (See exercise 8.) What is the wingspan of the Giant Swallowtail?

★ 10. Which butterfly pictures are larger than full-scale? How much greater are the wingspans in the picture than the actual wingspans?

Dividing a decimal by a whole number

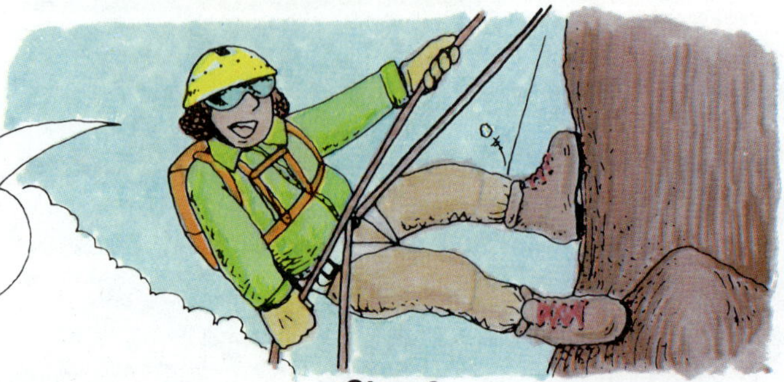

Study these examples.

EXAMPLE 1.

$2\overline{)9.78}$

Step 1. Divide ones.

```
    4
2)9.78
 -8
  1
```

Step 2. Regroup.

```
    4
2)9.78
 -8
  17
```

Be sure to place the decimal point in the quotient.

Step 3. Divide tenths.

```
    4.8
2)9.78
 -8
  17
 -16
   1
```

Step 4. Regroup.

```
    4.8
2)9.78
 -8
  17
 -16
   18
```

Step 5. Divide hundredths.

```
    4.89
2)9.78
 -8
  17
 -16
   18
  -18
    0
```

EXAMPLE 2.

```
     1.48
53)78.44
  -53
   254
  -212
    424
   -424
      0
```

EXAMPLE 3.

```
   0.75
4)3.00
 -28
   20
  -20
    0
```

EXERCISES

Divide.

1. $5\overline{)8}$ 2. $8\overline{)1}$ 3. $4\overline{)1}$ 4. $5\overline{)3}$ 5. $4\overline{)30}$

6. $4\overline{)0.3}$ 7. $8\overline{)30}$ 8. $8\overline{)0.3}$ 9. $5\overline{)0.07}$ 10. $2\overline{)0.5}$

11. $6\overline{)3.12}$ 12. $3\overline{)0.72}$ 13. $4\overline{)26.3}$ 14. $9\overline{)0.459}$ 15. $7\overline{)2.954}$

16. $8\overline{)4.368}$ 17. $5\overline{)39.65}$ 18. $9\overline{)4.815}$ 19. $7\overline{)34.23}$ 20. $6\overline{)537.6}$

21. $12\overline{)44.4}$ 22. $18\overline{)651.6}$ 23. $24\overline{)85.92}$ 24. $37\overline{)239.39}$ 25. $64\overline{)128.64}$

26. $47\overline{)143.82}$ 27. $83\overline{)42.33}$ 28. $91\overline{)278.46}$ 29. $75\overline{)243}$ 30. $86\overline{)3225}$

★ 31. $123\overline{)801.96}$ ★ 32. $204\overline{)73.032}$ ★ 33. $418\overline{)685.52}$ ★ 34. $528\overline{)331.848}$

Solve.

35. Four mountain climbers have to carry a total of 58.2 kilograms of supplies. How much should each climber carry if they divide the supplies evenly?

36. The climbers began at an elevation of 1858.5 meters above sea level. Twelve hours later, they were at an elevation of 3172.5 meters.

a. How many meters did they climb?
b. How many meters did they average per hour?

37. During the next day, they climbed for 6.5 hours. They averaged 168.2 meters per hour. How far did they climb?

38. Arrange the digits in the boxes to get the quotient.

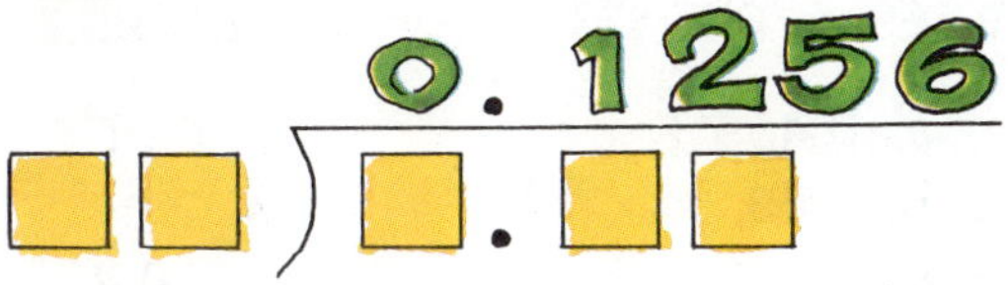

More about division

Before dividing, we will use a shortcut for estimating the quotient.

$39\overline{)9265.1}$

Step 1. Round the divisor to the nearest ten.

40

$39\overline{)9265.1}$

Step 2. Write the first digit of the quotient in the correct place.

$$\begin{array}{r} 2 \\ 39\overline{)9265.1} \end{array}$$

Step 3. Since the 2 is in the hundreds place, fill in two zeros.

$$\begin{array}{r} 200 \\ 39\overline{)9265.1} \end{array}$$

Notice that the division will never come out evenly. In such cases we usually round the quotient.

$$\begin{array}{r} 237.566 \\ 39\overline{)9265.100} \\ -78 \\ \hline 146 \\ -117 \\ \hline 295 \\ -273 \\ \hline 221 \\ -195 \\ \hline 260 \\ -234 \\ \hline 260 \\ -234 \\ \hline 26 \end{array}$$

Round to the nearest hundredth → **237.57**

Is the quotient between 200 and 300?

EXERCISES

Choose the quotient by estimating.

1. $38\overline{)2002.6}$
   a. 0.527
   b. 5.27
   c. 52.7
   d. 527

2. $42\overline{)153.3}$
   a. 0.365
   b. 3.65
   c. 36.5
   d. 365

3. $59\overline{)270.81}$
   a. 0.459
   b. 4.59
   c. 45.9
   d. 459

4. $96\overline{)195.84}$
   a. 0.204
   b. 2.04
   c. 20.4
   d. 204

5. 26)9.36

a. 0.36
b. 3.6
c. 36
d. 360

6. 42)2.184

a. 0.052
b. 0.52
c. 5.2
d. 52

7. 53)397.5

a. 0.75
b. 7.5
c. 75
d. 750

Divide. Round each quotient to the nearest hundredth.

8. 8)1
9. 6)1
10. 9)6
11. 3)22
12. 6)74
13. 8)35
14. 4)32.1
15. 9)4.38
16. 6)59.2
17. 6)5.97
18. 2)0.3956
19. 7)1.5931
20. 12)$3.97
21. 16)$5.63
22. 32)$8.10
23. 27)$.98
24. 23)$6.04
25. 48)$5.21
26. 17)$52.64
27. 25)$42.34
28. 83)$97.76
★ 29. 105)59.6
★ 30. 148)78.34
★ 31. 179)9.706
★ 32. 215)765.38
★ 33. 308)96.748
★ 34. 492)100.563

35. SLIP 'N' SLIDE

A frog fell to the bottom of a well that was 4 meters deep. During the first hour, the frog climbed up 17 centimeters and slid back 10 centimeters. At this rate, how long did it take the frog to climb out of the well?

KEEPING SKILLS SHARP

< or >?

1. 0.6 ● 0.7
2. 0.8 ● 0.08
3. 0.38 ● 0.39
4. 0.7 ● 0.69
5. 0.276 ● 0.277
6. 0.29 ● 0.3
7. 1 ● 0.99
8. 0.99 ● 0.989
9. 32.68 ● 32.65
10. 7.5123 ● 7.5132
11. 3.895 ● 4
12. 736.981 ● 736.975
13. 2.17 ● 21.7
14. 0.73 ● 0.728
15. 8.4 ● 8.36
16. 5.02 ● 5.2
17. 7.01 ● 7.001
18. 16.2 ● 9.14

Problem solving

To find the average number of points that Beal scored during the first four games, we add the four numbers and divide by 4.

$$3 + 4 + 3 + 5 = 15$$

$$\begin{array}{r} 3.75 \\ 4\overline{)15.00} \\ -12 \\ \hline 30 \\ -28 \\ \hline 20 \\ -20 \\ \hline 0 \end{array}$$

Player	Height in cm	Weight in kg	Points Scored					
			Game 1	Game 2	Game 3	Game 4	Game 5	Game 6
Beal	160	48	3	4	3	5	9	7
Davis	150	41.5	8	5	7	8	7	6
Franklin	162	49.5	5	7	8	3	4	3
Lopez	155	45.5	2	5	0	4	5	4
Manning	153	44	4	6	7	6	0	8
Porter	147	39.5	6	4	2	5	3	6
Rogers	157	46.5	0	3	4	4	3	5
Wilson	158	48.5	8	5	6	8	6	7

Rounded to the nearest tenth, her scoring average for the first four games was 3.8 points.

EXERCISES

Round each decimal answer to the nearest tenth.

1. How many points did the team score in their first game?
2. How many more points did the team score in their sixth game than in their first game?
3. What is the average height of the team members?
4. What is the average weight of the team members?
5. How many points did Davis average during the first five games?
6. In which games did Franklin score below her average?
7. A former player who holds the school scoring record averaged 9.5 points per game for 12 games. How many points did she score?
8. Who had the greater scoring average, Wilson or Lopez? How much greater?
9. What was the team's average score for the first three games?
10. What was the team's average score for the first six games?

11. Washington played on another team. She averaged 8.5 points for the first 6 games and 9.2 points for the last 5 games.
    a. How many points did she score in all 11 games?
    b. What was her 11-game average? Round your answer to the nearest tenth.

12. Carter averaged 8.4 points for the first 5 games and 10.5 points for the next 7 games. What was her 12-game average? Round your answer to the nearest tenth.

★ 13. King averaged 7.5 points for the first 8 games. What would she have to average in the next 4 games to have a 12-game average of 9 points per game?

KEEPING SKILLS SHARP

Round to the nearest whole number.

1. 56.3
2. 1.829
3. 35.50
4. 64.832
5. 42.059
6. 39.51

Round to the nearest tenth.

7. 3.13
8. 12.06
9. 25.093
10. 62.702
11. 45.63
12. 59.95

Round to the nearest hundredth.

13. 25.068
14. 7.005
15. 36.324
16. 125.3619
17. 82.7628
18. 49.995

Dividing by a decimal

To divide by a decimal, we can first divide as if the divisor were a whole number. Then we can estimate to place the decimal point in the quotient.

EXAMPLE 1. **1.8)97.74**

Step 1. Divide 9774 by 18.

```
     543
1.8)97.74
   -90
     77
    -72
     54
    -54
      0
```

Step 2. Estimate the quotient. The quotient will be "near" 50.

```
1.8)97.74

   50
 2)100
```

Step 3. Place the decimal point in the quotient.

```
     54.3
1.8)97.74
   -90
     77
    -72
     54
    -54
      0
```

If we multiply *both* the divisor and the dividend by the same nonzero number, the quotient remains the same.

```
  4                    4
3)12   Multiply by 10.  30)120
```

```
  17                      17
2)34   Multiply by 100.  200)3400
```

This fact gives us a shortcut for dividing by a decimal.

EXAMPLE 2. Divide 74.76 by 1.2.

Step 1. Make the divisor a whole number.

```
1.2)74.76
```

Multiply by 10.

Step 2. Move the decimal point the same number of places in the dividend.

```
12.)74.76
```

Multiply by 10.

Step 3. Divide 747.6 by 12.

```
     62.3
12.)747.6
   -72
     27
    -24
     36
    -36
      0
```

EXERCISES

Divide.

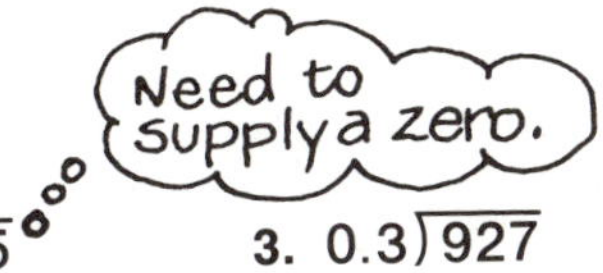

1. $0.5\overline{)36.5}$ 2. $0.5\overline{)365}$ 3. $0.3\overline{)927}$ 4. $0.02\overline{)0.642}$ 5. $0.07\overline{)0.91}$

6. $0.04\overline{)13.20}$ 7. $0.003\overline{)0.1701}$ 8. $0.9\overline{)1.287}$ 9. $0.08\overline{)56}$ 10. $0.005\overline{)7.5}$

11. $3.5\overline{)1.05}$ 12. $5.2\overline{)19.24}$ 13. $6.3\overline{)3.654}$ 14. $0.28\overline{)20.16}$ 15. $0.31\overline{)0.0961}$

★ 16. $25.3\overline{)397.21}$ ★ 17. $1.28\overline{)59.776}$ ★ 18. $0.235\overline{)5.1324}$ ★ 19. $46.2\overline{)1186.878}$

Divide. Round each quotient to the nearest tenth.

20. $0.8\overline{)29.7}$ 21. $0.06\overline{)368}$ 22. $0.14\overline{)5.297}$ 23. $2.6\overline{)7.438}$

24. $7.5\overline{)62.91}$ 25. $3.9\overline{)7.503}$ 26. $0.78\overline{)68.2}$ 27. $1.23\overline{)49.37}$

★ 28. $4.56\overline{)68.214}$ ★ 29. $0.387\overline{)5.2837}$ ★ 30. $1.06\overline{)0.3577}$ ★ 31. $98.7\overline{)62.357}$

Solve.

32. Janice brought hot dogs to the picnic. She spent $8.64. How many pounds of hot dogs did Janice buy?

33. Leonard brought the drink mix. The mix cost about 5¢ an ounce. What did he bring, lemonade or iced tea?

★ 34. Amy brought a jar of relish and some packs of hot dog buns. She spent a total of $4.79. How many packs of hot dog buns did she bring?

Practice exercises

Multiply.

1. $\begin{array}{r} 3.84 \\ \times 1.9 \\ \hline \end{array}$
2. $\begin{array}{r} 56.5 \\ \times 5.3 \\ \hline \end{array}$
3. $\begin{array}{r} 81.3 \\ \times 0.27 \\ \hline \end{array}$
4. $\begin{array}{r} 7.02 \\ \times 5.8 \\ \hline \end{array}$
5. $\begin{array}{r} 51.8 \\ \times 6.4 \\ \hline \end{array}$
6. $\begin{array}{r} 5.36 \\ \times 4.02 \\ \hline \end{array}$
7. $\begin{array}{r} 38.5 \\ \times 5.63 \\ \hline \end{array}$
8. $\begin{array}{r} 4.93 \\ \times 2.81 \\ \hline \end{array}$
9. $\begin{array}{r} 6.35 \\ \times 34.6 \\ \hline \end{array}$
10. $\begin{array}{r} 71.5 \\ \times 11.5 \\ \hline \end{array}$

Divide. Round each quotient to the nearest hundredth.

11. $5.7\overline{)0.548}$
12. $0.68\overline{)4.82}$
13. $2.6\overline{)2.964}$
14. $0.53\overline{)0.8349}$
15. $4.8\overline{)0.7529}$
16. $0.26\overline{)75}$
17. $3.12\overline{)3.906}$
18. $4.08\overline{)52.831}$
19. $0.225\overline{)729.84}$
20. $0.826\overline{)3.0715}$

Compute. [You should be able to do these problems without pencil and paper.]

21. $(58.7 + 3.9) - 3.9$
22. $(358.62 - 16.94) + 16.94$
23. $(8.6 \times 5.3) \div 5.3$
24. $(2.25 \div 0.15) \times 0.15$
25. $(24.6 \times 7.8) \div 24.6$
26. $(5.76 \div 0.18) \times 0.18$
27. $(9.283 + 4.295) - 9.283$
28. $(3.719 \times 1.065) \div 3.719$

Find each end number.

29.

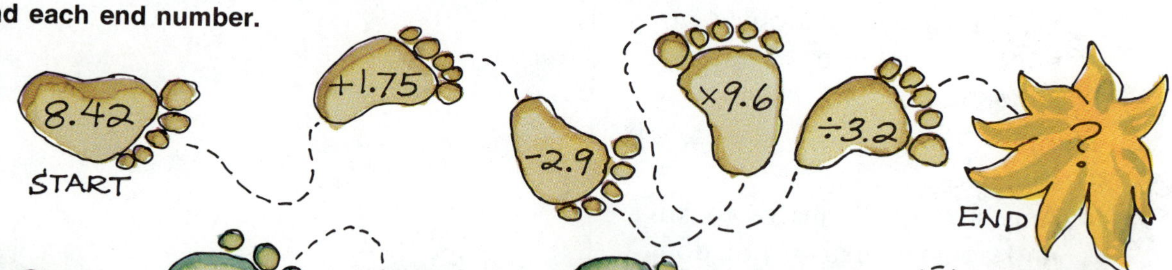

30.

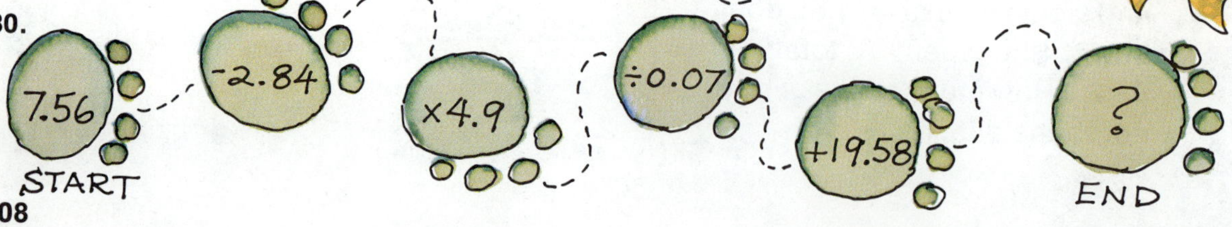

SKILL GAME

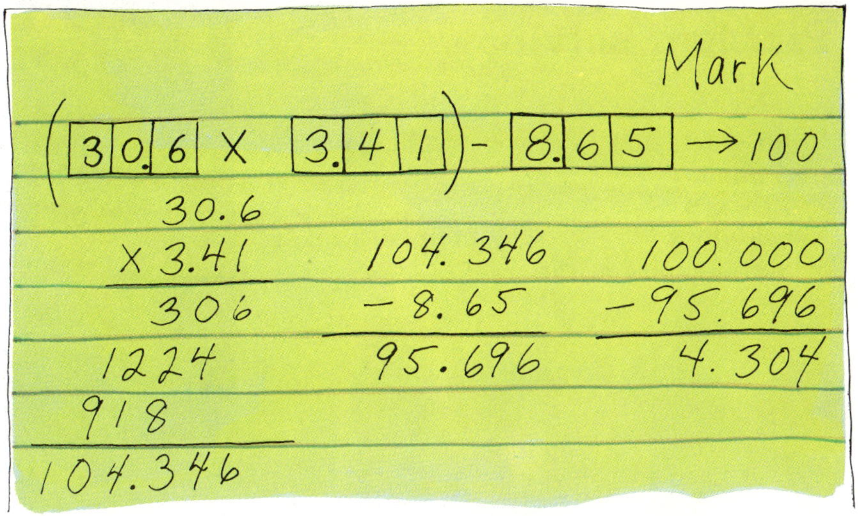

Notice that Mark missed the target number (100) by only 4.304.

How close did these players come to the target number? Round each quotient to the nearest hundredth.

1. (64.1 ÷ 30.8) × 3.56 → 100

2. (863. × .160) − 45.3 → 100

3. (13.6 × 5.03) ÷ .846 → 100

4. (5.36 + 13.0) − 6.84 → 100

Play the game.

1. Choose a game leader. Make two cards for each of the digits 0 through 9.
2. Draw a table like this on your paper.

   (□□□ □□□) □□□ → 100
3. The game leader displays nine digit cards.
4. Each player has five minutes to fill in the digits in his/her table, write in any two operation signs (+, −, ×, ÷), and place three decimal points.
5. The player getting closest to the target number, 100, wins the game.

Problem solving

In an emergency, a driver might have to stop quickly.

The graph below shows stopping distances for automobiles at different speeds.

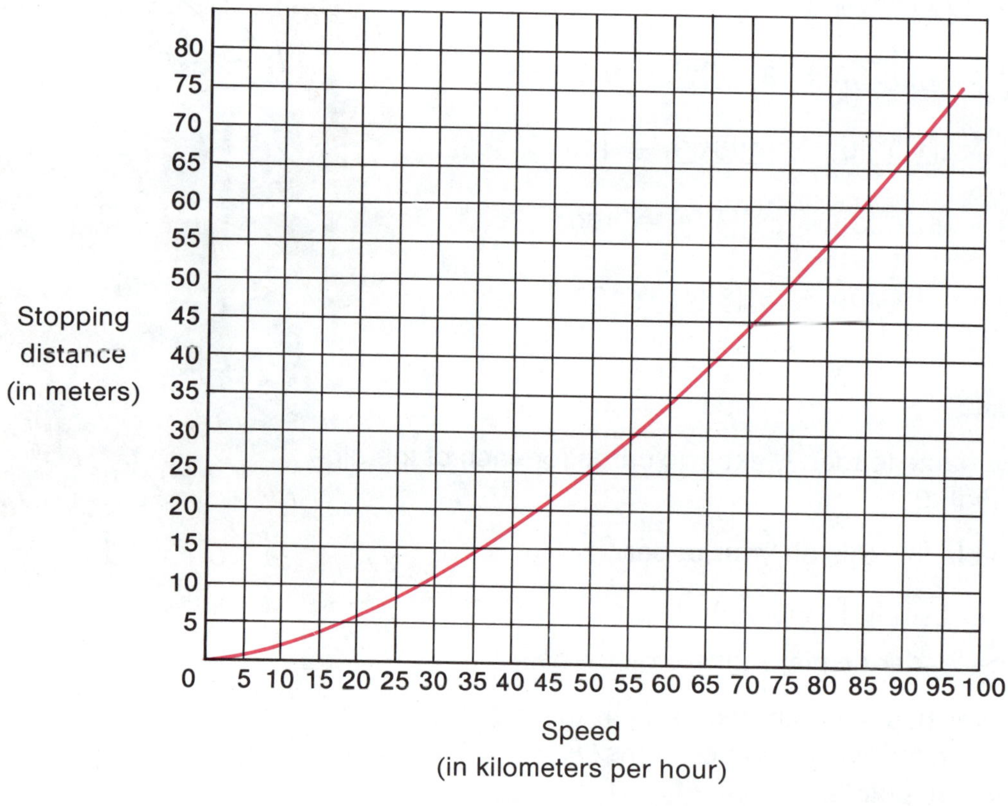

Note: In getting your facts from the graph, you must estimate.

Solve.

1. About how many meters will it take you to stop your car if you are traveling
   a. 10 kilometers per hour?
   b. 50 kilometers per hour?
   c. 75 kilometers per hour?
   d. 96 kilometers per hour?

2. At about what speed would you be traveling if your stopping distance was
   a. 20 meters?
   b. 40 meters?
   c. 65 meters?
   d. 75 meters?

3. What is the difference in the stopping distances between cars traveling at 45 kilometers per hour and 55 kilometers per hour?

4. If you change your speed from 30 kilometers per hour to 60 kilometers per hour (double your speed), do you double your stopping distance?

5. At 64 kilometers per hour, a safe following distance is 21.9 meters. Suppose that your car is 5.5 meters long. How many car lengths should you be from the car in front? Round your answer to the nearest whole number.

6. A recent study was made of 1500 fatal accidents. It was found that 0.65 of the drivers responsible for the accidents had been drinking. In how many of these accidents was drinking a factor?

7. One year the overall accident rate for female drivers was 12.02 accidents per 1 million kilometers driven. Suppose that in one week females drove 550 million kilometers. How many accidents would they be expected to have?

8. The overall accident rate for male drivers is 1.09 times the overall accident rate for females. What is the accident rate for males? (See exercise 7.)

CHAPTER CHECKUP

Multiply. [pages 92–94, 108–109]

1. $\begin{array}{r} 3.9 \\ \times 0.5 \\ \hline \end{array}$ 2. $\begin{array}{r} 4.2 \\ \times 2.7 \\ \hline \end{array}$ 3. $\begin{array}{r} 13.8 \\ \times 3.8 \\ \hline \end{array}$ 4. $\begin{array}{r} 50.7 \\ \times 0.21 \\ \hline \end{array}$ 5. $\begin{array}{r} 40.3 \\ \times 31.4 \\ \hline \end{array}$ 6. $\begin{array}{r} 8.07 \\ \times 1.04 \\ \hline \end{array}$

Multiply. [pages 96–97]

7. 5.8×10 8. 4.3×100 9. 0.5671×100

10. 3.2×0.1 11. 5.6×0.01 12. 3214.6×0.001

Divide. [pages 100–103]

13. $3\overline{)13.68}$ 14. $5\overline{)3.9}$ 15. $9\overline{)3.483}$ 16. $10\overline{)67.34}$

17. $21\overline{)37.8}$ 18. $82\overline{)323.08}$ 19. $134\overline{)3457.2}$ 20. $503\overline{)3274.53}$

Divide. Round each quotient to the nearest hundredth.
[pages 106–109]

21. $0.6\overline{)0.58}$ 22. $0.07\overline{)39.7}$ 23. $0.009\overline{)6.73}$ 24. $3.2\overline{)67.65}$

25. $0.12\overline{)1.397}$ 26. $0.62\overline{)4.18}$ 27. $12.4\overline{)58.369}$ 28. $2.79\overline{)493.752}$

Solve. [pages 95, 98–99, 104–105, 110–111]

29. To buy a stereo set, Mark has been saving $3.75 a week for the last 12 weeks. How much money has he saved during the 12 weeks?

30. The stereo that Mark wants costs $176.25. Mark now has $45. If he can save $3.75 a week, how much longer does he have to save to buy the stereo set?

31. Paula wants to buy stereo headphones that cost $29.95 and 2 eight-track tapes that cost $8.99 each. How much money does she need?

CHAPTER PROJECT

For these projects, you may wish to work with a classmate.

1. See how many pieces of notebook paper there are in a stack that is 1 centimeter thick. Divide to find the thickness of 1 sheet. Round your answer to the nearest thousandth of a centimeter.
2. Use a balance to determine the weight of 1 paper clip. Round your answer to the nearest hundredth of a gram. *Hint:* First see how many paper clips it takes to make a gram.
3. Can you find out how far apart the grooves are on a record?

CHAPTER REVIEW

$$\begin{array}{r} 6.3 \\ \times 0.21 \\ \hline 63 \\ 126 \\ \hline 1.323 \end{array}$$

6.3 and × 0.21: 3 places; 1.323: 3 places

Multiply.

1. 2.7×8	**2.** 3.9×0.6	**3.** 0.57×9	**4.** 0.35×0.7
5. 0.02×0.3	**6.** 6.7×3.4	**7.** 9.28×3.6	**8.** 5.47×6.24

$3.75 \times 10 = 37.5$

$6.21 \times 0.01 = 0.0621$

Give each product.

9. 6.5×10 **10.** 1.23×10

11. 3.8×0.1 **12.** 623.4×0.01

13. 37.611×100 **14.** 0.48×1000

$$\begin{array}{r} 8.81 \\ 3\overline{)26.43} \\ -24 \\ \hline 2\,4 \\ -2\,4 \\ \hline 3 \\ -3 \\ \hline 0 \end{array}$$

Divide.

15. $5\overline{)23.25}$ **16.** $3\overline{)1.44}$ **17.** $9\overline{)321.3}$

18. $6\overline{)0.426}$ **19.** $5\overline{)3.1}$ **20.** $8\overline{)0.0584}$

21. $12\overline{)2.76}$ **22.** $25\overline{)3.2}$ **23.** $40\overline{)0.72}$

$$\begin{array}{r} 6.055 \\ 0.32\overline{)1.9376} \\ -1\,92 \\ \hline 1\,76 \\ -1\,60 \\ \hline 160 \\ -160 \\ \hline 0 \end{array}$$

Divide. Round each quotient to the nearest hundredth.

24. $0.9\overline{)5.762}$ **25.** $0.8\overline{)93.58}$

26. $0.05\overline{)3.711}$ **27.** $2.5\overline{)7.12}$

28. $0.19\overline{)46}$ **29.** $8.1\overline{)0.731}$

$6 + 4 + 3 + 8 = 21$ (4 numbers)

$$\begin{array}{r} 5.25 \\ 4\overline{)21.00} \end{array}$$

Compute the average.

30. 8, 8, 9, 5, 6, 3 **31.** 28, 31, 26, 25, 38

32. 126, 143, 182, 135

CHAPTER CHALLENGE

Scientists often write numbers in **scientific notation.** For example, the diameter of the sun, written in scientific notation, is

1.38×10^6 kilometers

A number in scientific notation is the product of a number between 1 and 10 and a power of 10. Here are some more examples of numbers written in scientific notation:

3.2×10^1 4.56×10^2 2.03×10^4

$3.2 \times 10 = 32$ $4.56 \times 100 = 456$ $2.03 \times 10{,}000 = 20{,}300$

EXERCISES

Give the standard numeral.

1. 3×10^3
2. 5.3×10^4
3. 6.04×10^5
4. 2.753×10^6

Express each of the following numbers in scientific notation.

5. 2600
6. 384,000
7. 5,930,000
8. 369,000,000,000
9. The sun is about 150,000,000 kilometers from the earth.
10. The sun is thought to be about 5,000,000,000 years old.
11. The mass of the sun is about 332,000 times greater than the mass of the earth.
12. The temperature at the center of the sun is about 20,000,000 degrees Celsius.

MAJOR CHECKUP

STANDARDIZED FORMAT

Choose the correct letter.

1. The standard numeral for six million, sixty thousand, six is

- **a.** 6,066,000
- **b.** 6,060,006
- **c.** 6,000,066
- **d.** none of these

2. Give the sum.

56 + 381 + 9

- **a.** 446
- **b.** 447
- **c.** 456
- **d.** none of these

3. Give the difference.

5002 − 1327

- **a.** 4775
- **b.** 4785
- **c.** 3675
- **d.** none of these

4. Multiply.

$$\begin{array}{r} 324 \\ \times 602 \\ \hline \end{array}$$

- **a.** 195,048
- **b.** 20,088
- **c.** 196,148
- **d.** none of these

5. Divide.

$38\overline{)7752}$

- **a.** 204
- **b.** 240
- **c.** 24
- **d.** none of these

6. 4.2031 rounded to the nearest hundredth is

- **a.** 4.21
- **b.** 4.20
- **c.** 4.203
- **d.** none of these

7. Add.

3.71 + 28.6

- **a.** 7.57
- **b.** 75.7
- **c.** 31.31
- **d.** none of these

8. Subtract.

70.03 − 2.56

- **a.** 78.47
- **b.** 78.57
- **c.** 67.47
- **d.** none of these

9. Multiply.

$$\begin{array}{r} 0.32 \\ \times 2.4 \\ \hline \end{array}$$

- **a.** 0.768
- **b.** 7.68
- **c.** 76.8
- **d.** none of these

10. Divide.

$0.08\overline{)1.72}$

- **a.** 0.215
- **b.** 2.15
- **c.** 21.5
- **d.** none of these

11. Doris ran 0.4 km 3 times. Linda ran 0.4 km 5 times. How much farther did Linda run?

- **a.** 2 km
- **b.** 0.08 km
- **c.** 8 km
- **d.** none of these

12. Records cost $5 each. Tapes cost $9 each. What is the total cost of 2 records and 1 tape?

- **a.** $14
- **b.** $28
- **c.** $19
- **d.** none of these

Measurement

4

Measuring length

The basic unit for measuring length in the metric system is the **meter (m)**. The dog is about 1 meter long.

These units of length are used in the metric system:

1 kilometer (km)	=	1000 meters
1 hectometer (hm)*	=	100 meters
1 dekameter (dam)*	=	10 meters
1 meter (m)	=	1 meter
1 decimeter (dm)*	=	0.1 meter
1 centimeter (cm)	=	0.01 meter
1 millimeter (mm)	=	0.001 meter

* These units are seldom used.

The metric system, like our place-value system, is based on the number 10. Because of this, decimals are used in writing metric measurements.

The length of the segment is 12.5 cm.
What is the length of the segment in mm?

EXERCISES

To the nearest centimeter, measure

1. your height.
2. your waist measure.
3. your arm span.
4. your foot length.
5. the width of your desk top.
6. the length of a chalk eraser.
7. the width of a book.
8. the height of a table.
9. the width of a door.
10. the length of your pencil.

Measure

11. the width of a chalkboard to the nearest decimeter.
12. the length of your classroom to the nearest meter.

Get a ruler marked off in millimeters. Draw segments having these lengths.

13. 1 mm	**14.** 1 cm	**15.** 1 dm	**16.** 15 mm	**17.** 5.1 cm
18. 2.5 cm	**19.** 3.6 cm	**20.** 128 mm	**21.** 12.8 cm	**22.** 1.28 dm

23. Draw a segment that is 21 cm long.
    a. How many mm long is it?
    b. How many dm long is it?

24. On the chalkboard, draw a segment that is 1.23 m long.
    a. How many dm long is it?
    b. How many cm long is it?
    c. How many mm long is it?

25. On the chalkboard, draw a segment that is 425 mm long.
    a. How many cm long is it?
    b. How many dm long is it?
    c. How many m long is it?

Changing units in the metric system

Changing units in the metric system is easy because you need only multiply or divide by a power of 10.

This feather is 12.7 centimeters long.

EXAMPLE 1. What is the length of the feather in millimeters?

1 cm = 10 mm

12.7 cm = (12.7 × 10) mm
= 127 mm

EXAMPLE 2. What is the length of the feather in meters?

100 cm = 1 m

12.7 cm = (12.7 ÷ 100) m
= 0.127 m

EXAMPLE 3. 392 m = __?__ km

To change to a larger unit, divide.

1000 m = 1 km

392 m = (392 ÷ 1000) km
= 0.392 km

EXAMPLE 4. 49.3 m = __?__ cm

To change to a smaller unit, multiply.

1 m = 100 cm

49.3 m = (49.3 × 100) cm
= 4930 cm

EXERCISES

Copy and complete each table. You can change from one unit to another by moving the decimal point.

1. smaller units, larger numbers

	m	dm	cm	mm
a.	4.2	42	420	
b.	0.52	5.2		
c.	1.68			
d.	43.6			
e.	6.0			

2. larger units, smaller numbers

	m	dam	hm	km
a.	325	32.5		
b.	54.3			
c.	8.6			
d.	1.67			
e.	27.0			

Complete.

3. 2 m = __?__ dm

4. 4 m = __?__ cm

5. 5 m = __?__ mm

6. 1245 m = __?__ km

7. 1500 m = __?__ km

8. 354 m = __?__ km

9. 6.7 m = __?__ cm

10. 59.2 m = __?__ dm

11. 63.5 m = __?__ km

12. 83 cm = __?__ m

13. 57.2 cm = __?__ mm

14. 63.8 cm = __?__ dm

15. 63 km = __?__ m

16. 58.9 dm = __?__ mm

17. 173 mm = __?__ cm

18. 173 mm = __?__ dm

19. 173 mm = __?__ m

20. 675.41 mm = __?__ m

21. 52.3 km = __?__ m

22. 52.3 mm = __?__ cm

23. 42.77 m = __?__ cm

24. 42.77 cm = __?__ m

25. 56.761 m = __?__ km

26. 35.92 km = __?__ m

27. DECIMAL DEBATE

Decide who is right.

Liquid volume

A unit for measuring liquid volume in the metric system is the **liter (L)**.

The dog dish holds about 1 L of water.

The **milliliter (mL)** is used to measure small liquid volumes.

$$1 \text{ L} = 1000 \text{ mL}$$

The liter and the milliliter are the only metric units that are commonly used to measure liquid volume.

EXERCISES

Choose the correct liquid volume.

1. Bath water

a. 30 L **b.** 30 mL

2. Vitamin syrup

a. 395 L **b.** 395 mL

3. Medicine

a. 10 L **b.** 10 mL

4. Coat conditioner

a. 0.47 L **b.** 0.47 mL

Complete.

5. 1 L = ___?___ mL
6. 2 L = ___?___ mL
7. 1.5 L = ___?___ mL
8. 2.4 L = ___?___ mL
9. 0.25 L = ___?___ mL
10. 0.5 L = ___?___ mL
11. 1.35 L = ___?___ mL
12. 2.25 L = ___?___ mL
13. 1000 mL = ___?___ L
14. 3000 mL = ___?___ L
15. 1800 mL = ___?___ L
16. 2600 mL = ___?___ L
17. 750 mL = ___?___ L
18. 560 mL = ___?___ L
19. 125 mL = ___?___ L
20. 75 mL = ___?___ L

Solve.

21. A veterinarian told Judy to put 0.75 milliliter (15 drops) of medicine into her dog's food each day for a week. How many milliliters of medicine did the dog get?

22. Judy's dog drinks about 1.25 liters of water each day. About how many liters does her dog drink in a week?

23. Judy gives her dog 7.5 milliliters of liquid vitamins each day. How many days will a bottle that contains 0.3 liter last?

24. Judy gives her dog 15 milliliters of coat conditioner each day. A bottle of coat conditioner contains 945 milliliters of liquid and costs $9.45. What is the cost of coat conditioner each week?

KEEPING SKILLS SHARP

Estimate to find the two incorrect answers.

1. a. 343.8 − 96.9 = 146.9
   b. 703.41 − 29.7 = 673.71
   c. 81.44 + 58.84 = 140.28
   d. 429.3 + 8.94 = 518.7

Add or subtract.

2. 5.07 + 0.265
3. 0.29 − 0.163
4. 0.841 + 78.9
5. 3.25 + 0.325
6. 40 − 36.3
7. 2.03 − 1.99
8. 26.9 + 9.28
9. 5.47 − 0.8
10. 0.059 + 4.62
11. 7.03 − 2.59
12. 80.3 − 4.75
13. 27.6 + 9.38

Weight

A unit for measuring weight (mass) in the metric system is the **gram (g)**. The gram is a small unit. Here are some other units for measuring weight:

1 kilogram (kg) = 1000 g

1 milligram (mg) = 0.001 g

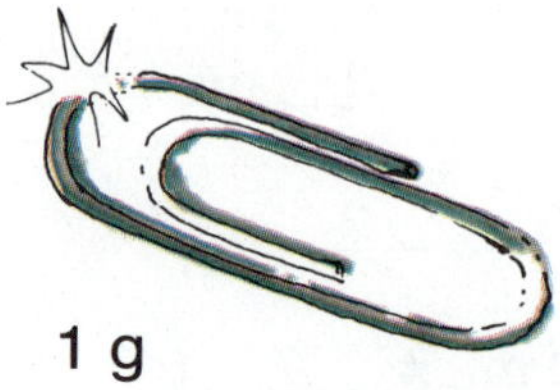

1 g

400 mg

5 g

2 kg

EXERCISES

Milligram, gram, or kilogram? Which unit would be used to measure the weight of

1. a brick?

2. an aspirin?

3. a sunflower seed?

4. a softball?

5. a bicycle?

6. a straight pin?

7. a grain of sand?

8. a package of notebook paper?

Choose the correct weight.

9.

a. 44 mg

b. 44 g

c. 44 kg

10.

a. 250 mg

b. 250 g

c. 250 kg

11. 

a. 45 mg

b. 45 g

c. 45 kg

12. 

a. 125 mg

b. 125 g

c. 125 kg

Complete.

13. 1 g = ___?___ mg
14. 8 g = ___?___ mg
15. 15 g = ___?___ mg
16. 2000 mg = ___?___ g
17. 1500 mg = ___?___ g
18. 800 mg = ___?___ g
19. 1 kg = ___?___ g
20. 2.1 kg = ___?___ g
21. 0.845 kg = ___?___ g
22. 3000 g = ___?___ kg
23. 2400 g = ___?___ kg
24. 560 g = ___?___ kg

3.1 grams

5.0 grams

2.3 grams

5.7 grams

11.2 grams

Use the given weights to solve.

25. What is the total weight of a penny, a nickel, a dime, a quarter, and a half-dollar?

26. What is the total weight of two half-dollars and three quarters?

★ 27. Gary has a collection of coins. Their value is 60¢ and their weight is 21.4 grams. What are the coins?

★ 28. Barbara has a collection of coins. Their value is 50¢ and their weight is 23 grams. What are the coins?

Temperature

The Celsius thermometer is used to measure temperature. Here are some sample temperatures.

Water boils	Body temperature	Hot summer day

100°C	37°C	35°C
Summer day	Room temperature	Cool day

25°C	20°C	10°C
Water freezes	Winter day	Very cold day

0°C	−5°C	−15°C

EXERCISES

Match each exercise with a temperature range.

1. an ice-cream sundae
2. a cup of hot chocolate
3. a glass of cold orange juice
4. a pizza oven
5. a swimming pool
6. a refrigerator
7. a campfire
8. warm bath water
9. January 1 in the Arctic

a. above 100°C

b. between 37°C and 100°C

c. between 20°C and 37°C

d. between 0°C and 20°C

e. below 0°C

Solve.

10. The body temperature of a hamster can be as high as 38.3°C and as low as 3.5°C when it hibernates. How many degrees' difference is that?

11. The highest shade temperature recorded on the earth was 58°C in Libya. The highest shade temperature recorded in the United States was 56.7°C in Death Valley, California. How many degrees' difference is that?

12. The lowest temperature recorded in Honolulu, Hawaii, was 11.7°C. The highest temperature recorded in Honolulu, Hawaii, was 22.2°C warmer. What was that temperature?

★ 13.

How warm is it inside?

Problem solving

Use the fact to solve.

Fact: A \$1 bill, a \$10 bill, and a \$100 bill each weigh 1 gram.

1. Which is heavier, a champion Yorkshire hog that weighs 94 kilograms or \$1,000,000 in \$1 bills? How many kilograms heavier?
2. What is the weight in kilograms of \$1,000,000 in \$10 bills?
3. What is the weight in kilograms of \$1,000,000 in \$100 bills?
4. Suppose you inherited 2.5 kilograms of \$1 bills. How much money would you get?

Fact: A \$1 bill, a \$10 bill, and a \$100 bill each have a thickness of 0.1 millimeter.

5. Which is taller, the Sears Tower, which is 443.2 meters tall, or a stack of \$1,000,000 in \$1 bills? How many meters taller?
6. How many meters tall is a stack of \$10 bills that is worth \$1,000,000?
7. How many meters tall is a stack of \$100 bills that is worth \$1,000,000?
8. Suppose you found a stack of \$10 bills that was 3 centimeters high. How much money would you find?

Fact: A \$1 bill, a \$10 bill, and a \$100 bill each have a length of 15.6 centimeters.

9. Which is the greater distance, the 155 kilometers from New York to Philadelphia or \$1,000,000 in \$1 bills laid end-to-end? How many kilometers greater?

10. How many kilometers long is \$1,000,000 in \$10 bills laid end-to-end?

11. How many kilometers long is \$1,000,000 in \$100 bills laid end-to-end?

★ 12. Suppose you won a contest. The prize is five of your giant steps in \$100 bills laid end-to-end. How much money would you win?

Length—customary units

Here are some customary units for measuring length:

12 inches (in. or ″) = 1 foot (ft)
3 ft (′) = 1 yard (yd)
1760 yd = 1 mile (mi)

These examples show how to change from one unit to another.

EXAMPLE 1. 2 mi = __?__ yd

There are 1760 yards in 1 mile. So, multiply 1760 by 2.

$$\begin{array}{r} 1760 \\ \times 2 \\ \hline 3520 \end{array}$$

2 mi = 3520 yd

EXAMPLE 2. 57 in. = __?__ ft __?__ in.

There are 12 inches in 1 foot. So, divide 57 by 12.

$$\begin{array}{r} 4 \\ 12\overline{)57} \\ -48 \\ \hline 9 \end{array}$$

57 in. = 4 ft 9 in.

EXERCISES

Complete.

1. 9 ft = __?__ in.
2. 6 yd = __?__ ft
3. 6 yd = __?__ in.
4. 3 mi = __?__ yd
5. 1 mi = __?__ ft
6. 8 yd = __?__ in.
7. 108 in. = __?__ ft
8. 140 in. = __?__ ft __?__ in.
9. 48 ft = __?__ yd
10. 76 ft = __?__ yd __?__ ft
11. 92 in. = __?__ ft __?__ in.
12. 150 in. = __?__ ft __?__ in.
13. 2 ft 3 in. = __?__ in.
14. 5 yd 1 ft = __?__ ft
15. 2 yd 5 in. = __?__ in.

Add. Regroup whenever possible.

EXAMPLE.

$$\begin{array}{r} 4 \text{ ft } 9 \text{ in.} \\ +2 \text{ ft } 5 \text{ in.} \\ \hline \end{array}$$

Step 1. Add.

$$\begin{array}{r} 4 \text{ ft } 9 \text{ in.} \\ +2 \text{ ft } 5 \text{ in.} \\ \hline 6 \text{ ft } 14 \text{ in.} \end{array}$$

Step 2. Regroup 12 inches for 1 foot.

$$\begin{array}{r} 4 \text{ ft } 9 \text{ in.} \\ +2 \text{ ft } 5 \text{ in.} \\ \hline \not{6} \text{ ft } \not{14} \text{ in.} \\ 7 \quad\; 2 \quad\;\; \end{array}$$

16. $\begin{array}{r} 2 \text{ ft } 10 \text{ in.} \\ +1 \text{ ft } \;\; 7 \text{ in.} \\ \hline \end{array}$ **17.** $\begin{array}{r} 7 \text{ ft } 8 \text{ in.} \\ +3 \text{ ft } 7 \text{ in.} \\ \hline \end{array}$ **18.** $\begin{array}{r} 9 \text{ ft } 6 \text{ in.} \\ +5 \text{ ft } 9 \text{ in.} \\ \hline \end{array}$ **19.** $\begin{array}{r} 5 \text{ yd } 2 \text{ ft} \\ +2 \text{ yd } 1 \text{ ft} \\ \hline \end{array}$

20. $\begin{array}{r} 3 \text{ yd } 2 \text{ ft} \\ +5 \text{ yd } 2 \text{ ft} \\ \hline \end{array}$ **21.** $\begin{array}{r} 4 \text{ yd } 2 \text{ ft } 8 \text{ in.} \\ +3 \text{ yd } 2 \text{ ft } 2 \text{ in.} \\ \hline \end{array}$ **22.** $\begin{array}{r} 3 \text{ yd } 1 \text{ ft } 8 \text{ in.} \\ +2 \text{ yd } 2 \text{ ft } 5 \text{ in.} \\ \hline \end{array}$ **23.** $\begin{array}{r} 2 \text{ yd } 2 \text{ ft } 7 \text{ in.} \\ +1 \text{ yd } 1 \text{ ft } 6 \text{ in.} \\ \hline \end{array}$

Subtract.

EXAMPLE.

$$\begin{array}{r} 5 \text{ ft } 1 \text{ in.} \\ -1 \text{ ft } 4 \text{ in.} \\ \hline \end{array}$$

Step 1. Regroup 1 foot for 12 inches.

$$\begin{array}{r} 4 \quad\; 13 \quad\;\; \\ \not{5} \text{ ft } \not{1} \text{ in.} \\ -1 \text{ ft } 4 \text{ in.} \\ \hline \end{array}$$

Step 2. Subtract.

$$\begin{array}{r} 4 \quad\; 13 \quad\;\; \\ \not{5} \text{ ft } \not{1} \text{ in.} \\ -1 \text{ ft } 4 \text{ in.} \\ \hline 3 \text{ ft } 9 \text{ in.} \end{array}$$

24. $\begin{array}{r} 7 \text{ ft } 2 \text{ in.} \\ -4 \text{ ft } 9 \text{ in.} \\ \hline \end{array}$ **25.** $\begin{array}{r} 8 \text{ ft } 5 \text{ in.} \\ -3 \text{ ft } 9 \text{ in.} \\ \hline \end{array}$ **26.** $\begin{array}{r} 3 \text{ ft} \quad\quad\;\; \\ -1 \text{ ft } 8 \text{ in.} \\ \hline \end{array}$ **27.** $\begin{array}{r} 6 \text{ yd } 2 \text{ ft} \\ -4 \text{ yd } 1 \text{ ft} \\ \hline \end{array}$

28. $\begin{array}{r} 8 \text{ yd } 1 \text{ ft} \\ -3 \text{ yd } 2 \text{ ft} \\ \hline \end{array}$ **29.** $\begin{array}{r} 5 \text{ yd} \quad\quad \\ -2 \text{ yd } 1 \text{ ft} \\ \hline \end{array}$ **30.** $\begin{array}{r} 3 \text{ yd } 2 \text{ ft } 8 \text{ in.} \\ -1 \text{ yd } 1 \text{ ft } 9 \text{ in.} \\ \hline \end{array}$ **31.** $\begin{array}{r} 11 \text{ yd } 2 \text{ ft } 4 \text{ in.} \\ -6 \text{ yd} \quad\quad\; 8 \text{ in.} \\ \hline \end{array}$

Solve.

32. How many yards long is a quarter-mile running track?

33. A marathon is a race that is 26 miles 385 yards long. How many feet is that?

★ **34.** The fastest marathon was run in about 2 hours 9 minutes. About what was the average time per mile?

★ **35.** The world record for the 100-yard dash is about 9.1 seconds. If a runner could keep up that rate, how long would it take to run a mile?

Liquid volume and weight—customary units

2 cups (c) = 1 pint (pt)

2 pints = 1 quart (qt)

2 quarts = 1 half-gallon

2 half-gallons = 1 gallon (gal)

EXERCISES

Complete.

1. 24 pt = __?__ qt
2. 8 c = __?__ pt
3. 4 qt = __?__ gal
4. 3 qt = __?__ c
5. 2 gal = __?__ qt
6. 6 pt = __?__ c
7. 1 gal = __?__ pt
8. 2 qt = __?__ half-gallon
9. 12 c = __?__ qt
10. 16 pt = __?__ qt
11. 2 qt = __?__ c
12. 3 gal = __?__ half-gallons
13. 3 pt 1 c = __?__ c
14. 3 gal 1 qt = __?__ qt
15. 8 qt 1 pt = __?__ pt

In the customary system, dry material such as flour and sugar is measured with these units:

2 pints (pt) = 1 quart (qt)
8 quarts = 1 peck (pk)
4 pecks = 1 bushel (bu)

Complete.

16. 6 pt = __?__ qt
17. 16 qt = __?__ pk
18. 3 bu = __?__ pk
19. 5 pk = __?__ qt

20. GOOD TO THE LAST DROP!

Suppose that you have a 3-qt container and a 5-qt container. How could you use the containers to measure out 4 qt of water?

Here are some units used for measuring weight:

16 ounces (oz) = 1 pound (lb)

2000 pounds (lb) = 1 ton (T)

Complete.

21. 4 lb = ___?___ oz

22. 3 T = ___?___ lb

23. 256 oz = ___?___ lb

24. 8000 lb = ___?___ T

25. 24 lb = ___?___ oz

26. 288 oz = ___?___ lb

27. 3 lb 6 oz = ___?___ oz

28. 4 lb 9 oz = ___?___ oz

29. 12 lb 3 oz = ___?___ oz

Solve.

30. A chili recipe calls for 2 lb of ground beef. If you had 1 lb 3 oz of ground beef, how much more would you need?

31. A truck weighing 21,400 lb came to a bridge with this sign:

Closed to traffic
over
10 tons

Should the truck cross the bridge?

CHAPTER CHECKUP

Complete. [pages 118–121]

1. 30 mm = ___?___ cm
2. 200 cm = ___?___ m
3. 135 cm = ___?___ dm
4. 68 cm = ___?___ m
5. 2.1 km = ___?___ m
6. 82.4 mm = ___?___ dm
7. 34.9 m = ___?___ km
8. 426 mm = ___?___ m
9. 85.2 cm = ___?___ mm

Complete. [pages 122–123]

10. 1 L = ___?___ mL
11. 2 L = ___?___ mL
12. 3.6 L = ___?___ mL
13. 250 mL = ___?___ L
14. 500 mL = ___?___ L
15. 1747.2 mL = ___?___ L

Complete. [pages 124–125]

16. 1 kg = ___?___ g
17. 1 g = ___?___ mg
18. 475 g = ___?___ kg
19. 3.9 kg = ___?___ g
20. 3.9 g = ___?___ mg
21. 3.9 mg = ___?___ g

Complete. [pages 130–133]

22. 48 ft = ___?___ in.
23. 48 in. = ___?___ ft
24. 48 ft = ___?___ yd
25. 48 yd = ___?___ ft
26. 6 yd = ___?___ in.
27. 1 mi = ___?___ ft
28. 6 c = ___?___ pt
29. 8 qt = ___?___ pt
30. 4 gal = ___?___ qt
31. 4 T = ___?___ lb
32. 3 lb = ___?___ oz
33. 2400 lb = ___?___ T

Solve. [pages 128–129]

34. How many laps around a 440-yard track would you have to run to run a mile?

35. Jennifer bought 48 cups of milk for the cross-country team. How many gallons was that?

CHAPTER PROJECT

1. Hold your hand as if to shake hands with someone. Keep your thumb straight up. Have a classmate hold a meterstick with the zero mark just above your hand.
2. When the stick is dropped, catch it.
3. Record the number of centimeters that your thumb is on.
4. Repeat steps 1 through 3 until you have done them 10 times. Then find your average "catching distance."
5. Repeat the experiment using your other hand.
6. Are you quicker with your left hand or your right hand?

CHAPTER REVIEW

15.35 cm = ___?___ m

1 m = 100 cm

Divide by 100.

15.35 cm = 0.1535 m

Complete.

1. 723.1 m = ___?___ km
2. 52.7 m = ___?___ cm
3. 423 mm = ___?___ m
4. 53 cm = ___?___ m
5. 74.3 cm = ___?___ mm
6. 8.4 mm = ___?___ cm
7. 3.5 km = ___?___ m
8. 3.5 m = ___?___ km
9. 7.8 m = ___?___ mm
10. 6.7 dm = ___?___ m

1 L = 1000 mL

Complete.

11. 2.7 L = ___?___ mL
12. 250 mL = ___?___ L
13. 342.6 mL = ___?___ L
14. 342.6 L = ___?___ mL

1 kg = 1000 g

1 g = 1000 mg

Complete.

15. 2.6 g = ___?___ mg
16. 473 g = ___?___ kg
17. 750 mg = ___?___ g
18. 25 kg = ___?___ g
19. 750 g = ___?___ kg
20. 25 g = ___?___ mg

12 in. = 1 ft

3 ft = 1 yd

1760 yd = 1 mi

Complete.

21. 3 ft = ___?___ in.
22. 7 ft = ___?___ in.
23. 4 yd = ___?___ ft
24. 4 yd = ___?___ in.
25. 2 mi = ___?___ yd
26. 1 mi = ___?___ ft

2 c = 1 pt

2 pt = 1 qt

4 qt = 1 gal

16 oz = 1 lb

2000 lb = 1 T

Complete.

27. 8 c = ___?___ pt
28. 8 pt = ___?___ c
29. 3 qt = ___?___ pt
30. 8 gal = ___?___ qt
31. 1 gal = ___?___ pt
32. 4 qt = ___?___ c
33. 2.4 lb = ___?___ oz
34. 4.2 T = ___?___ lb

CHAPTER CHALLENGE

Error of measurement

Suppose that you measured each of these chains to the nearest centimeter.

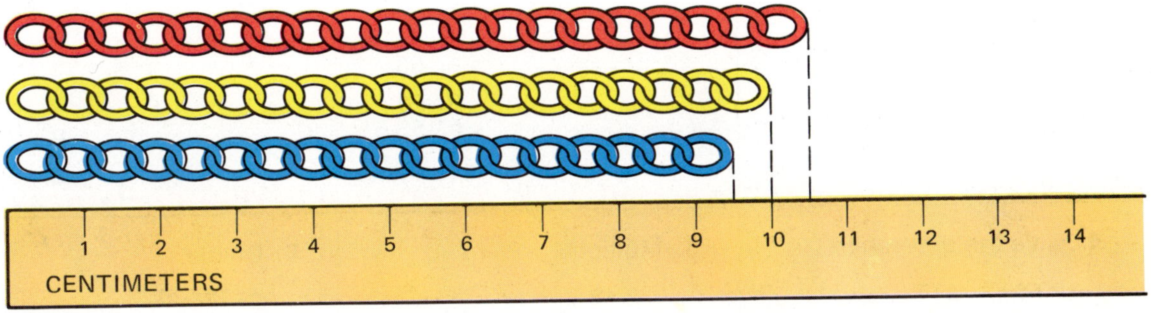

Each length would be recorded as 10 cm. In each case there is some error in the measurement—the difference between the true length and the recorded length. The **greatest possible error** (g.p.e.) occurs in the measurements of the blue chain and the red chain. The greatest possible error is always half of the unit used. In this case the g.p.e. is 0.5 cm.

If we use a smaller unit, we have less error. Compare these two measurements:

recorded measure: 7 m
unit used: meter
g.p.e.: 0.5 m

shortest possible true length: 6.5 m
longest possible true length: 7.5 m

recorded measure: 7.00 m
unit used: 0.01 m (1 cm)
g.p.e.: 0.005 m (0.5 cm)

shortest possible true length: 6.995 m
longest possible true length: 7.005 m

Give the g.p.e., the shortest possible true length, and the longest possible true length for each of these recorded measurements.

1. 15 m
2. 8 cm
3. 3.5 m
4. 20.45 m
5. 5.2 km
6. 6.0 cm
7. 18.5 mm
8. 15.00 m
9. A measurement is given as 8 m ± 0.5 m. The g.p.e. is 0.5 m.
   a. What is the shortest possible true length?
   b. What is the longest possible true length?

1. a b c d 2. a b c d 5. a b c d 6. a b
7. a b c d 8. a b c d 9. a b c d 10. a b c d 11. a b c d 12. a b

MAJOR CHECKUP
STANDARDIZED FORMAT

Choose the correct letter.

1. Give the standard numeral for fifty-seven tenths.

a. 0.57
b. 0.057
c. 57
d. none of these

2. Round 53.499 to the nearest whole number.

a. 53
b. 54
c. 53.5
d. none of these

3. Give the sum.

6.2 + 3.8 + 0.57

a. 1.57
b. 15.7
c. 10.57
d. none of these

4. Give the difference.

20.04 − 6.77

a. 13.27
b. 14.37
c. 23.27
d. none of these

5. Multiply.

$$\begin{array}{r} 2.48 \\ \times 0.13 \\ \hline \end{array}$$

a. 3.2240
b. 0.3224
c. 32.2400
d. none of these

6. Divide.

$25\overline{)75.25}$

a. 3.1
b. 31
c. 30.1
d. none of these

7. What is the average of 8, 6, 5, 9, and 8?

a. 7
b. 7.1
c. 7.2
d. none of these

8. Divide.

$0.08\overline{)3.216}$

a. 40.2
b. 4.02
c. 0.402
d. none of these

9. 324 mm = ___?___ m

a. 324,000
b. 3.24
c. 0.324
d. none of these

10. 256 kg = ___?___ g

a. 256,000
b. 0.256
c. 2.56
d. none of these

11. Avon is 39 km from Beta. Coro is 2.8 km closer to Beta than Avon is. How far is Coro from Beta?

a. 36.2 km **b.** 37.8 km
c. 11 km **d.** none of these

12. A tree is 58 m tall. A building is 0.61 km tall. How much taller is the building?

a. 3 m **b.** 300 m
c. 552 m **d.** none of these

Number Theory

5

Whole-number patterns

The set of whole numbers can be represented this way:

0, 1, 2, 3, . . .

Ancient Greek mathematicians were fascinated by the patterns in this set of numbers. Their explorations began a branch of mathematics called **number theory.**

The first pattern we consider is a **triangular number** pattern. The first four triangular numbers can be pictured this way:

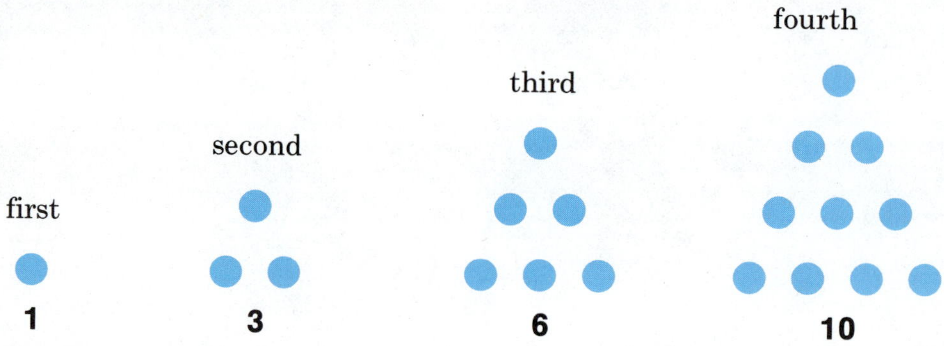

EXAMPLE. What is the fifth triangular number?

The bottom row of dots would have 5 dots, the row above it would have 4 dots, and so on. The top row would have 1 dot.

The fifth triangular number is $5 + 4 + 3 + 2 + 1 = 15$.

EXERCISES

1. Complete the chart. Draw pictures if you need to.

Triangular numbers							
1st	2nd	3rd	4th	5th	6th	7th	8th
1	3	6	10	15	?	?	?

2. Find the 12th triangular number.

3. Tell how you would find the 15th triangular number.

A **square number** pattern can be pictured in a similar way.

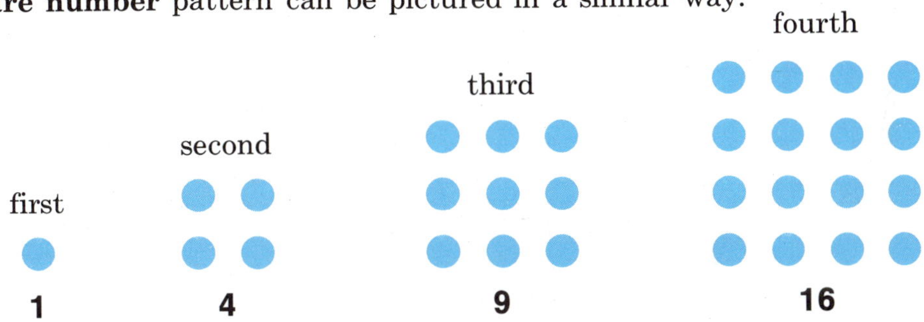

4. Complete the chart. Draw pictures if you need to.

Square numbers						
1st	2nd	3rd	4th	5th	10th	20th
1	4	9	16	25	?	?

The **odd numbers** 1, 3, 5, 7, . . . , can also be used to find the square numbers.

$1 = 1$

$3 + 1 = 4$

$5 + 3 + 1 = 9$

$7 + 5 + 3 + 1 = 16$

5. What odd numbers would you use to find the 10th square number?

6. Give two ways to find the 100th square number.

Factors

What rectangles can be made with 12 squares?

1 × 12

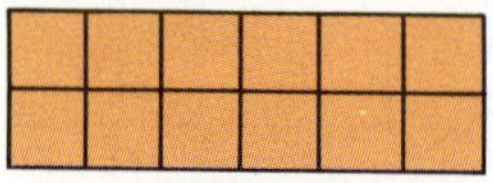

2 × 6

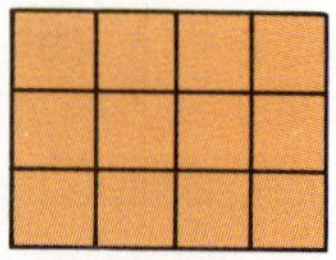

3 × 4

Each rectangle shows a product equal to 12. Since $3 \times 4 = 12$, both 3 and 4 are **factors,** or **divisors,** of 12. The other factors of 12 are 1, 2, 6, and 12.

Notice that 5 is not a factor of 12, because no whole number times 5 will give 12.

What rectangles can be made with 18 squares?

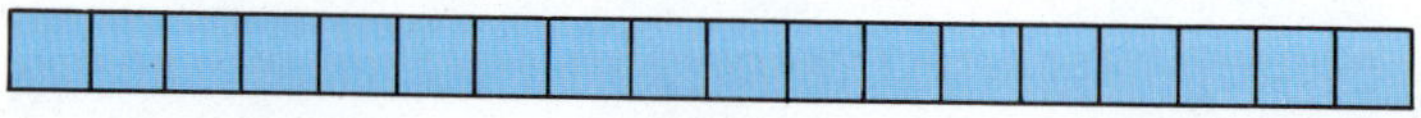

1 × 18

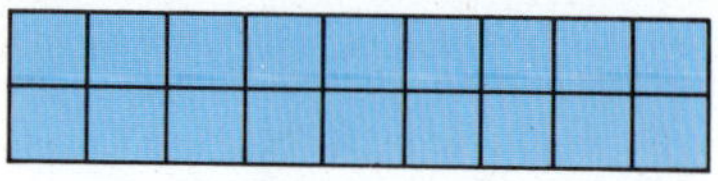

2 × 9

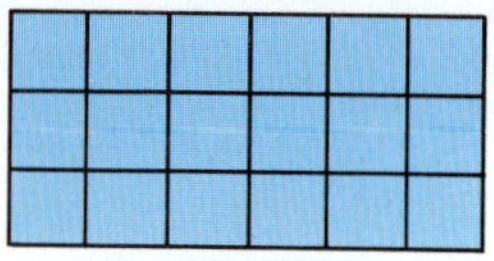

3 × 6

The factors of 18 are 1, 2, 3, 6, 9, and 18.

Notice that 1 and the number itself are always factors of the number.

factors of 12: 1, 2, 3, 4, 6, 12

factors of 18: 1, 2, 3, 6, 9, 18

1, 2, 3, and 6 are factors of both 12 and 18. They are **common factors** of 12 and 18.

The **greatest common factor (GCF)** of 12 and 18 is 6.

EXERCISES

1. Copy and complete the list of factors.

Number	*Factors*
2	1, 2
3	
4	
6	
8	
9	
10	
16	1, 2, ?, ?, 16

Number	*Factors*
17	
24	1, ?, ?, ?, 6, ?, ?, ?
30	
34	
36	1, 2, ?, ?, ?, 9, ?, ?, ?
40	
48	
50	

Give the GCF of each pair of numbers. Use the list to help you.

2. 6, 9
3. 12, 16
4. 10, 24
5. 16, 40
6. 4, 36
7. 10, 40
8. 24, 50
9. 24, 48
10. 17, 34
11. 36, 48
12. 24, 36
13. 16, 40
14. What number is a factor of every number?
15. What number is a common factor of every pair of numbers?
16. What is the smallest number that has exactly 3 factors?
17. What is the smallest number that has exactly 4 factors?

18. **DON'T BE FLOORED!**

There are fewer than 200 floor tiles.
2, 3, 7, and 9 are factors of the number.
How many floor tiles are there?

Divisibility

Dividing 225 by 3 leaves no remainder. We say that 225 is **divisible** by 3. Dividing 225 by 4 leaves a remainder. We say that 225 is not divisible by 4.

We can tell whether a whole number is divisible by 2, 3, 4, 5, 9, or 10 without actually dividing. The following rules tell whether a number is divisible by 2, 3, 4, 5, 9, or 10.

A whole number is divisible by

2

if its last digit is divisible by 2.

These numbers are divisible by 2:

36 120 148

372 514

3

if the sum of its digits is divisible by 3.

These numbers are divisible by 3:

1326 17,421

$1 + 3 + 2 + 6 = 12$ $1 + 7 + 4 + 2 + 1 = 15$

4

if its last two digits are divisible by 4.

These numbers are divisible by 4:

128 172 2908 3100

5

if its last digit is 5 or 0.

These numbers are divisible by 5:

95 170 2405

9

if the sum of its digits is divisible by 9.

These numbers are divisible by 9:

207 71,928

$2 + 0 + 7 = 9$ $7 + 1 + 9 + 2 + 8 = 27$

10

if its last digit is 0.

These numbers are divisible by 10:

830 4170

EXERCISES

Copy and complete.

	Number	*Divisible by*					
		2?	3?	4?	5?	9?	10?
1.	720	yes	?	?	?	?	?
2.	525	?	?	?	?	?	?
3.	724	?	?	?	?	?	?
4.	3612	?	?	?	?	?	?
5.	5634	?	?	?	?	?	?
6.	81,720	?	?	?	?	?	?
7.	38,955	?	?	?	?	?	?
8.	60,288	?	?	?	?	?	?

Use the divisibility rules to find the numbers.

9. What is the smallest 2-digit number divisible by 4?

10. What is the largest 2-digit number divisible by 10?

11. What is the smallest 3-digit number divisible by 3?

12. What is the largest 3-digit number divisible by 5?

13. What is the smallest 4-digit number divisible by 2?

14. What is the largest 4-digit number divisible by 9?

15. What is the smallest 5-digit number divisible by 2 and 3?

16. What is the largest 5-digit number divisible by 3 and 5?

17. Can you find and write a rule for divisibility by 6? (*Hint:* Think about the divisibility rules for 2 and 3.)

18. MYSTERY AGE

My age this year is divisible by 4.
Next year my age will be divisible by 3.
My older sister is now 19. How old am I?

Prime numbers and composite numbers

A whole number that has exactly two factors is called a **prime number.**

A whole number (other than 0) that has more than two factors is called a **composite number.**

The whole numbers 0 and 1 are neither prime nor composite.

$7 = 1 \times 7$

7 is a prime number.

$6 = 1 \times 6 \qquad 6 = 2 \times 3$

6 is a composite number.

Eratosthenes, an ancient Greek mathematician, invented a way to find prime numbers. It is called the Sieve of Eratosthenes.

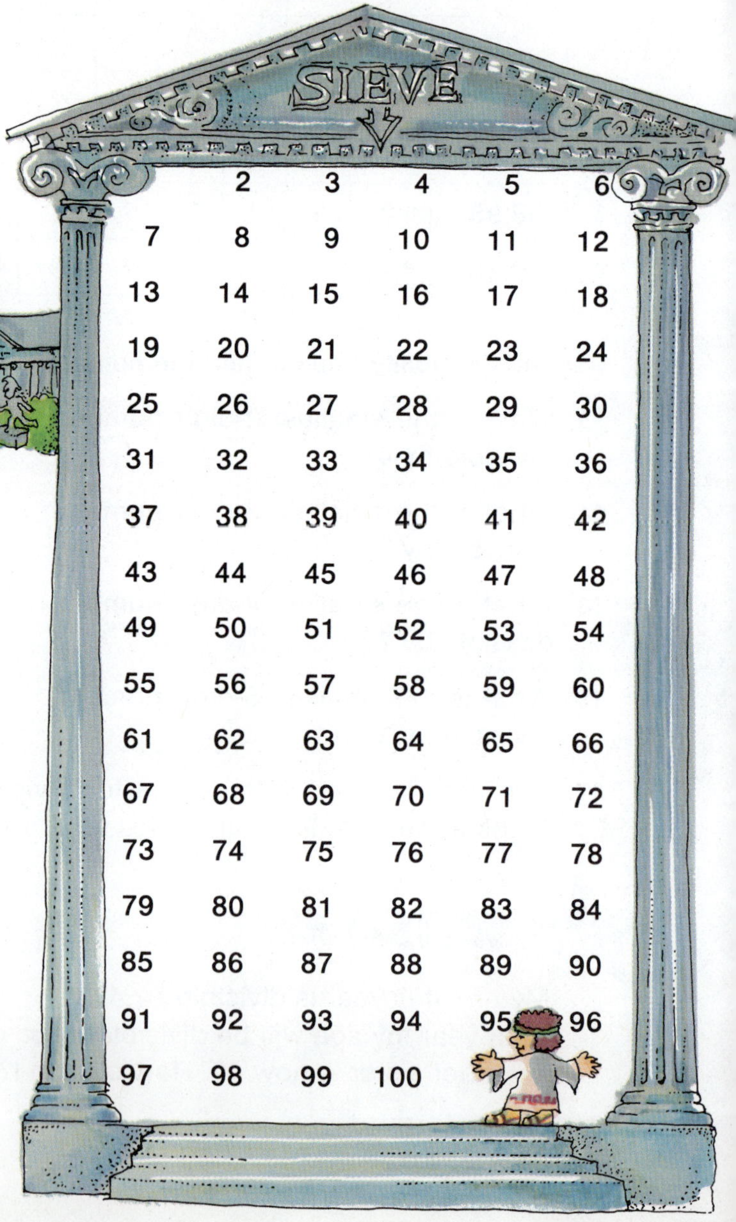

	2	3	4	5	6
7	8	9	10	11	12
13	14	15	16	17	18
19	20	21	22	23	24
25	26	27	28	29	30
31	32	33	34	35	36
37	38	39	40	41	42
43	44	45	46	47	48
49	50	51	52	53	54
55	56	57	58	59	60
61	62	63	64	65	66
67	68	69	70	71	72
73	74	75	76	77	78
79	80	81	82	83	84
85	86	87	88	89	90
91	92	93	94	95	96
97	98	99	100		

EXERCISES

Follow these steps to make your own sieve.

1. Make a number table as shown.
2. Cross off each number with a factor of 2 except 2.
3. The next number that is not crossed off is 3. Cross off each number with a factor of 3 except 3.
4. What is the next number not crossed off? Cross off each number with a factor of that number except the number itself.
5. Repeat step 4 until no more numbers can be crossed off. The numbers not crossed off are prime numbers.

Use your sieve to answer these questions.

6. How many numbers in your sieve are prime?

7. How many numbers in your sieve are composite?

8. How many primes in your sieve have a
   a. 3 in the ones place?
   b. 5 in the ones place?
   c. 7 in the ones place?
   d. 9 in the ones place?

9. The primes appear in only two columns after the first row. Why? Would this be true if you continued the sieve through 200?

10. How many prime numbers are even?

11. Are all composite numbers even?

12. What is the smallest composite number?

13. What is the greatest 2-digit prime number?

14. Find two pairs of primes that differ by
   a. 4 **b.** 14 **c.** 18

15. Twin primes are pairs of primes that differ by 2. How many twin primes are in your sieve?

16. Pairs of "mirror" primes such as 13 and 31 have their digits reversed. Find the pairs of mirror primes in your sieve.

17. True or false? "The difference of two primes, each greater than 2, is always an even number." Give a reason for your answer.

18. PRIME OF LIFE

a. My age is a prime number. The tens digit is 3 more than the ones digit. How old am I?

b. My age is a prime number. If you increase the tens digit by 1 and divide by 3, you get the ones digit. How old am I?

Prime factorization

Every composite number can be factored into a product of prime numbers. Notice that the numbers in the bottom row of each **factor tree** are prime numbers.

30 ← composite number → **30**

5×6 $\qquad$ 3×10

$5 \times 2 \times 3$ ← prime numbers → $3 \times 2 \times 5$

18

2×9

$2 \times 3 \times 3$

2×3^2

56

7×8

$7 \times 2 \times 4$

$7 \times 2 \times 2 \times 2$

$2^3 \times 7$

To express a composite number as a product of prime numbers is to give the **prime factorization** of the number.

$15 = 3 \times 5$

$$\begin{aligned}60 &= 6 \times 10\\ &= 2 \times 3 \times 2 \times 5\\ &= 2^2 \times 3 \times 5\end{aligned}$$

$$\begin{aligned}66 &= 6 \times 11\\ &= 2 \times 3 \times 11\end{aligned}$$

EXERCISES

Copy and complete these factor trees.

1.
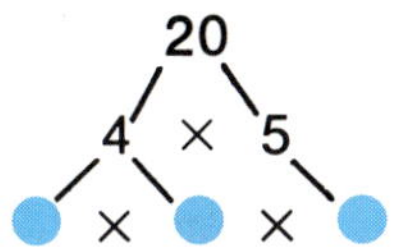

2.
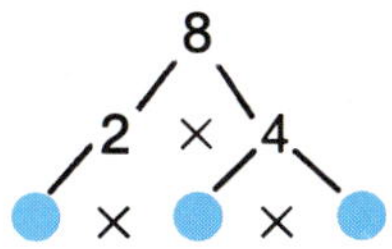

3.
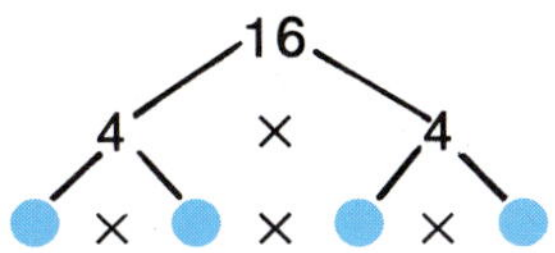

4.
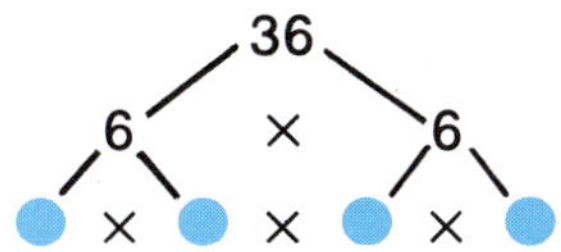

Give the prime factorization. Use exponents for factors that occur more than once.

5. 6	**6.** 14	**7.** 21	**8.** 28
9. 32	**10.** 38	**11.** 46	**12.** 49
13. 50	**14.** 63	**15.** 64	**16.** 72
17. 78	**18.** 84	**19.** 96	**20.** 100
21. 112	**22.** 128	**23.** 132	**24.** 144
25. 153	**26.** 165	**27.** 176	**28.** 195

29. GOLDBACH SUMS

Any even number greater than 2 can be expressed as the sum of two primes. (This was first proposed by a mathematician called Goldbach.)

$$4 = 2 + 2$$

$$6 = 3 + 3$$

$$8 = 3 + 5$$

Express other even numbers as the sum of two primes.

KEEPING SKILLS SHARP

Give each sum or difference.

1. 3.86 + 2.59

2. 7.09 − 3.14

3. 2.678 + 3.259

4. 9.1783 − 2.9999

5. 15.86 + 123.4

6. 32.785 − 2.96

7. 252 − 1.96

8. 58.03 + 26.529

9. 741.03 + 26.975

10. 28.1 − 6.942

11. 8.21 + 53.9

12. 15.06 + 218.4

13. 64.9 − 7.36

14. 302.8 − 24.07

15. 6.5 − 0.613

16. 8.7 + 0.372

17. 0.8 − 0.24

18. 57 − 0.62

19. 0.58 + 0.9

20. 29 + 101.6

Multiples

If you begin at 1 and multiply each whole number by 3, you get the numbers shown in red. These are the **multiples** of 3.

1 2 3 4 5 6 7 8 9 10 11 12 13 14 15 16 17 18 19 20 21 22 23 24

multiples of 3: 3, 6, 9, 12, . . .

If you begin at 1 and multiply each whole number by 4, you get the multiples of 4.

1 2 3 4 5 6 7 8 9 10 11 12 13 14 15 16 17 18 19 20 21 22 23 24 25 26 27 28

multiples of 4: 4, 8, 12, 16, . . .

The numbers that are multiples of both 3 and 4 are called **common multiples** of 3 and 4:

12, 24, 36, . . .

The **least common multiple (LCM)** of 3 and 4 is 12.

EXERCISES

Give four multiples of each number.

1. 2 **2.** 3 **3.** 5 **4.** 7 **5.** 9 **6.** 12

Give two common multiples for each pair of numbers.

7. 2, 3 **8.** 3, 4 **9.** 2, 4 **10.** 5, 7 **11.** 10, 6 **12.** 3, 6

13. Give the LCM for each pair of numbers in exercises 7–12.

True or false?

14. Every number is a multiple of itself.

15. Every number is a factor of itself.

16. 1 is a multiple of every number.

17. 1 is a factor of every number.

18. Every number is a multiple of 1.

19. Every number is a factor of 1.

20. 2 is a factor of 6.

21. 6 is a multiple of 2.

22. 5 is a factor of 10.

23. 10 is a multiple of 5.

24. 6 is a factor of 9.

25. 9 is a multiple of 6.

26. If a first number is a factor of a second number, then the second number is a multiple of the first number.

Copy and complete.

27. 

28. 

29. 

30. 

31. In exercises 27–30, compare the product of the "green" numbers with the product of the "yellow" numbers. What do you find?

32. Zero and multiples of 2 are called **even numbers.** What can you say about the digit in the ones place of any even number?

33. Whole numbers that are not even numbers are called **odd numbers.** What can you say about the digit in the ones place of any odd number?

34. **AGE OLD MYSTERY**

The least common multiple of our ages is 60. The greatest common factor of our ages is 3. I am 12 years old. What is her age?

Variables and expressions

If we let n be the regular price of any radio, then the sale price can be given by the expression

$$n - 5$$

The letter n is called a **variable.**

To find the sale price of each radio shown, we substitute the regular price for n and simplify the expression.

	Pocket model	Power-pack model	Deluxe model
Regular price in dollars (n)	39	59	83
Substitute in $n - 5$	39 − 5	59 − 5	83 − 5
Simplify	34	?	?

The pocket model is on sale for $34. What is the sale price of the power-pack model? The deluxe model?

Each telescope is on sale for half price. If we use n for the regular price, then the sale price can be given by the expression

$$n \div 2$$

We can substitute and simplify to find each sale price.

	Beginner's model	Junior model	Expert model
Regular price in dollars (n)	88	128	416
Substitute in $n \div 2$	88 ÷ 2	128 ÷ 2	416 ÷ 2
Simplify	44	?	?

The beginner's model is on sale for $44. What is the sale price of the junior model? The expert model?

EXERCISES

Copy and complete.

1.

$9-off sale					
Regular price in dollars (n)	15	24	16	21	38
Substitute in $n - 9$	?	?	?	?	?
Simplify	?	?	?	?	?

2.

Half-price sale					
Regular price in dollars (n)	60	38	46	51	72
Substitute in $n \div 2$	?	?	?	?	?
Simplify	?	?	?	?	?

3.

n	15	32	9	44	88	101
$n + 7$	?	?	?	?	?	?
Simplify	?	?	?	?	?	?

4.

n	10	15	22	39	50	11
$6 \times n$	?	?	?	?	?	?
Simplify	?	?	?	?	?	?

Substitute the prices and simplify. Give your answers in dollars.

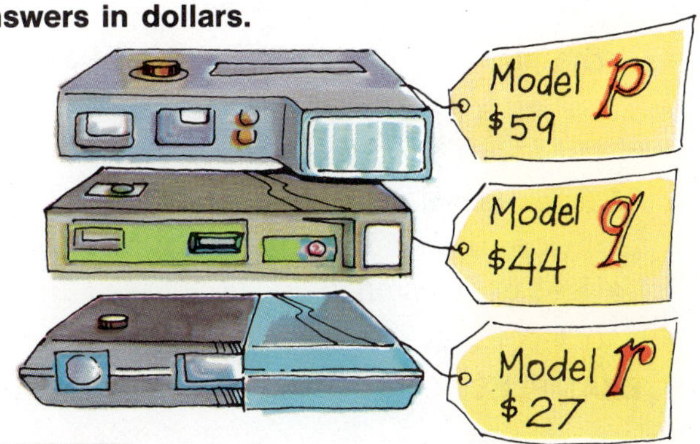

5. $p + q$
6. $3 \times r$
7. $q \div 2$
8. $p - r$
9. $p + r$
10. $q - r$
11. $2 \times p$
12. $r \div 3$
13. $r + q$

Substitute and simplify.

a	b	D	d	S	T
9	6	7	18	5	27

14. $a \times b$
15. $D \times d$
16. $D + d$
17. $d \div b$
18. $6 \times S$
19. $a \div 3$
20. $T \div a$
21. $d - D$
22. $a + D + S$
23. $2 \times a \times S$
24. $(D + d) - a$
25. $(T - D) \div S$

Writing expressions

Look at the expression cards below. They show mathematical expressions for different word expressions.

$n + 3$

a number added to 3
the sum of a number and 3
a number plus 3
3 more than a number
a number increased by 3
3 added to a number

$n - 3$

3 subtracted from a number
a number minus 3
3 less than a number
a number decreased by 3

$3 \times n$

3 times a number
a number times 3
3 multiplied by a number
a number multiplied by 3
the product of 3 and a number

$n \div 3$

a number divided by 3
one third of a number

Mathematical expressions are short ways of writing word expressions that refer to addition, subtraction, multiplication, or division.

EXAMPLES.

a number increased by 5 $\longrightarrow g + 5$

8 less than a number $\longrightarrow e - 8$

6 times a number $\longrightarrow 6 \times d$

one fourth of a number $\longrightarrow r \div 4$

EXERCISES

Write a mathematical expression for

1. 4 more than a number n.
2. 6 more than a number x.
3. 7 less than a number y.
4. 12 less than a number y.
5. a number x increased by 3.
6. a number y increased by 5.
7. a number r decreased by 1.
8. a number z decreased by 3.
9. 5 times a number b.
10. 6 times a number x.
11. the sum of a number m and 9.
12. the product of 5 and a number y.
13. 7 subtracted from a number c.
14. a number a divided by 2.
15. 6 divided by a number d.
16. 3 less than 6 times a number m.

Let c be the number of telephone calls you made last week. Write a mathematical expression for the number of calls that is

17. 5 telephone calls less.
18. 7 telephone calls more.
19. 2 times as many.
20. your number of calls decreased by 4 calls.
21. your number of calls divided by 6.
22. your number of calls increased by 8 calls.

Let y be your age in years. Write a mathematical expression for the age that is

23. your age increased by 10 years.
24. your age decreased by 3 years.
25. 4 times as old.
26. one half as old.
27. ★ 3 years less than 6 times as old.
28. ★ 4 years more than 2 times as old.
29. ★ 3 times as old plus 5 years.
30. ★ 4 times as old minus 15 years.

Solving addition and subtraction equations

If we let n be the number of marbles in the box that is closed, we get the equation

$$n + 3 = 11$$

$n + 3 = 11$

To **solve** the equation is to find the number that we can substitute for n to make the equation true.

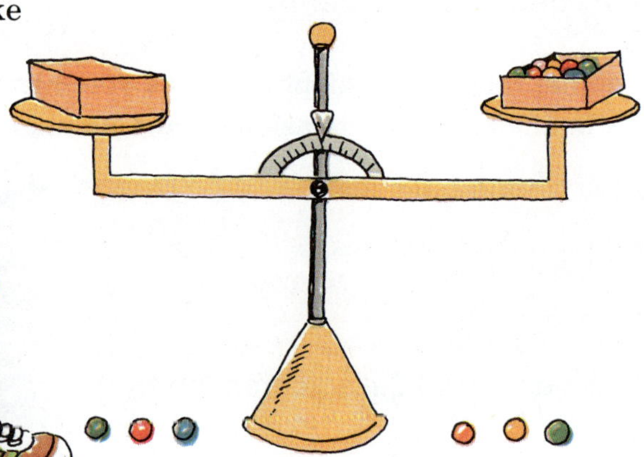

To find the solution, we can subtract 3 from *both sides* and then simplify both sides.

$$n + 3 - 3 = 11 - 3$$

$$n = 8$$

To check the solution, we substitute 8 for n in the first equation. Since the equation is true, it checks!

Check: $8 + 3 = 11$

To solve this equation, we can add 6 to *both sides* and then simplify both sides.

$$m - 6 = 17$$

$$m - 6 + 6 = 17 + 6$$

$$m = 23$$

Check: $23 - 6 = 17$

EXERCISES

Solve and check.

1. $n + 6 = 23$
   Hint: Subtract 6 from both sides.
2. $8 + n = 37$
   Hint: Subtract 8 from both sides.
3. $x + 7 = 19$
4. $11 + n = 27$
5. $10 + h = 38$
6. $z + 19 = 29$
7. $y + 18 = 46$
8. $15 + t = 31$
9. $17 + y = 35$
10. $s + 23 = 29$
11. $r + 37 = 43$
12. $27 + w = 52$
13. $42 + n = 73$
14. $38 + k = 95$
15. $14 + x = 29$
16. $43 + y = 102$
17. $n - 28 = 35$
    Hint: Add 28 to both sides.
18. $k - 42 = 39$
    Hint: Add 42 to both sides.
19. $a - 18 = 19$
20. $h - 23 = 16$
21. $e - 27 = 23$
22. $r - 15 = 25$
23. $k - 32 = 28$
24. $b - 35 = 73$
25. $d - 42 = 32$
26. $g - 33 = 26$
27. $x - 30 = 53$
28. $s - 39 = 61$
29. $f - 28 = 45$
30. $c - 46 = 58$
31. $y - 47 = 40$
32. $j - 52 = 27$
33. $x + 17 = 32$
34. $a - 62 = 41$
35. $19 = n - 58$
36. $53 = x - 72$
37. $81 = 19 + y$
38. $69 = r - 15$
39. ★ $14 - a = 9$
40. ★ $65 - x = 27$

KEEPING SKILLS SHARP

Estimate to find the two incorrect products.

1. a. $\begin{array}{r} 2.05 \\ \times 6.8 \\ \hline 1.394 \end{array}$

   b. $\begin{array}{r} 0.84 \\ \times 6.5 \\ \hline 5.46 \end{array}$

   c. $\begin{array}{r} 6.8 \\ \times 3.5 \\ \hline 23.8 \end{array}$

   d. $\begin{array}{r} 90.4 \\ \times 0.95 \\ \hline 95.88 \end{array}$

Multiply.

2. $\begin{array}{r} 3.8 \\ \times 1.6 \\ \hline \end{array}$

3. $\begin{array}{r} 38.2 \\ \times 3.8 \\ \hline \end{array}$

4. $\begin{array}{r} 6.75 \\ \times 0.14 \\ \hline \end{array}$

5. $\begin{array}{r} 396 \\ \times 0.25 \\ \hline \end{array}$

6. $\begin{array}{r} 25.35 \\ \times 1.05 \\ \hline \end{array}$

Solving multiplication and division equations

Each box contains the same number of marbles. If we let n be the number of marbles in each box, we get the equation

$3 \times n$

$$3n = 12$$

To find the solution, we *divide* both sides by 3 and then simplify both sides.

$$\frac{3n}{3} = \frac{12}{3}$$

$$n = 4$$

The solution is 4. To check the solution, we substitute 4 for n in the original equation.

Check: $3 \times 4 = 12$

We get a true equation, so the solution is 4.

In this equation, n is divided by 5. To find the solution, we multiply both sides by 5 and simplify both sides.

$$\frac{n}{5} = 7$$

$$\frac{5 \times n}{5} = 5 \times 7$$

$$n = 35$$

Check: $\frac{35}{5} = 7$

The solution is 35.

EXERCISES

Solve and check.

1. $4n = 24$
2. $5n = 30$
3. $6n = 42$
4. $3y = 21$
5. $2n = 12$
6. $8k = 88$
7. $5z = 50$
8. $7n = 70$
9. $4x = 28$
10. $8w = 120$
11. $4j = 48$
12. $9y = 108$
13. $2n = 18$
14. $6n = 72$
15. $7n = 91$
16. $3n = 51$
17. $3x = 45$
18. $6n = 120$
19. $\frac{w}{6} = 4$
20. $\frac{k}{7} = 5$
21. $\frac{r}{4} = 7$

Hint: Multiply by 6.

22. $\frac{t}{8} = 6$
23. $\frac{t}{12} = 4$
24. $\frac{n}{15} = 3$
25. $\frac{k}{20} = 8$
26. $\frac{y}{4} = 13$
27. $\frac{r}{25} = 7$
28. $\frac{n}{17} = 9$
29. $\frac{x}{21} = 40$
30. $\frac{w}{34} = 6$
31. $\frac{c}{11} = 12$
32. $\frac{n}{13} = 20$
33. $\frac{x}{5} = 22$

Find each number by writing and solving an equation. Check the solution in the problem.

I AM THINKING OF A NUMBER

34. If I multiply it by 3, I get 42.

★ 35. If I multiply it by 4, I get 39.

★ 36. If I add 19 to it, I get 57.

★ 37. If I subtract 37 from it, I get 74.

KEEPING SKILLS SHARP

Estimate to find the incorrect answer.

1. a. $4.6\overline{)232.3}$ (answer: 50.5)

   b. $0.73\overline{)175.2}$ (answer: 24)

Divide.

2. $2.1\overline{)20.223}$
3. $1.1\overline{)3916}$
4. $0.30\overline{)16.110}$
5. $3.2\overline{)2.1312}$
6. $50\overline{)458.50}$
7. $6.3\overline{)505.89}$
8. $5.8\overline{)25.346}$
9. $3.2\overline{)14}$
10. $36\overline{)26.568}$
11. $0.046\overline{)3.1326}$

Formulas

By the time you finish reading this sentence, the earth will have traveled about 200 kilometers around the sun.

The table shows the distance traveled in terms of time.

In 1 second, the earth travels 30 kilometers.

In 2 seconds, the earth travels 60 kilometers.

Notice that the distance traveled in kilometers is 30 times the time in seconds.

Time (seconds)	*Distance (kilometers)*
1	30
2	60
3	90
4	120
5	150
6	180
7	210
8	240
9	270

The relationship can be expressed as a **formula:**

$$d = 30t$$

where d is the distance in kilometers and t is the time in seconds.

EXAMPLE. How far does the earth travel in 20 seconds?

Substitute 20 for t in the formula:

$$\begin{aligned} d &= 30 \times 20 \\ &= 600 \end{aligned}$$

The earth travels 600 kilometers in 20 seconds.

EXERCISES

Use the formula. How far does the earth travel in the following times?

1. 15 seconds **2.** 45 seconds **3.** 60 seconds **4.** 1 hour (3600 seconds)

The formula below gives the distance, m, that you could long-jump on the moon in terms of the distance, e, you could long-jump on the earth:

$$m = 6.02e$$

Use the formula to find how far these long-jumps on the earth would be on the moon.

5. 5 meters

6. 6.5 meters

7. 3.91 meters

8. 5.67 meters

9. The record on the earth for the long-jump is 7.09 meters. Would a 42.14-meter long-jump on the moon set a new record?

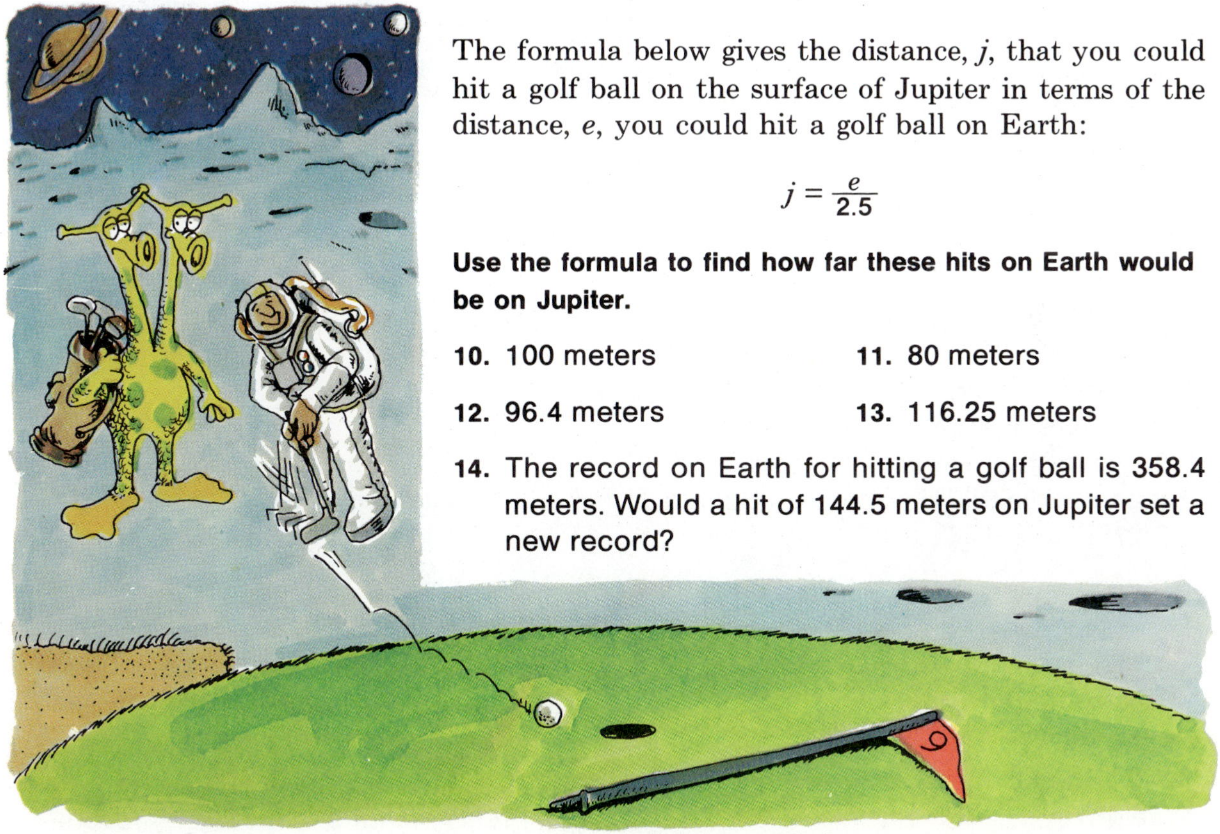

The formula below gives the distance, j, that you could hit a golf ball on the surface of Jupiter in terms of the distance, e, you could hit a golf ball on Earth:

$$j = \frac{e}{2.5}$$

Use the formula to find how far these hits on Earth would be on Jupiter.

10. 100 meters

11. 80 meters

12. 96.4 meters

13. 116.25 meters

14. The record on Earth for hitting a golf ball is 358.4 meters. Would a hit of 144.5 meters on Jupiter set a new record?

Equations in problem solving

The pony express carried mail from St. Joseph, Missouri, to Sacramento, California, a distance of 1966 miles. In 1860 this ad appeared in a newspaper:

WANTED - Young, skinny, wiry fellows not over 18. Must be expert riders willing to risk death.

To solve each problem:

- Write an equation.
- Solve the equation.
- Check the solution in the problem.

Solve.

1. Two of the most famous men who worked for the pony express were James B. "Wild Bill" Hickok and William F. Cody, "Buffalo Bill." When they were hired, Wild Bill Hickok was 23 years old. He was 8 years older than Buffalo Bill. How old was Buffalo Bill?

   Hint: $n + 8 = 23$

   (n = Buffalo Bill's age; $n + 8$ = Wild Bill's age; 23 = Wild Bill's age)

2. The pony express had a weight limit of 125 pounds for their riders. Wild Bill Hickok weighed about 33 pounds more than the limit. How much did he weigh?

3. At one time a total of 477 men worked for the pony express. Of these, 79 were riders and the rest worked at the relay stations. How many worked at the relay stations?

4. Pony express riders rode the same route as the Overland Mail stagecoaches. The Overland Mail drivers were paid $350 per month. This was 7 times the salary the pony express riders were paid. How much were the pony express riders paid per month?

Pony express rider's salary

Hint: $\underbrace{7 \times n}$ = 350

Overland Mail driver's salary — Overland Mail driver's salary

5. By Overland Mail, a letter from St. Joseph, Missouri, to Sacramento took 21 days. This was 9 days longer than by pony express. How many days did it take by pony express?

6. The fee for carrying letters by the pony express was $10 per ounce. This was 5 times the fee charged by the Overland Mail. How much did the Overland Mail charge per ounce?

7. The distance between relay stations was about 11 miles. After a rider reached the first relay station, he still had about 64 miles left to ride. About how far did each rider ride?

8. The record time of 185 hours was set when the pony express carried Lincoln's inaugural address. It took 55 hours less than the first trip. How long did it take to make the first trip?

9. The pony express ended its career 18 months after it began. During this time, 11,397 pieces of mail were carried west to Sacramento. This was 11,959 pieces less than the amount of mail carried east to St. Joseph. How many pieces of mail were carried east by the pony express?

CHAPTER CHECKUP

Give the greatest common factor of the two numbers. [pages 142–143]

1. 12, 18 **2.** 8, 10 **3.** 6, 12 **4.** 14, 15

Give the prime factorization of each number. [pages 146–149].

5. 6 **6.** 18 **7.** 36 **8.** 42 **9.** 72 **10.** 78

Give the least common multiple of the two numbers. [pages 150–151]

11. 3, 5 **12.** 4, 6 **13.** 5, 10 **14.** 8, 9

Substitute and simplify. [pages 152–153]

b	c	P	Q
3	2	36	9

15. $b + 5$ **16.** $7 \times c$ **17.** $15 - Q$ **18.** $18 \div b$

19. $P + c$ **20.** $P \div Q$ **21.** $20 - (b + Q)$ **22.** $5 \times (Q - b)$

Let h be your height in centimeters. Write a mathematical expression for the height that is [pages 154–155]

23. 3 centimeters less. **24.** 2 times as much. **25.** 5 centimeters more.

Solve each equation. [pages 156–159]

26. $n + 18 = 21$ **27.** $n - 6 = 19$ **28.** $14 + n = 42$ **29.** $37 = n - 10$

30. $3n = 21$ **31.** $5n = 65$ **32.** $\frac{n}{4} = 8$ **33.** $\frac{n}{3} = 30$

Solve. [pages 162–163]

34. Judy had 12 records. Frank had 3 times as many. How many records did Frank have?

35. George read 13 books. This was 5 fewer books than Barbara read. How many books did Barbara read?

CHAPTER PROJECT

1. The dot diagram above shows the factors of the numbers 1 through 6. On grid paper make a similar dot diagram for the numbers 1 through 12.
2. Find a way to use your dot diagram to find the common factors of 6 and 12.
3. How can you use your dot diagram to find the GCF of 6 and 12?
4. Use your dot diagram to find the GCF of 4, 8, and 10.
5. How can you use your dot diagram to find common multiples of 2 and 3?
6. Use your dot diagram to find the LCM of 2, 3, and 4.

CHAPTER REVIEW

factors of 8: 1, 2, 4, 8
factors of 12: 1, 2, 3, 4, 6, 12
The greatest common factor is 4.

Give the greatest common factor (GCF).

1.	2.	3.	4.
6	12	18	27
9	16	24	45

Whole numbers with exactly 2 factors are called **prime numbers.**

(1, 7)
7 is a prime number

Which of these numbers are prime numbers?

5. 9 6. 11 7. 19 8. 27

9. 39 10. 43 11. 49 12. 51

multiples of 6: 6, 12, 18, . . .
multiples of 9: 9, 18, 27, . . .
The least common multiple is 18.

Give the least common multiple (LCM).

13.	14.	15.	16.
5	8	12	15
6	12	16	20

$8 + e$
$8 + 7$
15

Substitute and simplify.

e	f	N	T
7	10	12	3

17. $N - 7$
18. $f \div 2$
19. $e \times T$
20. $6 \times (f + T)$

$n + 1 = 17$
$n + 1 - 1 = 17 - 1$
$n = 16$
Check: $16 + 1 = 17$

$5n = 15$
$\frac{5n}{5} = \frac{15}{5}$
$n = 3$
Check: $5 \times 3 = 15$

Solve and check.

21. $n + 13 = 23$
22. $n - 18 = 30$
23. $9n = 27$
24. $15n = 165$
25. $\frac{n}{2} = 8$
26. $12 = n - 7$

CHAPTER CHALLENGE

Study the clues. Solve.

1. How many A's must be placed in the empty pan to balance the C ?

2. How many B's are needed to balance the two C's and the D?

3. How many C's are needed to balance the B and the D ?

MAJOR CHECKUP
STANDARDIZED FORMAT

Choose the correct letter.

1. 56,752 rounded to the nearest hundred is
   - a. 56,700
   - b. 56,800
   - c. 57,000
   - d. none of these

2. Add.

$$\begin{array}{r} 624 \\ 38 \\ +577 \\ \hline \end{array}$$

   - a. 1238
   - b. 1139
   - c. 1239
   - d. none of these

3. Divide.

$43\overline{)8944}$

   - a. 208
   - b. 280
   - c. 28
   - d. none of these

4. The standard numeral for five hundred and six thousandths is
   - a. 0.506
   - b. 600.006
   - c. 506,000
   - d. none of these

5. Give the difference.

$9.6 - 0.38$

   - a. 9.22
   - b. 5.8
   - c. 0.58
   - d. none of these

6. Multiply.

$$\begin{array}{r} 0.13 \\ \times 0.04 \\ \hline \end{array}$$

   - a. 0.52
   - b. 0.052
   - c. 0.0052
   - d. none of these

7. Divide.

$0.9\overline{)3.771}$

   - a. 4.19
   - b. 41.9
   - c. 0.419
   - d. none of these

8. 35 mL = ___?___ L
   - a. 0.350
   - b. 0.035
   - c. 35,000
   - d. none of these

9. The greatest common factor of 8 and 12 is
   - a. 24
   - b. 4
   - c. 1
   - d. none of these

10. The least common multiple of 4 and 6 is
    - a. 12
    - b. 24
    - c. 2
    - d none of these

11. Anne spent $20 for 1 book and 2 tapes. She spent $8 for the book. How much did she spend for the 2 tapes?
    - a. $12
    - b. $6
    - c. $28
    - d. none of these

12. Frank had 24 tapes. This was 4 times as many tapes as Ira had. How many tapes did Ira have?
    - a. 96
    - b. 20
    - c. 6
    - d. none of these

Fractions—Addition and Subtraction

6

READY OR NOT?

What fraction of the region is colored?

1. 

2. 

3. 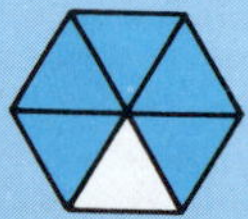

4. 

5. 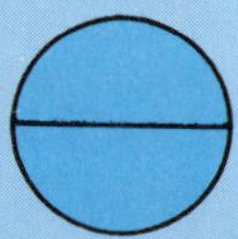

What fraction of the objects are colored?

6. 

7. 

8. 

9. 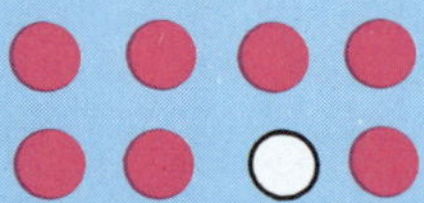

10. 

Equivalent fractions

Fractions that name the same number are called **equivalent fractions.** Multiplying both the numerator and the denominator by the same nonzero number gives an equivalent fraction.

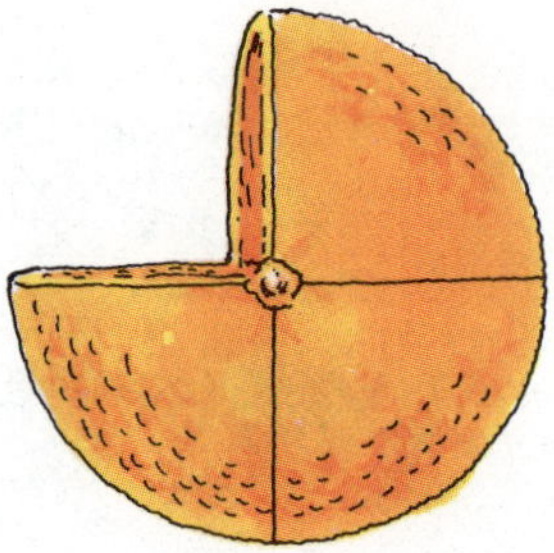

3 fourths of an orange

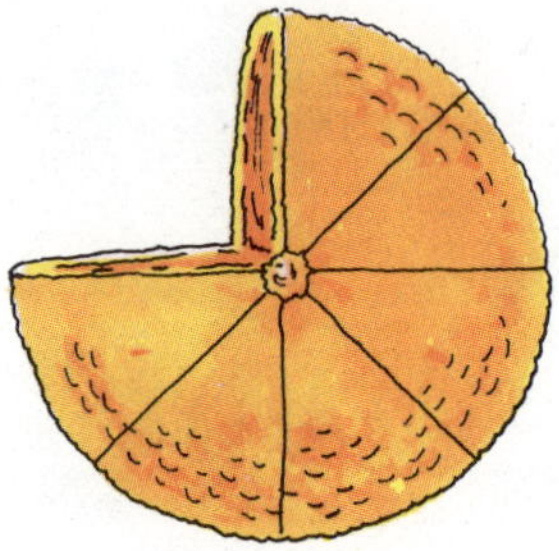

6 eighths of an orange

numerator → $\frac{3}{4} = \frac{6}{8}$ ← denominator (×2 on the numerator, ×2 on the denominator)

EXERCISES

Give the equivalent fraction.

1. $\frac{1}{3} \overset{\times 2}{\underset{\times 2}{=}} ?$
2. $\frac{1}{2} \overset{\times 3}{\underset{\times 3}{=}} ?$
3. $\frac{1}{4} \overset{\times 2}{\underset{\times 2}{=}} ?$
4. $\frac{3}{4} \overset{\times 4}{\underset{\times 4}{=}} ?$
5. $\frac{3}{8} \overset{\times 5}{\underset{\times 5}{=}} ?$
6. $\frac{3}{5} \overset{\times 3}{\underset{\times 3}{=}} ?$
7. $\frac{4}{3} \overset{\times 2}{\underset{\times 2}{=}} ?$
8. $\frac{3}{2} \overset{\times 6}{\underset{\times 6}{=}} ?$
9. $\frac{5}{4} \overset{\times 4}{\underset{\times 4}{=}} ?$
10. $\frac{5}{6} \overset{\times 3}{\underset{\times 3}{=}} ?$

Copy and complete. ***Hint:*** **First decide what the denominator was multiplied by.**

11. $\frac{1}{2} = \frac{?}{8}$
12. $\frac{2}{2} = \frac{?}{8}$
13. $\frac{1}{5} = \frac{?}{20}$
14. $\frac{3}{2} = \frac{?}{10}$
15. $\frac{1}{4} = \frac{?}{16}$
16. $\frac{2}{3} = \frac{?}{9}$
17. $\frac{4}{3} = \frac{?}{12}$
18. $\frac{1}{3} = \frac{?}{6}$
19. $\frac{5}{2} = \frac{?}{10}$
20. $\frac{1}{2} = \frac{?}{12}$
21. $\frac{1}{3} = \frac{?}{12}$
22. $\frac{8}{3} = \frac{?}{6}$
23. $\frac{5}{2} = \frac{?}{6}$
24. $\frac{2}{3} = \frac{?}{12}$
25. $\frac{1}{6} = \frac{?}{18}$
26. $\frac{3}{2} = \frac{?}{6}$
27. $\frac{2}{3} = \frac{?}{6}$
28. $\frac{1}{2} = \frac{?}{6}$
29. $\frac{3}{3} = \frac{?}{9}$
30. $\frac{1}{3} = \frac{?}{15}$
31. $\frac{1}{2} = \frac{?}{10}$
32. $\frac{0}{2} = \frac{?}{8}$
33. $\frac{4}{3} = \frac{?}{6}$
34. $\frac{1}{4} = \frac{?}{20}$
35. $\frac{8}{7} = \frac{?}{14}$
36. $\frac{5}{2} = \frac{?}{8}$
37. $\frac{1}{3} = \frac{?}{9}$
38. $\frac{1}{5} = \frac{?}{10}$
39. $\frac{1}{4} = \frac{?}{8}$
40. $\frac{1}{2} = \frac{?}{4}$
41. $\frac{1}{4} = \frac{?}{12}$
42. $\frac{3}{2} = \frac{?}{4}$
43. $\frac{3}{3} = \frac{?}{6}$
44. $\frac{1}{6} = \frac{?}{24}$
45. $\frac{0}{2} = \frac{?}{6}$

Solve.

46. Which is more, $\frac{5}{12}$ of the oranges or $\frac{7}{12}$ of the oranges?
47. Which is less, $\frac{4}{6}$ of the oranges or $\frac{5}{6}$ of the oranges?
48. Which is more, $\frac{1}{2}$ or $\frac{3}{4}$?
49. Which is less, $\frac{5}{6}$ or $\frac{3}{4}$?

Fractions in lowest terms

Dividing both the numerator and the denominator of a fraction by their greatest common factor gives the equivalent fraction that is in **lowest terms.**

$\frac{6}{9}$ of the pencils are blue.

$$\frac{6 \div 3}{9 \div 3} = \frac{2}{3}$$

$\frac{6}{9} = \frac{2}{3}$ ← fraction in lowest terms

$\frac{2}{3}$ of the pencils are blue.

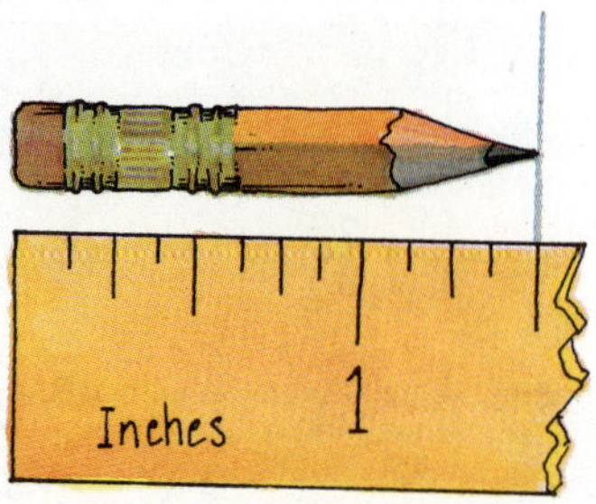

The pencil is $\frac{12}{8}$, $\frac{6}{4}$, or $\frac{3}{2}$ inches long.

$$\frac{12 \div 4}{8 \div 4} = \frac{3}{2}$$

$\frac{12}{8} = \frac{3}{2}$ ← fraction in lowest terms

Here is a shortcut that can be used to write a fraction in lowest terms.

$$\frac{\overset{2}{\cancel{6}}}{\underset{3}{\cancel{9}}} \qquad \frac{\overset{3}{\cancel{12}}}{\underset{2}{\cancel{8}}}$$

A fraction is in lowest terms if the greatest common factor of the numerator and the denominator is 1.

EXERCISES

Give the greatest common factor of the numerator and the denominator.

1. $\frac{2}{6}$ 2. $\frac{14}{6}$ 3. $\frac{10}{5}$ 4. $\frac{12}{18}$ 5. $\frac{6}{8}$

6. $\frac{18}{15}$ 7. $\frac{12}{4}$ 8. $\frac{6}{4}$ 9. $\frac{3}{9}$ 10. $\frac{4}{16}$

Write each fraction in lowest terms.

11. $\frac{6}{10}$ 12. $\frac{6}{9}$ 13. $\frac{6}{12}$ 14. $\frac{2}{4}$ 15. $\frac{10}{8}$

16. $\frac{10}{12}$ 17. $\frac{4}{16}$ 18. $\frac{2}{8}$ 19. $\frac{2}{6}$ 20. $\frac{6}{10}$

21. $\frac{15}{18}$ 22. $\frac{6}{4}$ 23. $\frac{3}{9}$ 24. $\frac{15}{10}$ 25. $\frac{3}{6}$

26. $\frac{5}{15}$ 27. $\frac{4}{6}$ 28. $\frac{3}{12}$ 29. $\frac{4}{8}$ 30. $\frac{18}{15}$

31. $\frac{9}{6}$ 32. $\frac{10}{4}$ 33. $\frac{4}{12}$ 34. $\frac{16}{32}$ 35. $\frac{10}{15}$

36. $\frac{10}{20}$ 37. $\frac{5}{10}$ 38. $\frac{12}{24}$ 39. $\frac{24}{18}$ 40. $\frac{32}{12}$

Who am I?

41. I am a fraction. My numerator is 3 and my denominator is 7.
42. I am equivalent to $\frac{4}{8}$. My denominator is 2.
43. I am equivalent to $\frac{3}{2}$. My numerator is 6.
44. I am equivalent to $\frac{3}{2}$. My denominator is 6.

45. Write in lowest terms. $\frac{9{,}702}{14{,}553}$

KEEPING SKILLS SHARP

Multiply.

1. 8.2×2.6
2. 14.3×1.8
3. 2.07×6.2
4. 3.19×3.9
5. 0.453×5.6
6. 7.29×0.81

Divide. Give each quotient to the nearest tenth.

7. $0.35\overline{)4.771}$
8. $4.2\overline{)1.234}$
9. $0.27\overline{)1.8}$
10. $0.51\overline{)396}$
11. $2.4\overline{)0.166}$
12. $25\overline{)4.03}$

Comparing fractions

Some students decided to build a birdhouse for a shop project. To build the birdhouse, they had to compare fractions. They found fractions with the same denominator easy to compare. For example:

$$\frac{1}{8} < \frac{3}{8}$$

To compare fractions with different denominators, they used equivalent fractions that had the same (common) denominator. For example:

$$\left(\frac{3}{4}, \frac{6}{8}, \frac{9}{12}, \frac{12}{16}\right) \quad \frac{3}{4} > \frac{11}{16}$$

The least common multiple of the denominators is called the **least common denominator.**

Here is another example:

Step 1. Find the least common denominator.

$$\frac{5}{2} \bullet \frac{7}{3} \quad (6)$$

Step 2. Change to equivalent fractions.

$$\left(\frac{15}{6}\right) \frac{5}{2} \bullet \frac{7}{3} \left(\frac{14}{6}\right)$$

Step 3. Compare.

$$\left(\frac{15}{6}\right) \frac{5}{2} > \frac{7}{3} \left(\frac{14}{6}\right)$$

EXERCISES

< or >?

1. $\frac{4}{6}$ ● $\frac{5}{6}$ **2.** $\frac{6}{10}$ ● $\frac{3}{10}$ **3.** $\frac{9}{4}$ ● $\frac{7}{4}$ **4.** $\frac{5}{7}$ ● $\frac{2}{7}$ **5.** $\frac{5}{9}$ ● $\frac{4}{9}$

6. $\frac{2}{2}$ ● $\frac{3}{2}$ **7.** $\frac{5}{3}$ ● $\frac{4}{3}$ **8.** $\frac{7}{5}$ ● $\frac{5}{5}$ **9.** $\frac{7}{8}$ ● $\frac{5}{8}$ **10.** $\frac{0}{8}$ ● $\frac{1}{8}$

Give the least common denominator.

11. $\frac{1}{2}, \frac{1}{3}$ 12. $\frac{2}{3}, \frac{3}{4}$ 13. $\frac{5}{2}, \frac{8}{5}$ 14. $\frac{4}{3}, \frac{6}{5}$ 15. $\frac{3}{4}, \frac{5}{6}$ 16. $\frac{7}{6}, \frac{11}{9}$

17. $\frac{7}{5}, \frac{8}{7}$ 18. $\frac{2}{3}, \frac{5}{8}$ 19. $\frac{5}{6}, \frac{7}{8}$ 20. $\frac{5}{12}, \frac{1}{3}$ 21. $\frac{7}{9}, \frac{4}{8}$ 22. $\frac{5}{6}, \frac{2}{1}$

<, =, or >? *Hint:* **First find the least common denominator.**

23. $\frac{2}{3}$ ● $\frac{5}{6}$ 24. $\frac{1}{4}$ ● $\frac{3}{8}$ 25. $\frac{2}{3}$ ● $\frac{5}{9}$ 26. $\frac{1}{3}$ ● $\frac{2}{7}$ 27. $\frac{3}{2}$ ● $\frac{5}{3}$

28. $\frac{3}{8}$ ● $\frac{1}{2}$ 29. $\frac{1}{3}$ ● $\frac{3}{10}$ 30. $\frac{1}{2}$ ● $\frac{5}{8}$ 31. $\frac{1}{6}$ ● $\frac{1}{8}$ 32. $\frac{5}{2}$ ● $\frac{5}{3}$

33. $\frac{2}{9}$ ● $\frac{1}{3}$ 34. $\frac{2}{5}$ ● $\frac{3}{8}$ 35. $\frac{6}{8}$ ● $\frac{3}{4}$ 36. $\frac{3}{5}$ ● $\frac{1}{2}$ 37. $\frac{0}{3}$ ● $\frac{0}{6}$

38. $\frac{15}{12}$ ● $\frac{5}{4}$ 39. $\frac{3}{10}$ ● $\frac{1}{4}$ 40. $\frac{1}{5}$ ● $\frac{1}{4}$ 41. $\frac{4}{4}$ ● $\frac{2}{2}$ 42. $\frac{5}{6}$ ● $\frac{7}{8}$

43. $\frac{7}{8}$ ● $\frac{3}{5}$ 44. $\frac{4}{3}$ ● $\frac{3}{4}$ 45. $\frac{5}{9}$ ● $\frac{7}{12}$ 46. $\frac{9}{16}$ ● $\frac{5}{8}$ 47. $\frac{7}{5}$ ● $\frac{9}{7}$

Solve.

48. Will a $\frac{3}{4}$-inch nail go through a $\frac{1}{2}$-inch-thick board?

49. The small hole for the perch stick had to be $\frac{1}{4}$ inch in diameter. Would a $\frac{3}{8}$-inch drill bit be too large or too small?

50. The hole for the birds to enter had to be $\frac{7}{8}$ inch or larger. Would a $\frac{13}{16}$-inch drill bit make a large enough hole?

Adding fractions

To add two fractions having the same denominator, add the numerators and use the common denominator.

$$\frac{1}{8} + \frac{5}{8} = \frac{6}{8}$$
$$= \frac{3}{4} \leftarrow \text{lowest terms}$$

When the fractions have different denominators, you must change to equivalent fractions with common denominators.

Step 1. Find the least common denominator.

$$\frac{2}{5} + \frac{1}{3} = \frac{\quad}{15} + \frac{\quad}{15}$$

Step 2. Change to equivalent fractions.

$$\frac{2}{5} + \frac{1}{3} = \frac{6}{15} + \frac{5}{15}$$

Step 3. Add.

$$\frac{2}{5} + \frac{1}{3} = \frac{6}{15} + \frac{5}{15} = \frac{11}{15}$$

EXERCISES

Add. Give each sum in lowest terms. *Hint:* First find the least common denominator.

1. $\frac{1}{3} + \frac{1}{3}$ **2.** $\frac{3}{8} + \frac{1}{8}$ **3.** $\frac{1}{6} + \frac{1}{6}$ **4.** $\frac{3}{5} + \frac{1}{5}$ **5.** $\frac{1}{4} + \frac{1}{4}$

6. $\frac{1}{4} + \frac{1}{2}$ **7.** $\frac{1}{2} + \frac{1}{6}$ **8.** $\frac{1}{4} + \frac{1}{3}$ **9.** $\frac{5}{6} + \frac{1}{8}$ **10.** $\frac{1}{4} + \frac{2}{5}$

11. $\frac{1}{6} + \frac{1}{5}$ **12.** $\frac{1}{6} + \frac{3}{4}$ **13.** $\frac{0}{4} + \frac{1}{2}$ **14.** $\frac{1}{15} + \frac{2}{5}$ **15.** $\frac{3}{5} + \frac{1}{10}$

16. $\frac{2}{3} + \frac{1}{9}$ **17.** $\frac{1}{10} + \frac{2}{3}$ **18.** $\frac{1}{2} + \frac{3}{8}$ **19.** $\frac{1}{3} + \frac{1}{6}$ **20.** $\frac{3}{10} + \frac{1}{2}$

Add. Give answers in lowest terms.

21. $\frac{3}{5} + \frac{1}{6}$ **22.** $\frac{1}{4} + \frac{1}{8}$ **23.** $\frac{0}{6} + \frac{2}{3}$ **24.** $\frac{5}{9} + \frac{1}{3}$ **25.** $\frac{1}{6} + \frac{4}{9}$ **26.** $\frac{5}{12} + \frac{1}{3}$

27. $\frac{3}{8} + \frac{1}{6}$ **28.** $\frac{3}{5} + \frac{1}{3}$ **29.** $\frac{2}{7} + \frac{2}{3}$ **30.** $\frac{1}{10} + \frac{2}{5}$ **31.** $\frac{3}{4} + \frac{1}{12}$ **32.** $\frac{2}{3} + \frac{1}{4}$

This map shows float times between stopping points on a White Water River trip.

33. How long does it take to go from Rapid View to Deep Canyon?

34. Which takes longer: to float from Rapid View to Deep Canyon or to float from Deep Canyon to Beaver Lake?

35. Can you make the entire trip in 3 hours?

★ **36.** If you started from Rapid View at 10:30 A.M., would you be at Sandy Bottom by noon?

Subtracting fractions

To subtract fractions having the same denominator, subtract the numerators for the numerator of the difference and use the common denominator for the denominator of the difference.

$$\frac{3}{4} - \frac{1}{4} = \frac{2}{4}$$
$$= \frac{1}{2}$$

lowest terms

When the two denominators are different, find the least common denominator and change to equivalent fractions.

$$\frac{1}{2} - \frac{1}{3} = \frac{3}{6} - \frac{2}{6}$$
$$= \frac{1}{6}$$

EXERCISES

Subtract. Give each difference in lowest terms.

1. $\frac{5}{8} - \frac{1}{8}$	**2.** $\frac{7}{6} - \frac{5}{6}$	**3.** $\frac{2}{3} - \frac{1}{3}$	**4.** $\frac{7}{5} - \frac{3}{5}$	**5.** $\frac{6}{7} - \frac{1}{7}$
6. $\frac{4}{3} - \frac{3}{3}$	**7.** $\frac{4}{8} - \frac{1}{8}$	**8.** $\frac{2}{5} - \frac{1}{5}$	**9.** $\frac{5}{9} - \frac{4}{9}$	**10.** $\frac{5}{4} - \frac{2}{4}$
11. $\frac{7}{5} - \frac{4}{5}$	**12.** $\frac{7}{9} - \frac{1}{9}$	**13.** $\frac{8}{8} - \frac{2}{8}$	**14.** $\frac{5}{3} - \frac{4}{3}$	**15.** $\frac{7}{4} - \frac{4}{4}$

Subtract. Give answers in lowest terms. *Hint:* First find the least common denominator.

16. $\frac{1}{3} - \frac{2}{9}$ **17.** $\frac{5}{6} - \frac{5}{9}$ **18.** $\frac{5}{7} - \frac{1}{3}$ **19.** $\frac{5}{8} - \frac{1}{4}$ **20.** $\frac{2}{3} - \frac{1}{6}$

21. $\frac{5}{3} - \frac{6}{5}$ **22.** $\frac{7}{6} - \frac{2}{3}$ **23.** $\frac{5}{4} - \frac{2}{3}$ **24.** $\frac{3}{4} - \frac{1}{3}$ **25.** $\frac{5}{6} - \frac{1}{3}$

26. $\frac{3}{4} - \frac{2}{3}$ **27.** $\frac{3}{4} - \frac{5}{8}$ **28.** $\frac{3}{8} - \frac{1}{4}$ **29.** $\frac{1}{2} - \frac{1}{6}$ **30.** $\frac{2}{3} - \frac{1}{2}$

31. $\frac{4}{5} - \frac{2}{3}$ **32.** $\frac{3}{4} - \frac{1}{2}$ **33.** $\frac{8}{8} - \frac{3}{4}$ **34.** $\frac{1}{2} - \frac{0}{8}$ **35.** $\frac{5}{4} - \frac{2}{2}$

Subtract. Give answers in lowest terms.

36. $\frac{8}{9} - \frac{3}{4}$ **37.** $\frac{3}{4} - \frac{5}{8}$ **38.** $\frac{3}{4} - \frac{3}{10}$ **39.** $\frac{3}{4} - \frac{1}{6}$ **40.** $\frac{2}{5} - \frac{3}{8}$ **41.** $\frac{5}{6} - \frac{3}{8}$

42. $\frac{7}{5} - \frac{2}{3}$ **43.** $\frac{5}{9} - \frac{1}{2}$ **44.** $\frac{2}{3} - \frac{1}{4}$ **45.** $\frac{5}{4} - \frac{5}{8}$ **46.** $\frac{5}{6} - \frac{0}{4}$ **47.** $\frac{9}{4} - \frac{3}{2}$

48. MEASURE UP

Using these 4 cups to measure with, Chef Louie measured 18 times to make the recipe for Crunchy Cereal. Find a way to make the recipe by measuring only 17 times with the cups. Keep a record of how you do it.

Practice exercises

Give each sum in lowest terms.

1. $\frac{1}{3} + \frac{1}{2}$
2. $\frac{1}{3} + \frac{5}{9}$
3. $\frac{2}{3} + \frac{1}{4}$
4. $\frac{1}{5} + \frac{3}{10}$
5. $\frac{2}{9} + \frac{1}{6}$
6. $\frac{1}{3} + \frac{3}{5}$
7. $\frac{5}{8} + \frac{1}{4}$
8. $\frac{1}{6} + \frac{1}{2}$
9. $\frac{1}{10} + \frac{3}{4}$
10. $\frac{2}{3} + \frac{2}{7}$
11. $\frac{4}{9} + \frac{0}{3}$
12. $\frac{3}{4} + \frac{1}{8}$
13. $\frac{1}{4} + \frac{3}{8}$
14. $\frac{2}{5} + \frac{1}{4}$
15. $\frac{1}{3} + \frac{5}{12}$
16. $\frac{3}{10} + \frac{1}{10}$
17. $\frac{1}{3} + \frac{3}{7}$
18. $\frac{1}{4} + \frac{5}{12}$
19. $\frac{1}{12} + \frac{5}{12}$
20. $\frac{3}{8} + \frac{1}{6}$
21. $\frac{1}{2} + \frac{3}{10}$
22. $\frac{1}{4} + \frac{1}{6}$
23. $\frac{2}{9} + \frac{4}{9}$
24. $\frac{1}{2} + \frac{1}{14}$

Give each difference in lowest terms.

25. $\frac{1}{2} - \frac{1}{4}$
26. $\frac{7}{8} - \frac{3}{4}$
27. $\frac{3}{4} - \frac{5}{8}$
28. $\frac{1}{2} - \frac{1}{3}$
29. $\frac{5}{6} - \frac{2}{3}$
30. $\frac{7}{4} - \frac{3}{2}$
31. $\frac{3}{10} - \frac{1}{5}$
32. $\frac{3}{4} - \frac{3}{8}$
33. $\frac{3}{4} - \frac{1}{4}$
34. $\frac{7}{9} - \frac{0}{2}$
35. $\frac{1}{2} - \frac{3}{8}$
36. $\frac{3}{4} - \frac{3}{10}$
37. $\frac{2}{3} - \frac{3}{8}$
38. $\frac{6}{6} - \frac{1}{3}$
39. $\frac{7}{6} - \frac{3}{4}$
40. $\frac{5}{8} - \frac{5}{8}$
41. $\frac{5}{9} - \frac{1}{3}$
42. $\frac{6}{5} - \frac{7}{10}$
43. $\frac{2}{3} - \frac{1}{3}$
44. $\frac{1}{2} - \frac{4}{9}$
45. $\frac{3}{4} - \frac{1}{2}$
46. $\frac{2}{3} - \frac{1}{4}$
47. $\frac{5}{6} - \frac{4}{9}$
48. $\frac{3}{4} - \frac{2}{3}$

Computers in industry

Many machines can do only one kind of task, but modern industrial robots are machines that can do many complicated tasks. These robots are controlled by computers that can be programmed to make the robots do such tasks as welding, lifting, and sorting.

The motions of a robot are similar to those of the human body. Some of these are shoulder, waist, and wrist rotations.

Solve.

1. One robot sorts telephone receiver caps by color into 12 different bins. It can sort 6500 caps in an hour. How many caps can it sort in 8 hours?

2. When sorting the 12 colors of receiver caps, the robot makes an average of one mistake for each 1000 caps sorted. How many mistakes does the robot make in 24 hours?

3. A human can sort one third as many caps as the robot and makes an average of 12 mistakes in sorting 10,000 caps. How many caps can a human sort in 8 hours? How many mistakes will be made?

4. The same number of vacuum cleaners were manufactured when a company in Japan replaced 120 workers by a computer, a team of robots, and 4 human monitors. How many fewer workers were there when the company used robots?

5. A heavy-duty robot can position a piece within 0.1 mm of a target point. A smaller, more flexible robot can position a piece within 0.04 mm of a target point. How many times as large is the variability of the larger robot?

6. A large manufacturing company is now using 120 robots and plans to increase the number in use by one half each year for the next three years. How many robots will be in use at the end of the three years?

Writing whole numbers and mixed numbers as fractions

Look at the pizzas. They show that there are 8 fourths in 2.

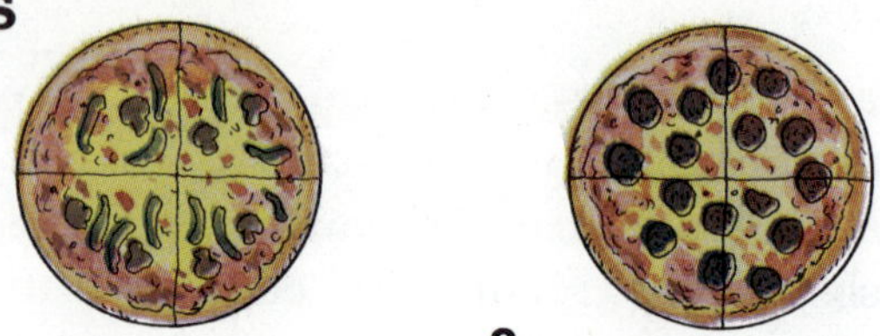

$$2 = \frac{8}{4}$$

To write a whole number as a fraction, you can write the whole number over a denominator of 1. Then, to change to an equivalent fraction, you multiply both numerator and denominator by the same whole number (not zero).

$$2 = \frac{2}{1}$$

$$\frac{2}{1} = \frac{8}{4} \quad (\times 4)$$

A **mixed number** has a whole-number part and a fraction part.

$2\frac{3}{4}$ is a mixed number.
It means $2 + \frac{3}{4}$.

To write a mixed number as a fraction, you can write the whole-number part as a fraction and then add.

$$2\frac{3}{4} = 2 + \frac{3}{4} = \frac{2}{1} + \frac{3}{4} = \frac{8}{4} + \frac{3}{4} = \frac{11}{4}$$

$$2\frac{3}{4} = \frac{11}{4}$$

$$(4 \times 2) + 3 = 11$$

EXERCISES

Copy and complete.

1. $4 = \frac{?}{2}$ 2. $4 = \frac{?}{3}$ 3. $4 = \frac{?}{4}$ 4. $4 = \frac{?}{5}$

5. $2 = \frac{?}{6}$ 6. $3 = \frac{?}{6}$ 7. $4 = \frac{?}{6}$ 8. $5 = \frac{?}{6}$

9. $1 = \frac{?}{8}$ 10. $6 = \frac{?}{4}$ 11. $7 = \frac{?}{2}$ 12. $3 = \frac{?}{10}$

Write each mixed number as a fraction.

13. $3\frac{1}{2}$ 14. $2\frac{1}{3}$ 15. $1\frac{1}{4}$ 16. $3\frac{1}{5}$ 17. $1\frac{3}{4}$

18. $1\frac{1}{2}$ 19. $3\frac{3}{4}$ 20. $3\frac{2}{3}$ 21. $1\frac{1}{3}$ 22. $4\frac{5}{6}$

23. $4\frac{1}{6}$ 24. $1\frac{2}{3}$ 25. $5\frac{1}{4}$ 26. $5\frac{2}{3}$ 27. $8\frac{1}{4}$

28. $5\frac{3}{4}$ 29. $2\frac{1}{4}$ 30. $4\frac{3}{4}$ 31. $4\frac{2}{3}$ 32. $6\frac{2}{5}$

33. $3\frac{7}{8}$ 34. $6\frac{3}{4}$ 35. $6\frac{3}{8}$ 36. $2\frac{5}{6}$ 37. $8\frac{3}{5}$

38. $6\frac{2}{3}$ 39. $2\frac{5}{8}$ 40. $3\frac{5}{8}$ 41. $4\frac{3}{8}$ 42. $5\frac{5}{8}$

Who am I?

43. I am a fraction equivalent to 3. My denominator is 4.

44. I am a fraction equivalent to $2\frac{3}{5}$. I am in lowest terms.

45. GO FOUR IT!

Using 4 fours, write expressions for the numbers 1 through 8. Here are expressions for 1 and 2:

$$\frac{44}{44} = 1 \qquad \frac{4}{4} + \frac{4}{4} = 2$$

KEEPING SKILLS SHARP

Give the greatest common factor (GCF) of each pair of numbers.

1. 4, 5 2. 4, 6
3. 4, 7 4. 4, 8
5. 6, 7 6. 6, 8
7. 6, 9 8. 6, 10
9. 8, 9 10. 8, 10
11. 8, 12 12. 12, 9
13. 12, 10 14. 12, 18
15. 10, 15 16. 9, 15

Give the least common multiple (LCM) of each pair of numbers.

17. 4, 5 18. 4, 6
19. 4, 7 20. 4, 8
21. 6, 7 22. 6, 8
23. 6, 9 24. 6, 10
25. 8, 9 26. 8, 10
27. 8, 12 28. 12, 9
29. 12, 10 30. 12, 18
31. 10, 15 32. 9, 15

Writing fractions as whole numbers or as mixed numbers

When the numerator is greater than or equal to the denominator, the fraction is greater than or equal to 1. You can write the fraction as either a whole number or a mixed number by dividing the numerator by the denominator.

$\frac{8}{4} = ?$

$\frac{10}{4} = ?$

$4\overline{)8}$ with quotient 2

number of fourths in a whole sandwich → 4; 2 ← number of whole sandwiches; 8 ← number of fourths

So, $\frac{8}{4} = 2$

$$\begin{array}{r} 2 \\ 4\overline{)10} \\ -8 \\ \hline 2 \end{array}$$

2 ← number of fourths left over

So, $\frac{10}{4} = 2\frac{2}{4}$

$= 2\frac{1}{2}$

Here is how to write your answers in **simplest form.**

Write a fraction that is less than 1 in lowest terms.

$\frac{5}{10} = \frac{1}{2}$

Write a mixed number so that its fraction part is less than 1 and is in lowest terms.

$2\frac{6}{9} = 2\frac{2}{3}$

Write a fraction that is greater than or equal to 1 as a whole number or as a mixed number in simplest form.

$\frac{12}{3} = 4$

$\frac{14}{6} = 2\frac{2}{6} = 2\frac{1}{3}$

EXERCISES

Write each fraction as either a whole number or a mixed number.

1. $\frac{20}{4}$ 2. $\frac{10}{3}$ 3. $\frac{12}{6}$ 4. $\frac{13}{4}$ 5. $\frac{13}{6}$ 6. $\frac{12}{3}$ 7. $\frac{8}{8}$

8. $\frac{18}{5}$ 9. $\frac{33}{10}$ 10. $\frac{11}{2}$ 11. $\frac{18}{6}$ 12. $\frac{23}{5}$ 13. $\frac{13}{7}$ 14. $\frac{18}{9}$

15. $\frac{22}{3}$ 16. $\frac{10}{10}$ 17. $\frac{27}{10}$ 18. $\frac{15}{5}$ 19. $\frac{6}{2}$ 20. $\frac{21}{5}$ 21. $\frac{16}{2}$

Write in simplest form.

22. $\frac{4}{6}$ 23. $\frac{5}{20}$ 24. $\frac{4}{8}$ 25. $\frac{8}{12}$ 26. $\frac{10}{12}$ 27. $\frac{9}{24}$

28. $3\frac{2}{4}$ 29. $4\frac{6}{8}$ 30. $1\frac{5}{10}$ 31. $2\frac{8}{10}$ 32. $1\frac{6}{9}$ 33. $4\frac{14}{16}$

34. $\frac{14}{2}$ 35. $\frac{18}{3}$ 36. $\frac{12}{4}$ 37. $\frac{24}{8}$ 38. $\frac{40}{5}$ 39. $\frac{24}{4}$

40. $\frac{7}{2}$ 41. $\frac{11}{4}$ 42. $\frac{12}{8}$ 43. $\frac{14}{6}$ 44. $\frac{18}{4}$ 45. $\frac{22}{6}$

46. $\frac{15}{20}$ 47. $\frac{20}{15}$ 48. $\frac{18}{24}$ 49. $\frac{24}{18}$ 50. $\frac{16}{12}$ 51. $\frac{12}{16}$

52. $9\frac{4}{12}$ 53. $\frac{32}{8}$ 54. $\frac{9}{3}$ 55. $8\frac{8}{10}$ 56. $5\frac{3}{6}$ 57. $\frac{8}{32}$

58. $\frac{22}{8}$ 59. $\frac{20}{5}$ 60. $\frac{15}{6}$ 61. $4\frac{9}{12}$ 62. $\frac{30}{5}$ 63. $\frac{15}{8}$

64. FRACTIONS FOR PHARAOHS

Ancient Egyptians expressed fractional numbers as sums of unit fractions with different denominators. (A unit fraction has a numerator of 1.) Here are some examples:

$$\frac{2}{5} = \frac{1}{3} + \frac{1}{15}$$

$$\frac{3}{5} = \frac{1}{3} + \frac{1}{15} + \frac{1}{5} \text{ or } \frac{1}{2} + \frac{1}{10}$$

How might they have expressed these fractions?

$$\frac{2}{7}, \frac{3}{7}, \frac{4}{9}, \frac{5}{9}$$

Practice exercises

Add. Give each sum in simplest form.

1. $\frac{1}{2} + \frac{3}{4}$
2. $\frac{1}{3} + \frac{4}{5}$
3. $\frac{2}{3} + \frac{3}{4}$
4. $\frac{1}{4} + \frac{7}{8}$
5. $\frac{3}{4} + \frac{5}{6}$
6. $\frac{9}{10} + \frac{3}{10}$
7. $\frac{8}{9} + \frac{2}{3}$
8. $\frac{5}{8} + \frac{1}{2}$
9. $\frac{7}{10} + \frac{4}{5}$
10. $\frac{13}{16} + \frac{1}{4}$
11. $\frac{9}{10} + \frac{3}{4}$
12. $\frac{4}{5} + \frac{1}{5}$
13. $\frac{3}{5} + \frac{2}{3}$
14. $\frac{1}{3} + \frac{9}{10}$
15. $\frac{7}{8} + \frac{1}{2}$
16. $\frac{2}{3} + \frac{1}{2}$
17. $\frac{5}{6} + \frac{7}{8}$
18. $\frac{2}{5} + \frac{1}{2}$
19. $\frac{11}{12} + \frac{1}{3}$
20. $\frac{7}{10} + \frac{1}{2}$
21. $\frac{5}{12} + \frac{1}{4}$
22. $\frac{2}{5} + \frac{3}{4}$
23. $\frac{8}{15} + \frac{4}{5}$
24. $\frac{3}{8} + \frac{5}{8}$

Subtract. Give each difference in simplest form.

25. $\frac{5}{4} - \frac{3}{4}$
26. $\frac{5}{3} - \frac{1}{4}$
27. $\frac{14}{9} - \frac{2}{3}$
28. $\frac{7}{6} - \frac{5}{6}$
29. $\frac{3}{2} - \frac{1}{4}$
30. $\frac{7}{4} - \frac{3}{4}$
31. $\frac{9}{8} - \frac{5}{6}$
32. $\frac{11}{6} - \frac{2}{3}$
33. $\frac{11}{8} - \frac{5}{8}$
34. $\frac{9}{10} - \frac{2}{5}$
35. $\frac{3}{2} - \frac{5}{6}$
36. $\frac{5}{4} - \frac{2}{3}$
37. $\frac{13}{8} - \frac{1}{4}$
38. $\frac{13}{10} - \frac{1}{2}$
39. $\frac{9}{5} - \frac{1}{2}$
40. $\frac{7}{4} - \frac{9}{10}$
41. $\frac{11}{6} - \frac{4}{9}$
42. $\frac{15}{8} - \frac{7}{8}$
43. $\frac{2}{3} - \frac{1}{9}$
44. $\frac{17}{12} - \frac{1}{4}$
45. $\frac{7}{5} - \frac{1}{2}$
46. $\frac{3}{2} - \frac{1}{3}$
47. $\frac{13}{12} - \frac{2}{3}$
48. $\frac{4}{3} - \frac{1}{5}$

Problem solving—not enough information

1. Study and understand.
2. Plan and do.
3. Answer and check.

A football field is 160 feet wide. How many yards is that?

Missing fact: the number of feet in a yard.

160 ÷ (missing fact)

Do you know the missing fact? If so, solve the problem.

What is the missing fact? If you know the missing fact, solve the problem.

1. A Canadian football team has 12 players. How many fewer players does an American football team have?
2. High school football games are 48 minutes long. How many minutes longer are college football games?
3. Between the goal lines on a football field there are yard lines that go across the field every 5 yards. How many yard lines are there on a football field?
4. The goalposts are 18 feet 6 inches apart in high school games and 23 feet 4 inches apart in college games. How much farther apart are the goalposts in college games?
5. At 3 minutes before the end of the game, 142 spectators left the stadium. How many spectators remained in the stadium?
6. The Lions scored 24 points. They made 2 field goals. What fraction of the Lions' points were for field goals?

Adding mixed numbers

Here is how to find the sum of two mixed numbers.

EXAMPLE 1.

The fractions already have a common denominator. So, add the fractions and add the whole numbers.

$$\begin{array}{r} 3\frac{5}{8} \\ +2\frac{1}{8} \\ \hline 5\frac{6}{8} = 5\frac{3}{4} \end{array}$$

Change to simplest form.

EXAMPLE 2.

Write the fractions with a common denominator.

$$\begin{array}{rcr} 5\frac{1}{3} & = & 5\frac{2}{6} \\ +4\frac{1}{6} & = & +4\frac{1}{6} \\ \hline \end{array}$$

Add the fractions and add the whole numbers.

$$\begin{array}{rcr} 5\frac{1}{3} & = & 5\frac{2}{6} \\ +4\frac{1}{6} & = & +4\frac{1}{6} \\ \hline & & 9\frac{3}{6} = 9\frac{1}{2} \end{array}$$

simplest form

EXAMPLE 3.

$$\begin{array}{rcr} 1\frac{2}{3} & = & 1\frac{8}{12} \\ +5\frac{3}{4} & = & +5\frac{9}{12} \\ \hline \end{array}$$

$$\begin{array}{rcr} 1\frac{2}{3} & = & 1\frac{8}{12} \\ +5\frac{3}{4} & = & +5\frac{9}{12} \\ \hline & & 6\frac{17}{12} = 7\frac{5}{12} \end{array}$$

$\frac{17}{12} = 1\frac{5}{12}$

EXERCISES

Add. Write the sum in simplest form.

1. $\begin{array}{r} 3\frac{1}{4} \\ +4\frac{1}{4} \\ \hline \end{array}$
2. $\begin{array}{r} 8\frac{1}{7} \\ +3\frac{4}{7} \\ \hline \end{array}$
3. $\begin{array}{r} 5\frac{1}{2} \\ +6\frac{1}{2} \\ \hline \end{array}$
4. $\begin{array}{r} 9\frac{3}{8} \\ +6\frac{1}{8} \\ \hline \end{array}$
5. $\begin{array}{r} 10\frac{1}{6} \\ +6\frac{5}{6} \\ \hline \end{array}$
6. $\begin{array}{r} 8\frac{3}{8} \\ +7\frac{3}{8} \\ \hline \end{array}$

7. $4\frac{1}{4} + 3\frac{1}{2}$

8. $3\frac{5}{6} + 3\frac{1}{3}$

9. $12\frac{1}{3} + 5\frac{2}{3}$

10. $6\frac{1}{2} + 7\frac{2}{3}$

11. $6\frac{3}{4} + 8\frac{1}{8}$

12. $13\frac{2}{5} + 6$

13. $6\frac{1}{2} + 4\frac{5}{6}$

14. $4\frac{5}{8} + 3\frac{1}{2}$

15. $2\frac{1}{2} + \frac{5}{8}$

16. $16\frac{3}{7} + 4\frac{1}{3}$

17. $12\frac{3}{4} + 5\frac{3}{8}$

18. $17\frac{2}{5} + 11\frac{3}{4}$

19. $15\frac{3}{5} + 8\frac{1}{2}$

20. $12\frac{2}{3} + 11\frac{3}{4}$

21. $16\frac{5}{8} + 13\frac{2}{3}$

22. $23\frac{4}{5} + 14$

23. $20\frac{5}{8} + 6\frac{5}{6}$

24. $24\frac{3}{4} + 19\frac{5}{8}$

Solve.

25. While helping to take an inventory at Collimore's Department Store, Antonio worked $2\frac{1}{2}$ hours on Monday and $1\frac{3}{4}$ hours on Tuesday. How many hours did he work in all?

26. Antonio counted $2\frac{3}{4}$ cartons of white tennis balls and $5\frac{1}{4}$ cartons of yellow tennis balls. How many cartons of tennis balls were there in all?

27. He counted $2\frac{3}{4}$ cartons of Tru-Fli golf balls and $3\frac{1}{3}$ cartons of Hi-Fli golf balls. How many cartons of golf balls were there in all?

★ 28. In the hardware section of the store there were $5\frac{1}{2}$ dozen, $6\frac{3}{4}$ dozen, and $4\frac{1}{3}$ dozen of three different kinds of bolts. How many dozen bolts were there in all?

★ 29.

Antonio said there were $14\frac{1}{3}$ dozen greeting cards. Check Antonio's addition. How many greeting cards were there?

Subtracting mixed numbers

Here is how to find the difference of two mixed numbers.

EXAMPLE 1.

$$\begin{array}{r} 4\frac{7}{10} \\ -1\frac{3}{10} \\ \hline 3\frac{4}{10} = 3\frac{2}{5} \end{array}$$

Change to simplest form.

EXAMPLE 2.

Write fractions with a common denominator.

$$\begin{array}{rl} 8\frac{5}{6} & = 8\frac{5}{6} \\ -7\frac{1}{3} & = -7\frac{2}{6} \\ \hline \end{array}$$

Subtract the fractions and subtract the whole numbers.

$$\begin{array}{rl} 8\frac{5}{6} & = 8\frac{5}{6} \\ -7\frac{1}{3} & = -7\frac{2}{6} \\ \hline & 1\frac{3}{6} = 1\frac{1}{2} \end{array}$$

simplest form

EXAMPLE 3. Sometimes you need to regroup before subtracting fractions.

$$\begin{array}{rl} 7\frac{1}{8} & = 6\frac{9}{8} \quad \left(\frac{1}{8}+\frac{8}{8}\right) \\ -2\frac{3}{8} & = -2\frac{3}{8} \\ \hline \end{array}$$

$$\begin{array}{rl} 7\frac{1}{8} & = 6\frac{9}{8} \\ -2\frac{3}{8} & = -2\frac{3}{8} \\ \hline & 4\frac{6}{8} = 4\frac{3}{4} \end{array}$$

EXERCISES

Subtract. Write the difference in simplest form.

1. $\begin{array}{r} 8\frac{3}{4} \\ -2\frac{1}{4} \\ \hline \end{array}$ **2.** $\begin{array}{r} 9\frac{7}{8} \\ -3\frac{3}{8} \\ \hline \end{array}$ **3.** $\begin{array}{r} 6\frac{1}{2} \\ -3\frac{1}{6} \\ \hline \end{array}$ **4.** $\begin{array}{r} 15\frac{3}{4} \\ -8\frac{3}{8} \\ \hline \end{array}$ **5.** $\begin{array}{r} 8\frac{5}{9} \\ -1\frac{1}{3} \\ \hline \end{array}$ **6.** $\begin{array}{r} 15\frac{5}{7} \\ -9\frac{1}{7} \\ \hline \end{array}$

7. $11\frac{3}{4} - 2\frac{1}{3}$

8. 9 ($8\frac{5}{5}$) $- 1\frac{3}{5}$

9. $12 - 3\frac{1}{4}$

10. $6 - 2\frac{5}{8}$

11. $8\frac{5}{6} - 2\frac{1}{6}$

12. $6\frac{1}{4} - 5\frac{3}{4}$

13. $16\frac{3}{8} - 5\frac{5}{8}$

14. $12\frac{3}{4} - 10$

15. $6\frac{3}{8} - 2\frac{3}{4}$

16. $11\frac{6}{7} - 7\frac{1}{2}$

17. $9\frac{2}{5} - 6\frac{1}{2}$

18. $24\frac{1}{3} - 18\frac{1}{2}$

19. $16 - 5\frac{7}{8}$

20. $26\frac{7}{8} - 4\frac{3}{4}$

21. $18 - 9\frac{1}{2}$

22. $23\frac{3}{8} - 12$

Solve.

23. One week Alvin exercised for $6\frac{1}{4}$ hours and Beth exercised for $3\frac{3}{4}$ hours. How many more hours did Alvin exercise?

24. John long-jumped $16\frac{1}{4}$ feet and Kevin long-jumped $14\frac{2}{3}$ feet. How many feet farther did John jump?

25. Susan had $78\frac{1}{2}$ feet of kite string. She gave $18\frac{3}{4}$ feet of it to a friend. How many feet of string did she have left?

26. HEEL AND TOE

Find a way to use these shoes to mark off a distance of $1\frac{1}{12}$ feet.

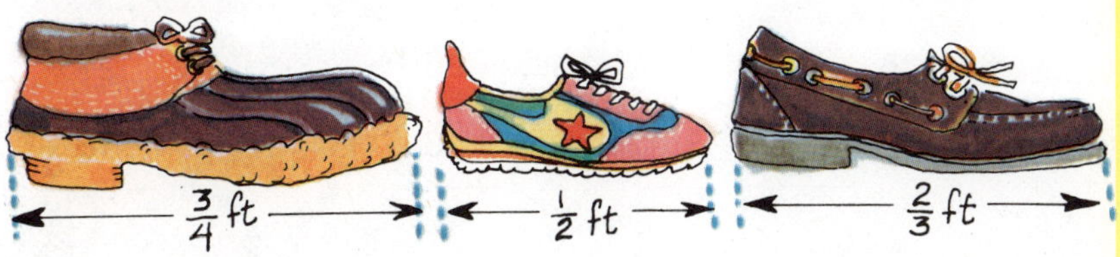

KEEPING SKILLS SHARP

$<$ or $>$?

1. $\frac{5}{8}$ ● $\frac{7}{8}$
2. $\frac{5}{9}$ ● $\frac{4}{9}$
3. $\frac{2}{9}$ ● $\frac{1}{3}$
4. $\frac{3}{4}$ ● $\frac{4}{5}$
5. $\frac{1}{4}$ ● $\frac{1}{6}$
6. $\frac{3}{4}$ ● $\frac{5}{8}$
7. $\frac{3}{4}$ ● $\frac{7}{10}$
8. $\frac{5}{8}$ ● $\frac{5}{6}$
9. $\frac{3}{8}$ ● $\frac{5}{12}$
10. $\frac{7}{9}$ ● $\frac{5}{6}$
11. $\frac{7}{10}$ ● $\frac{5}{6}$
12. $\frac{3}{10}$ ● $\frac{4}{15}$
13. $\frac{5}{8}$ ● $\frac{7}{10}$
14. $\frac{5}{8}$ ● $\frac{7}{12}$
15. $\frac{5}{12}$ ● $\frac{1}{2}$

Practice exercises

Add. Give the sum in simplest form.

1. $5\frac{2}{3} + 7\frac{1}{3}$
2. $6 + 8\frac{3}{5}$
3. $7\frac{3}{4} + 9\frac{1}{4}$
4. $7\frac{1}{6} + 8\frac{1}{8}$
5. $8\frac{3}{4} + 8\frac{1}{8}$
6. $5\frac{5}{9} + 9\frac{1}{3}$
7. $8\frac{1}{8} + 9\frac{1}{4}$
8. $6\frac{5}{8} + 6$
9. $9\frac{1}{2} + 9\frac{1}{3}$
10. $8\frac{2}{3} + 5\frac{1}{4}$
11. $12\frac{7}{8} + 8\frac{1}{4}$
12. $13\frac{5}{8} + 5\frac{1}{2}$
13. $19\frac{3}{4} + 6\frac{2}{5}$
14. $17\frac{3}{4} + 3\frac{3}{8}$
15. $18\frac{5}{6} + 7\frac{1}{3}$
16. $12\frac{2}{3} + 10\frac{5}{9}$
17. $15\frac{3}{8} + 13\frac{7}{8}$
18. $23\frac{3}{4} + 15\frac{1}{2}$
19. $25\frac{5}{8} + 18\frac{3}{4}$
20. $30\frac{2}{3} + 12\frac{3}{4}$

Subtract. Give the difference in simplest form.

21. $8\frac{4}{5} - 4\frac{1}{5}$
22. $9\frac{8}{9} - 3\frac{2}{9}$
23. $7\frac{2}{3} - 1\frac{1}{6}$
24. $9\frac{1}{2} - \frac{1}{3}$
25. $7\frac{5}{6} - 2\frac{3}{4}$
26. $6\frac{9}{10} - 3\frac{2}{5}$
27. $5\frac{3}{4} - 2\frac{1}{3}$
28. $8\frac{5}{6} - 3$
29. $5\frac{3}{4} - 2\frac{5}{8}$
30. $9\frac{3}{5} - 6$
31. $11 - 4\frac{2}{3}$
32. $13 - 8\frac{1}{4}$
33. $12\frac{1}{4} - 3\frac{1}{8}$
34. $15\frac{1}{3} - 9\frac{1}{2}$
35. $18\frac{1}{8} - 9\frac{3}{4}$
36. $22\frac{1}{2} - 8\frac{3}{4}$
37. $24 - 9\frac{3}{5}$
38. $25\frac{1}{4} - 11\frac{2}{5}$
39. $29\frac{1}{3} - 16\frac{5}{6}$
40. $27 - 18\frac{7}{8}$

Problem solving—drawing a picture

Use the information on the signs to solve these problems.

1. How many miles is it
   a. from the falls to the ledge?
   b. from the parking lot to the fire tower?
   c. from the ledge to the fire tower?

2. Suppose you are at the falls. How far are you
   a. from the fire tower?
   b. from the parking lot?
   c. from the ledge?

3. Suppose you started at the parking lot and hiked 6 miles toward the fire tower? How far are you
   a. from the fire tower?
   b. from the ledge?

★ 4. Suppose you are halfway between the parking lot and the ledge. How far are you
   a. from the falls?
   b. from the fire tower?

Draw another picture to solve these problems.

5. How many miles is it from the old oak to the exit?

6. Suppose you are at the cave. How far is it to the exit?

★ 7. Which is the cave closer to, the pond or the old oak? How much closer?

Problem solving

1. During a physical education class each student made 3 long jumps. The longest jump was recorded. Jan jumped $10' 2\frac{1}{2}''$, $11' 3\frac{1}{4}''$, and $10' 8\frac{1}{4}''$. What was her longest jump?

2. John's longest jump was $10'\ 8\frac{1}{2}''$. How much longer was Jan's longest jump? (See exercise 1.)

Regrouped 1 foot for 12 inches.

$$\begin{array}{rcr} 11'\ 3\frac{1}{4}'' & = & 10'\ 15\frac{1}{4}'' \\ -10'\ 8\frac{1}{2}'' & = & -10'\ \ 8\frac{1}{2}'' \\ \hline \end{array}$$

3. Mike's three jumps were $11'\ 1\frac{1}{2}''$, $10'\ 11\frac{3}{4}''$, and $11'\ 1\frac{1}{2}''$. Terry's three jumps were $10' \frac{1}{2}''$, $11' \frac{3}{4}''$, and $11' 4\frac{1}{4}''$. What was each boy's longest jump? How much longer was Terry's longest jump?

4. Bill's shortest jump was $9'\ 4\frac{1}{2}''$. His longest jump was $1'\ 7\frac{3}{4}''$ longer than his shortest jump. How long was his longest jump?

5. David's three jumps were $10'\ 8\frac{3}{4}''$, $10'\ 11\frac{1}{2}''$, and $10'\ 10''$. What was the difference between his longest and shortest jumps?

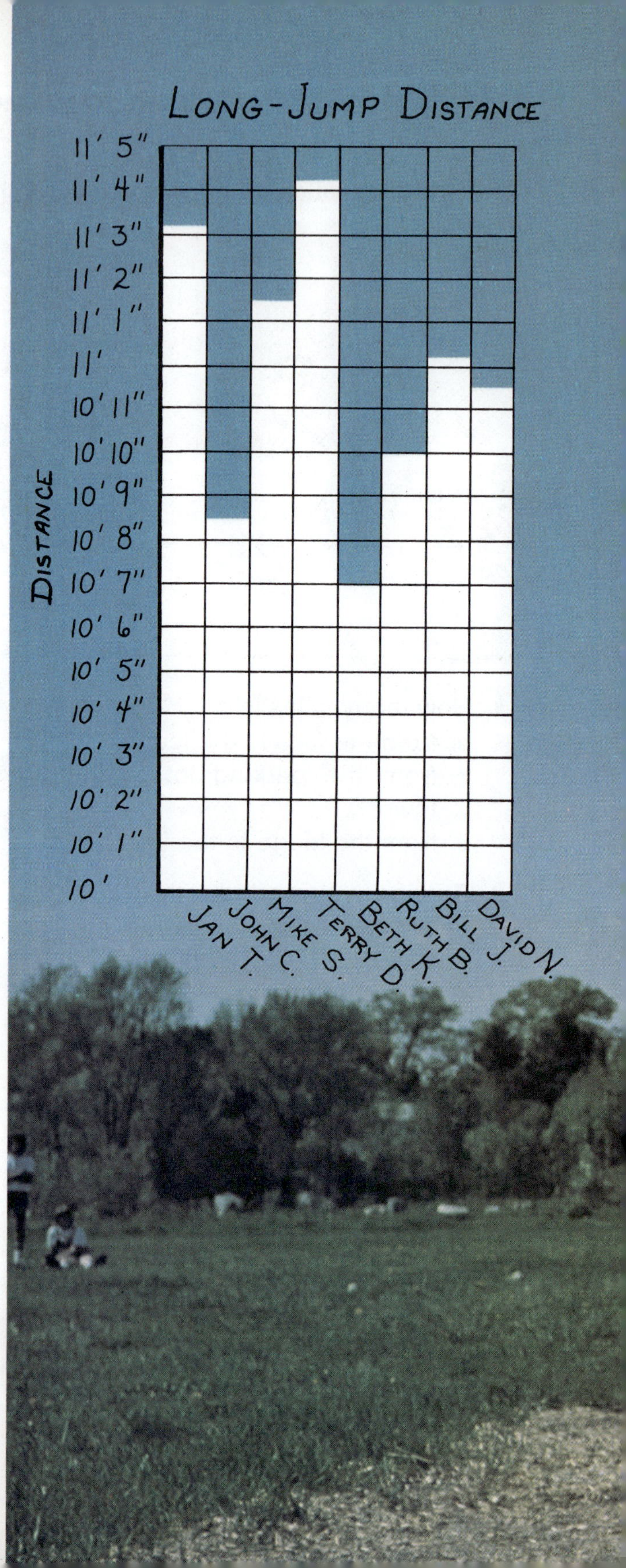

Refer to the graph to answer the following questions.

6. What was the longest jump?

7. Who had the shortest jump?

8. How many jumped more than 11 ft?

9. Who had the fifth longest jump?

10. How many jumped farther than Ruth?

11. How much farther did Bill jump than John?

12. Who jumped 2 in. farther than David?

13. What was the difference between the longest and shortest jumps?

★ 14. How much less than 12 ft was Terry's longest jump?

★ 15. How far in yards was Ruth's longest jump?

CHAPTER CHECKUP

Give each sum in lowest terms. [pages 172–173, 176–177, 180]

1. $\frac{1}{5} + \frac{3}{5}$ 2. $\frac{1}{4} + \frac{1}{2}$ 3. $\frac{3}{8} + \frac{1}{4}$ 4. $\frac{1}{3} + \frac{1}{4}$

5. $\frac{1}{6} + \frac{4}{9}$ 6. $\frac{2}{5} + \frac{3}{10}$ 7. $\frac{2}{9} + \frac{1}{3}$ 8. $\frac{0}{6} + \frac{3}{5}$

Give each difference in lowest terms. [pages 172–173, 178–180]

9. $\frac{5}{9} - \frac{2}{9}$ 10. $\frac{7}{8} - \frac{3}{8}$ 11. $\frac{5}{6} - \frac{1}{2}$ 12. $\frac{3}{2} - \frac{3}{4}$

13. $\frac{5}{7} - \frac{0}{3}$ 14. $\frac{1}{2} - \frac{1}{3}$ 15. $\frac{3}{4} - \frac{2}{3}$ 16. $\frac{5}{6} - \frac{5}{8}$

[pages 182–186]

Copy and complete.

	17.	18.	19.	20.	21.	22.	23.	24.	25.
Fraction	$\frac{3}{2}$	$\frac{5}{3}$	$\frac{11}{4}$	$\frac{9}{5}$	$\frac{15}{6}$				
Mixed number						$2\frac{1}{2}$	$3\frac{1}{4}$	$4\frac{2}{3}$	$1\frac{3}{5}$

Add. Write the sum in simplest form. [pages 188–189, 192]

26. $3\frac{1}{5} + 2\frac{2}{5}$ 27. $5\frac{1}{4} + 2\frac{1}{2}$ 28. $6\frac{5}{8} + 3\frac{3}{4}$ 29. $9\frac{5}{8} + 4\frac{3}{8}$ 30. $6\frac{2}{5} + 3\frac{3}{4}$ 31. $8 + 5\frac{2}{7}$

Subtract. Write the difference in simplest form. [pages 190–192]

32. $8\frac{5}{9} - 3\frac{4}{9}$ 33. $9\frac{3}{4} - 2\frac{1}{2}$ 34. $8\frac{1}{2} - 5\frac{1}{3}$ 35. $7\frac{1}{2} - 1\frac{3}{4}$ 36. $7\frac{2}{3} - 5\frac{5}{6}$ 37. $6 - 3\frac{7}{8}$

Solve. [pages 187, 193–195]

38. A bread recipe calls for $\frac{3}{4}$ cup of white flour and $\frac{1}{2}$ cup of whole wheat flour. How much flour does the recipe call for?

39. Alice had $\frac{5}{8}$ yard of ribbon. She used $\frac{1}{2}$ yard on a dress. What fraction of a yard did she have left?

CHAPTER PROJECT

1. Use pieces of graph paper to make strips of multiples for the numbers 1 through 10.
2. The examples show how you can use your strips of multiples to add and subtract fractions.

EXAMPLE 1.

$$\frac{1}{5} + \frac{2}{3} = \frac{13}{15}$$

EXAMPLE 2.

$$\frac{5}{6} - \frac{2}{9} = \frac{11}{18}$$

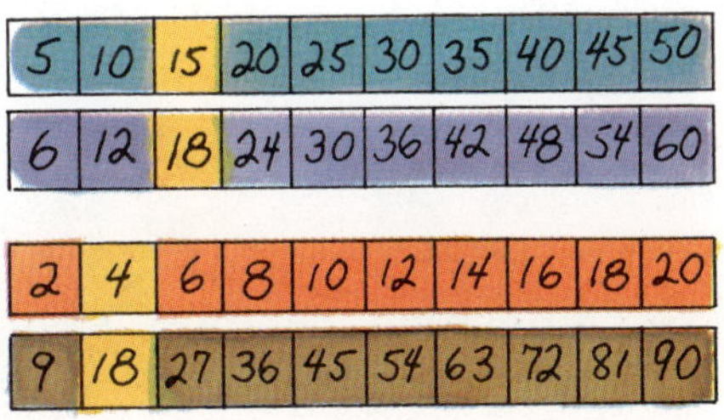

3. Use your strips of multiples to find these sums and differences in simplest form.

a. $\frac{2}{3} + \frac{3}{5}$ b. $\frac{5}{6} + \frac{9}{8}$ c. $\frac{5}{7} + \frac{2}{3}$ d. $\frac{7}{4} + \frac{5}{6}$

e. $\frac{8}{5} - \frac{5}{6}$ f. $\frac{7}{6} - \frac{4}{9}$ g. $\frac{9}{5} - \frac{3}{4}$ h. $\frac{7}{6} - \frac{3}{10}$

CHAPTER REVIEW

The least common multiple of two denominators is the least common denominator.

Give the least common denominator.

1. $\frac{1}{3}, \frac{1}{4}$
2. $\frac{2}{3}, \frac{1}{2}$
3. $\frac{3}{8}, \frac{3}{4}$
4. $\frac{2}{5}, \frac{1}{3}$
5. $\frac{5}{6}, \frac{3}{8}$
6. $\frac{4}{9}, \frac{2}{3}$
7. $\frac{5}{2}, \frac{3}{5}$
8. $\frac{7}{4}, \frac{5}{6}$

$$\frac{2}{3} + \frac{1}{5} = \frac{10}{15} + \frac{3}{15} = \frac{13}{15}$$

Add. Write the sum in simplest form.

9. $\frac{2}{9} + \frac{3}{9}$
10. $\frac{3}{8} + \frac{1}{8}$
11. $\frac{3}{4} + \frac{1}{2}$
12. $\frac{5}{6} + \frac{1}{3}$
13. $\frac{2}{3} + \frac{5}{4}$
14. $\frac{7}{6} + \frac{5}{8}$

$$\frac{3}{4} - \frac{2}{3} = \frac{9}{12} - \frac{8}{12} = \frac{1}{12}$$

Subtract. Write the difference in simplest form.

15. $\frac{5}{6} - \frac{1}{6}$
16. $\frac{1}{2} - \frac{1}{3}$
17. $\frac{5}{8} - \frac{1}{4}$
18. $\frac{11}{12} - \frac{3}{4}$
19. $\frac{5}{2} - \frac{5}{6}$
20. $\frac{7}{3} - \frac{0}{5}$

$$\begin{array}{rr} 3\frac{1}{2} = & 3\frac{4}{8} \\ +2\frac{5}{8} = & +2\frac{5}{8} \\ \hline & 6\frac{1}{8} \end{array}$$

Regroup $\frac{9}{8}$ for $1\frac{1}{8}$.

Add. Write the sum in simplest form.

21. $3\frac{1}{3} + 2\frac{1}{3}$
22. $4\frac{1}{4} + 3\frac{1}{2}$
23. $5\frac{1}{2} + 2\frac{1}{3}$
24. $6\frac{1}{2} + 3\frac{5}{8}$
25. $4\frac{5}{6} + 2\frac{1}{3}$
26. $5\frac{3}{4} + 3\frac{2}{3}$

Regroup 1 for $\frac{12}{12}$.

$$\begin{array}{rrr} 6\frac{1}{3} = & 6\frac{4}{12} = & 5\frac{16}{12} \\ -2\frac{3}{4} = & -2\frac{9}{12} = & -2\frac{9}{12} \\ \hline & & 3\frac{7}{12} \end{array}$$

Subtract. Write the difference in simplest form.

27. $8\frac{5}{9} - 2\frac{1}{9}$
28. $7\frac{3}{4} - 3\frac{1}{4}$
29. $9\frac{5}{8} - 4\frac{1}{4}$
30. $6\frac{1}{3} - 1\frac{1}{2}$
31. $8\frac{2}{3} - 2\frac{5}{6}$
32. $7\frac{1}{8} - 3\frac{5}{16}$

CHAPTER CHALLENGE

GOLD	PRICE
$\frac{1}{2}$ oz	$240
$\frac{1}{3}$ oz	$160
$\frac{1}{4}$ oz	$120
$\frac{1}{5}$ oz	$ 96

How many ounces?
How many dollars?

1.

2.

3.

4.

5.

MAJOR CHECKUP

STANDARDIZED FORMAT

Choose the correct letter.

1. 535,000,000 rounded to the nearest ten million is
   a. 530,000,000
   b. 535,000,000
   c. 500,000,000
   d. none of these

2. Divide.
   $136\overline{)14,282}$
   a. 15 R2
   b. 155 R2
   c. 105 R2
   d. none of these

3. Add.
   2.93
   0.08
   25.6
   +7.42
   a. 36.03
   b. 24.93
   c. 35.03
   d. none of these

4. Find this difference.
   $37.6 - 2.54$
   a. 1322
   b. 12.2
   c. 35.06
   d. none of these

5. Multiply.
   5.84
   ×6.5
   a. 37.96
   b. 63.24
   c. 3.796
   d. none of these

6. Divide.
   $2.6\overline{)7.93}$
   a. 3.05
   b. 0.35
   c. 3.5
   d. none of these

7. The greatest common factor of 9 and 12 is
   a. 36
   b. 9
   c. 3
   d. none of these

8. Which of these numbers is prime?
   a. 9
   b. 21
   c. 31
   d. 36

9. Subtract.
   $\frac{2}{5} - \frac{1}{3}$
   a. $\frac{1}{15}$
   b. $\frac{11}{15}$
   c. $\frac{1}{2}$
   d. none of these

10. Add.
   $2\frac{3}{5}$
   $+1\frac{2}{5}$
   a. $1\frac{1}{5}$
   b. $3\frac{4}{5}$
   c. 4
   d. none of these

11. Adele had 18 books. This was 3 more books than Vera had. How many books did Vera have?
   a. 21 b. 15
   c. 54 d. none of these

12. Leon weighs 97.4 kg. Alan weighs 138 kg. How many more kg does Alan weigh than Leon?
   a. 40.6 b. 41.4
   c. 235.4 d. none of these

Fractions—Multiplication and Division

READY OR NOT?

This is a unit square.

What fraction of a unit square is colored?

1. 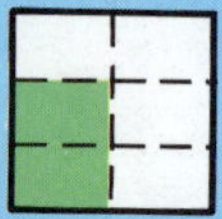

2. 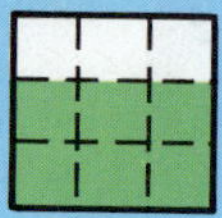

3. 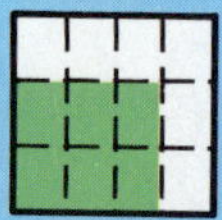

4. 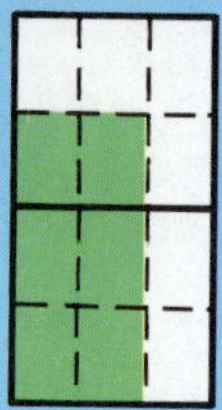

5. 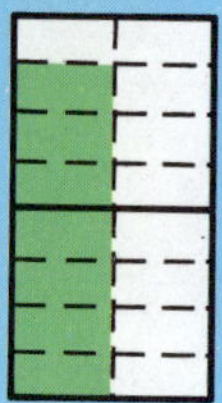

6. 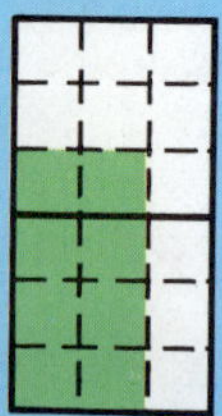

Multiplying fractions

You get the area of a rectangle by multiplying length and width.

$3 \times 2 = 6$

The area is 6 unit squares.

You can use the idea of area to help you multiply fractions.

$$\frac{2}{3} \times \frac{4}{5} = ?$$

(1) Pick a unit square. 

(2) Color a rectangle that is $\frac{4}{5}$ as long and $\frac{2}{3}$ as wide as the unit square.

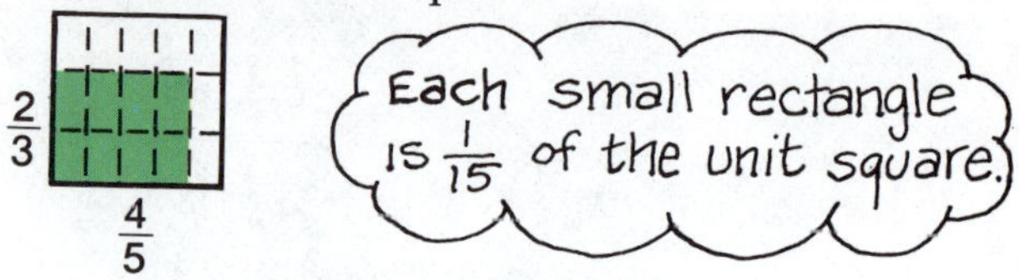

(3) The colored rectangle is $\frac{8}{15}$ of the unit square.

$$\frac{2}{3} \times \frac{4}{5} = \frac{8}{15}$$

To multiply two fractions:

Step 1. Multiply the numerators to get the numerator of the product.

Step 2. Multiply the denominators to get the denominator of the product.

$$\frac{2}{3} \times \frac{4}{5} = \frac{2 \times 4}{3 \times 5} = \frac{8}{15}$$

EXERCISES

Multiply.

1. 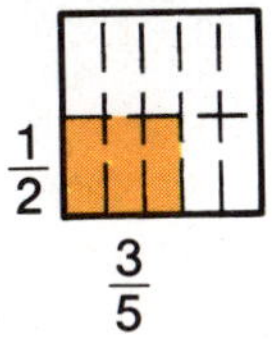

$\frac{1}{2} \times \frac{3}{5}$

2. 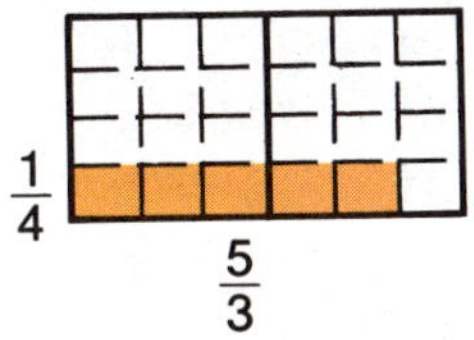

$\frac{1}{4} \times \frac{5}{3}$

3. 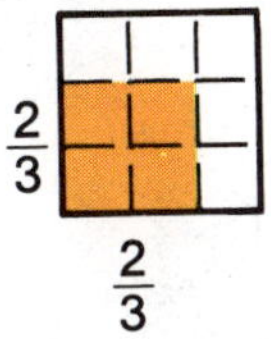

$\frac{2}{3} \times \frac{2}{3}$

4. 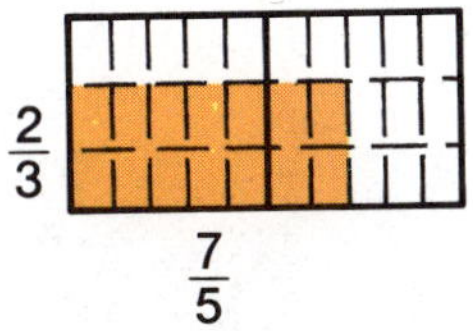

$\frac{2}{3} \times \frac{7}{5}$

Multiply. Give answers in simplest form.

5. $\frac{1}{4} \times \frac{1}{5}$

6. $\frac{1}{2} \times \frac{3}{8}$

7. $\frac{2}{5} \times \frac{1}{3}$

8. $\frac{3}{2} \times \frac{1}{4}$

9. $\frac{5}{6} \times \frac{2}{3}$

10. $\frac{7}{8} \times \frac{4}{5}$

11. $\frac{3}{8} \times \frac{2}{3}$

12. $\frac{3}{7} \times \frac{5}{6}$

13. $\frac{3}{2} \times \frac{3}{4}$

14. $\frac{4}{3} \times \frac{5}{3}$

15. $\frac{3}{2} \times \frac{5}{2}$

16. $\frac{3}{4} \times 5$

17. $\frac{5}{12} \times \frac{8}{3}$

18. $\frac{8}{5} \times \frac{3}{4}$

19. $\frac{3}{5} \times \frac{5}{3}$

20. $4 \times \frac{5}{6}$

21. $\frac{2}{3} \times \frac{3}{10}$

22. $\frac{4}{7} \times \frac{2}{5}$

23. $\frac{3}{4} \times \frac{8}{7}$

24. $\frac{5}{7} \times \frac{5}{6}$

25. $\frac{3}{4} \times \frac{4}{3}$

26. $\frac{3}{2} \times \frac{2}{5}$

27. $\frac{3}{5} \times \frac{1}{4}$

28. $\frac{5}{2} \times \frac{2}{3}$

29. $\frac{5}{9} \times 3$

30. $\frac{3}{10} \times \frac{5}{6}$

31. $\frac{4}{5} \times \frac{4}{5}$

32. $5 \times \frac{2}{3}$

33. $\frac{7}{8} \times 2$

34. $\frac{7}{4} \times 4$

35. $\frac{3}{8} \times \frac{5}{6}$

36. $\frac{2}{5} \times \frac{5}{2}$

37. $\frac{5}{12} \times \frac{3}{5}$

38. $\frac{7}{10} \times \frac{2}{5}$

39. $\frac{5}{3} \times \frac{6}{5}$

40. $\frac{2}{3} \times \frac{9}{8}$

41. $\frac{1}{8} \times 8$

42. $6 \times \frac{2}{3}$

43. $5 \times \frac{1}{5}$

44. $\frac{9}{4} \times \frac{2}{3}$

Fraction of a number

The Denim Den is having a sale on blue jeans. One style of jeans is regularly \$24. To find the sale price, we take $\frac{3}{4}$ of \$24. Finding a fraction of a number is a multiplication problem.

$$\frac{3}{4} \text{ of } \$24 = \frac{3}{4} \times \$24$$
$$= \frac{\$72}{4}$$
$$= \$18$$

$$\frac{3}{4} \text{ of } 20 = 15$$

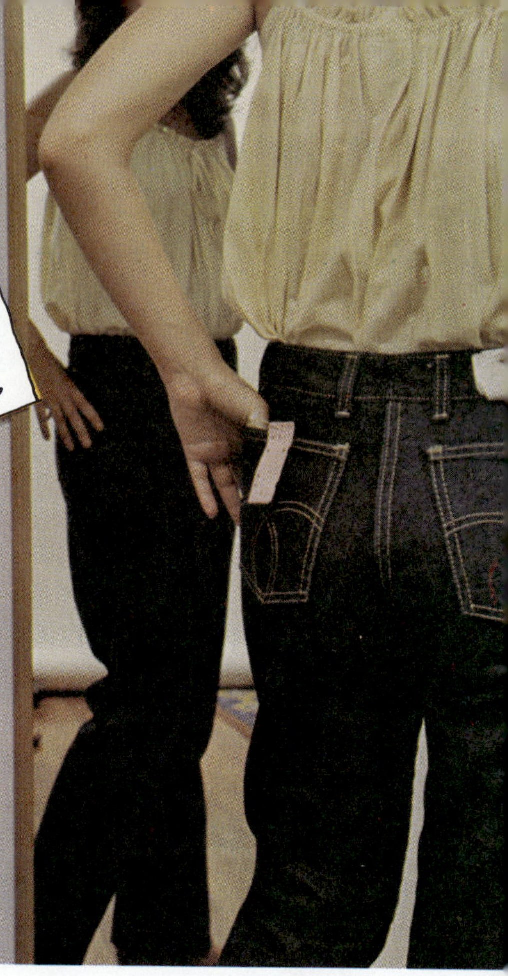

EXERCISES

Compute.

1. $\frac{1}{2}$ of 18	**2.** $\frac{1}{3}$ of 21	**3.** $\frac{1}{5}$ of 60	**4.** $\frac{1}{4}$ of 20
5. $\frac{1}{8}$ of 32	**6.** $\frac{1}{6}$ of 54	**7.** $\frac{1}{3}$ of 18.3	**8.** $\frac{2}{3}$ of 3.18
9. $\frac{4}{9}$ of 38.7	**10.** $\frac{1}{4}$ of 264	**11.** $\frac{2}{4}$ of 264	**12.** $\frac{3}{4}$ of 264
13. $\frac{5}{8}$ of 24	**14.** $\frac{3}{8}$ of 16	**15.** $\frac{5}{4}$ of 100	**16.** $\frac{3}{2}$ of 140
17. $\frac{5}{6}$ of 30	**18.** $\frac{7}{8}$ of 56	**19.** $\frac{2}{3}$ of 36	**20.** $\frac{3}{2}$ of 36
21. $\frac{6}{5}$ of 40	**22.** $\frac{3}{10}$ of 54.6	**23.** $\frac{4}{5}$ of 3.75	**24.** $\frac{2}{2}$ of 9.38

Here are some price tags from the Denim Den. Compute the sale prices. *Remember:* The sale price is $\frac{3}{4}$ of the regular price. Round each price to the nearest cent.

25.

26.

27.

28.

Solve.

29. Sarah wants to buy a new bicycle that cost \$96.99. Her father will pay $\frac{1}{3}$ of the price. How much money will Sarah need for her share?

30. Joe can throw a football $\frac{2}{5}$ of the length of a football field (100 yards). How many yards can he throw a football?

31. A club wants to sell 120 tickets for a chili supper. They have sold $\frac{5}{6}$ of their tickets. How many tickets do they have left to sell?

32. To get to school, Alex ran $\frac{3}{8}$ of a mile. How many yards is that?

33. There are 464 students in a school. Five-eighths of them are girls. How many boys are there in the school?

34. Mr. Adams earns \$1120 per month. How much money would he save in a year if he saved $\frac{1}{8}$ of his earnings?

★ 35. A furlong, $\frac{1}{8}$ of a mile, is a unit used in horse racing. How many yards long is a 7-furlong race?

★ 36. Carol earns \$1.75 per hour. She saves $\frac{5}{7}$ of that. How many hours must she work to save \$25?

Reciprocals

Two numbers are **reciprocals** when their product is 1.

$\frac{2}{5}$ and $\frac{5}{2}$ are reciprocals.

$$\frac{2}{5} \times \frac{5}{2} = \frac{10}{10} = 1$$

$2\frac{1}{3}$ and $\frac{3}{7}$ are reciprocals.

$$2\frac{1}{3} = \frac{7}{3}$$

$$\frac{7}{3} \times \frac{3}{7} = \frac{21}{21} = 1$$

6 and $\frac{1}{6}$ are reciprocals.

$$6 = \frac{6}{1}$$

$$\frac{6}{1} \times \frac{1}{6} = \frac{6}{6} = 1$$

EXERCISES

Copy and complete.

1. $\frac{2}{3} \times ? = 1$ **2.** $\frac{1}{2} \times ? = 1$ **3.** $\frac{3}{5} \times ? = 1$ **4.** $8 \times ? = 1$

5. $\frac{7}{4} \times ? = 1$ **6.** $3\frac{1}{2} \times ? = 1$ **7.** $1\frac{4}{5} \times ? = 1$ **8.** $10 \times ? = 1$

Give the reciprocal of each number.

9. $\frac{5}{8}$ **10.** 3 **11.** $\frac{7}{10}$ **12.** 5 **13.** $1\frac{1}{6}$ **14.** $\frac{4}{9}$ **15.** $\frac{6}{5}$ **16.** $\frac{1}{4}$

17. $\frac{1}{7}$ **18.** $\frac{10}{3}$ **19.** $2\frac{3}{5}$ **20.** $\frac{7}{6}$ **21.** 9 **22.** $1\frac{2}{9}$ **23.** $\frac{1}{12}$ **24.** $2\frac{1}{5}$

Who am I?

★ **25.** I am my own reciprocal.

★ **26.** I do not have a reciprocal.

Problem solving

Solve.

1. There are 250 species of sharks. One tenth of these species are dangerous to humans. How many dangerous species are there?

2. The largest shark is the whale shark, which grows up to 60 feet in length. A basking shark grows up to $\frac{2}{3}$ of that length. How many feet long is the largest basking shark?

3. The most dangerous shark is the white shark, which grows up to $1\frac{2}{3}$ times the length of a hammerhead shark. A hammerhead shark grows up to 15 feet in length. How many feet long is the largest white shark?

4. A thresher shark grows up to 21 feet in length. Half of a thresher shark's length is its tail. How long is the tail of the largest thresher shark?

★ 5. The largest white shark ever caught by rod and reel weighed 2664 pounds. The largest tiger shark ever caught by rod and reel weighed 4 pounds more than $\frac{2}{3}$ as much as the largest white shark caught. What was the weight of the largest tiger shark caught?

★ 6. The largest white shark ever caught by rod and reel was 16 feet 10 inches long. The largest thresher shark ever caught by rod and reel was 5 inches longer than $\frac{1}{2}$ as long as the largest white shark caught. How long was the largest thresher shark caught?

Dividing by a fraction

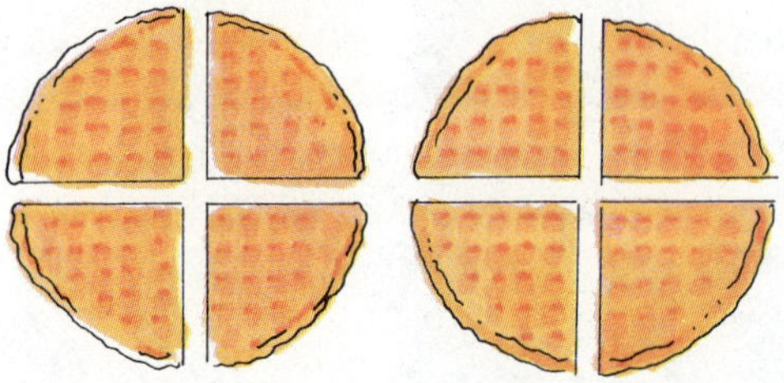

From the picture above, you can see that there are 8 fourths in 2.

$$2 \div \frac{1}{4} = 8$$

EXERCISES

1.

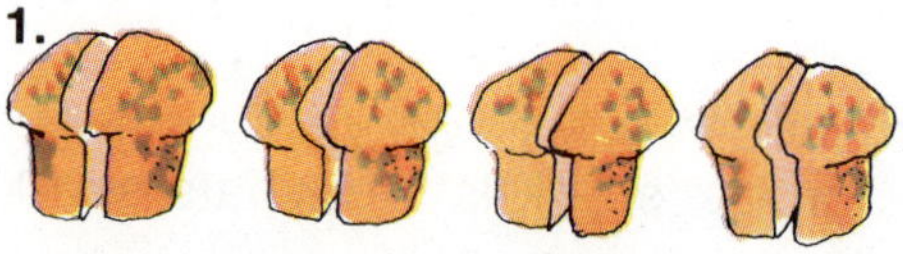

How many $\frac{1}{2}$'s are there in 4?

2.

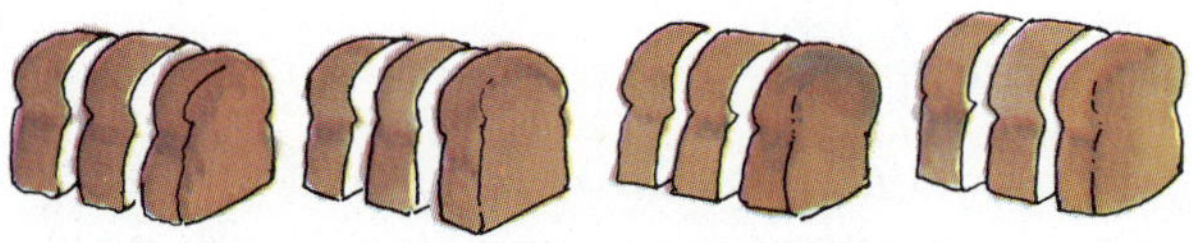

How many $\frac{1}{3}$'s are there in 4?

3.

How many $\frac{1}{4}$'s are there in 3?

4.

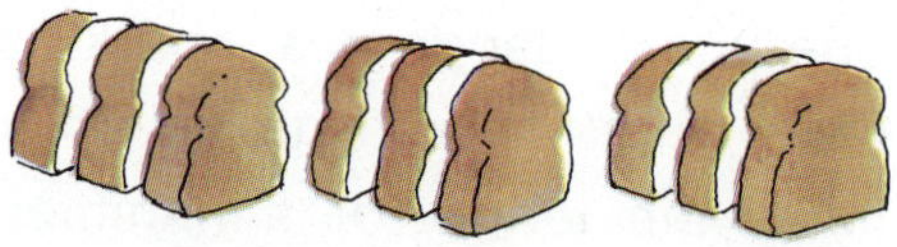

How many $\frac{1}{3}$'s are there in 3?

5.

How many $\frac{1}{3}$'s are there in $\frac{2}{3}$?

6.

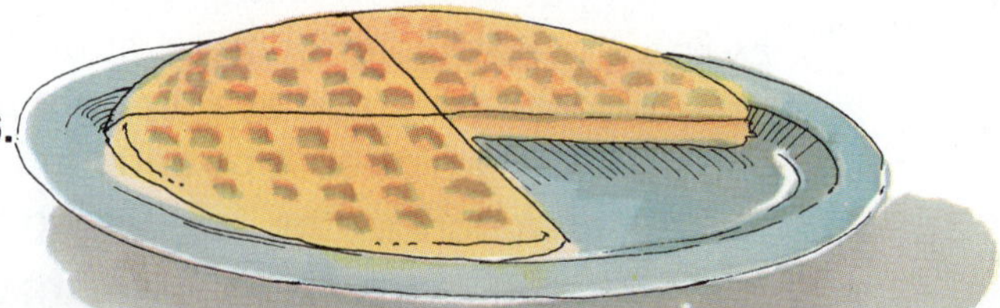

How many $\frac{1}{4}$'s are there in $\frac{3}{4}$?

First complete. Then compare your answers in each exercise.

7. 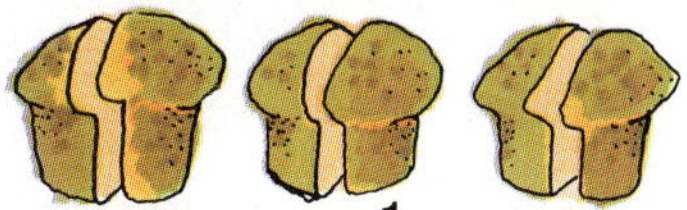

a. How many $\frac{1}{2}$'s are in 3?

b. $3 \div \frac{1}{2} = ?$

c. $3 \times 2 = ?$

8. 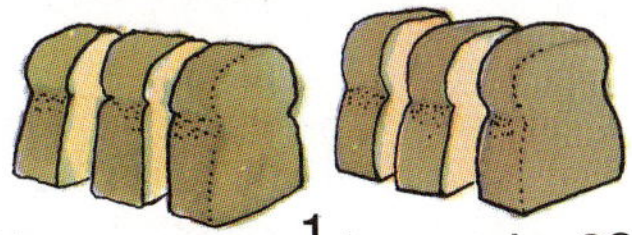

a. How many $\frac{1}{3}$'s are in 2?

b. $2 \div \frac{1}{3} = ?$

c. $2 \times 3 = ?$

9. 

a. How many $\frac{2}{3}$'s are in 2?

b. $2 \div \frac{2}{3} = ?$

c. $2 \times \frac{3}{2} = ?$

10. 

a. How many $\frac{3}{4}$'s are in 3?

b. $3 \div \frac{3}{4} = ?$

c. $3 \times \frac{4}{3} = ?$

11. 

a. How many $\frac{2}{5}$'s are in $\frac{4}{5}$?

b. $\frac{4}{5} \div \frac{2}{5} = ?$

c. $\frac{4}{5} \times \frac{5}{2} = ?$

12. 

a. How many $\frac{2}{3}$'s are in $\frac{8}{3}$?

b. $\frac{8}{3} \div \frac{2}{3} = ?$

c. $\frac{8}{3} \times \frac{3}{2} = ?$

Complete.

13. Look at exercise 9. Dividing by $\frac{2}{3}$ is the same as multiplying by __?__.

14. Look at exercise 10. Dividing by $\frac{3}{4}$ is the same as multiplying by __?__.

15. Look at exercise 11. Dividing by $\frac{2}{5}$ is the same as multiplying by the reciprocal of __?__.

16. Look at exercise 12. Dividing by $\frac{2}{3}$ is the same as multiplying by the reciprocal of __?__.

17. What is the reciprocal of 0.125? Of 0.0125?

Dividing by a fraction

Louisa and 4 friends planned to make some batches of cookies. The recipe required $\frac{3}{4}$ of a pound of butter for each batch of cookies. How many batches could be made if each of the 5 girls brought $\frac{1}{2}$ pound (2 quarter-pound sticks) of butter?

Here are the sticks of butter the 5 girls brought.

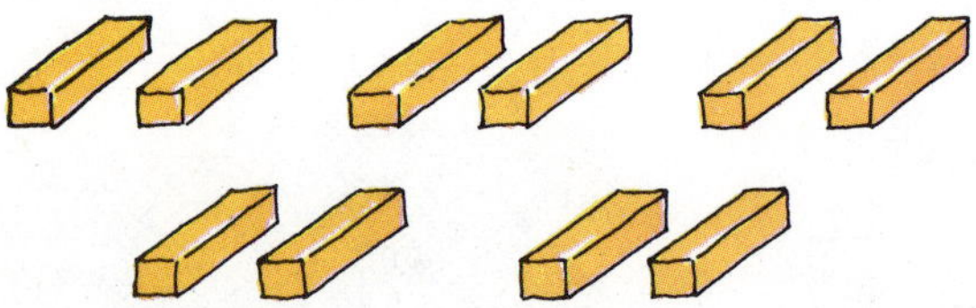

Louisa divided them into groups of 3 quarter-pound sticks.

There was enough butter for $3\frac{1}{3}$ batches of cookies.

Let's look at it this way.

The total number of pounds of butter is 5 $\frac{1}{2}$'s, or $\frac{5}{2}$.

The number of pounds of butter needed for each batch is $\frac{3}{4}$.

The number of batches of cookies is $3\frac{1}{3}$.

$$\frac{5}{2} \div \frac{3}{4} = 3\frac{1}{3}$$

To divide by a fraction, you can multiply by its reciprocal.

reciprocals

$$\frac{5}{2} \div \frac{3}{4} = \frac{5}{2} \times \frac{4}{3}$$
$$= \frac{20}{6}$$
$$= \frac{10}{3}$$
$$= 3\frac{1}{3}$$

EXERCISES

Complete.

1. To divide by $\frac{3}{5}$, you can multiply by the reciprocal of __?__.
2. Dividing by $\frac{4}{7}$ is the same as multiplying by __?__.
3. To divide by 3, you can multiply by the reciprocal of __?__.
4. Dividing by 5 is the same as multiplying by __?__.

Divide. Give each quotient in simplest form.

5. $\frac{2}{3} \div 2$
6. $\frac{3}{2} \div \frac{1}{4}$
7. $\frac{3}{4} \div \frac{1}{4}$
8. $\frac{3}{8} \div 2$
9. $\frac{3}{4} \div \frac{2}{3}$
10. $\frac{5}{8} \div 3$
11. $\frac{4}{5} \div \frac{2}{2}$
12. $8 \div \frac{3}{5}$
13. $5 \div \frac{2}{5}$
14. $\frac{5}{6} \div \frac{3}{8}$
15. $\frac{3}{5} \div 4$
16. $\frac{0}{4} \div \frac{1}{2}$
17. $\frac{5}{9} \div \frac{1}{3}$
18. $6 \div \frac{1}{2}$
19. $\frac{6}{5} \div \frac{3}{3}$
20. $\frac{7}{4} \div \frac{7}{4}$
21. $\frac{3}{8} \div \frac{3}{4}$
22. $\frac{3}{4} \div \frac{3}{8}$
23. $\frac{5}{9} \div \frac{2}{3}$
24. $\frac{2}{3} \div \frac{5}{9}$
25. $\frac{5}{6} \div \frac{2}{3}$
26. $\frac{2}{3} \div \frac{5}{6}$
27. $\frac{3}{8} \div \frac{3}{2}$
28. $\frac{3}{2} \div \frac{3}{8}$
29. $\frac{2}{5} \div \frac{1}{4}$
30. $\frac{0}{5} \div \frac{5}{6}$
31. $\frac{5}{9} \div \frac{4}{3}$
32. $\frac{7}{8} \div \frac{5}{4}$
33. $\frac{5}{4} \div 3$
34. $\frac{6}{5} \div \frac{1}{10}$
35. $\frac{8}{3} \div \frac{4}{3}$
36. $\frac{9}{2} \div \frac{3}{4}$

Solve.

37. A recipe requires $\frac{3}{4}$ pound of butter. How much butter is needed to make 2 recipes?

38. A recipe requires $\frac{3}{4}$ pound of butter. How many recipes can be made from 6 pounds of butter?

39. Gregory had 3 cups of flour. He used $2\frac{1}{4}$ cups of flour to make bread. How much flour did he have left?

40. Bernice used $\frac{2}{3}$ cup of chocolate to make 8 cookies. How much chocolate was in each cookie?

Practice exercises

Give each product in simplest form.

1. $\frac{2}{3} \times \frac{1}{4}$
2. $\frac{3}{10} \times \frac{2}{9}$
3. $\frac{2}{5} \times \frac{4}{7}$
4. $\frac{5}{8} \times \frac{0}{6}$
5. $\frac{1}{8} \times \frac{1}{6}$
6. $\frac{1}{4} \times \frac{1}{2}$
7. $\frac{1}{2} \times \frac{5}{8}$
8. $\frac{4}{4} \times \frac{3}{8}$
9. $\frac{5}{3} \times \frac{2}{2}$
10. $\frac{5}{6} \times \frac{3}{10}$
11. $6 \times \frac{5}{12}$
12. $\frac{5}{3} \times 2$
13. $\frac{7}{9} \times \frac{3}{7}$
14. $2 \times \frac{3}{4}$
15. $\frac{1}{4} \times 5$
16. $\frac{0}{3} \times \frac{5}{2}$
17. $\frac{3}{8} \times 5$
18. $\frac{5}{8} \times \frac{8}{5}$
19. $6 \times \frac{3}{2}$
20. $\frac{7}{10} \times \frac{4}{3}$
21. $\frac{5}{6} \times \frac{1}{5}$
22. $\frac{5}{12} \times \frac{3}{8}$
23. $\frac{5}{9} \times \frac{9}{5}$
24. $5 \times \frac{2}{5}$

Give each quotient in simplest form.

25. $\frac{1}{2} \div \frac{1}{3}$
26. $\frac{1}{3} \div \frac{1}{2}$
27. $\frac{2}{3} \div \frac{2}{3}$
28. $\frac{3}{4} \div \frac{1}{4}$
29. $3 \div \frac{4}{5}$
30. $\frac{3}{4} \div \frac{2}{2}$
31. $\frac{7}{3} \div 2$
32. $\frac{1}{8} \div \frac{9}{5}$
33. $\frac{8}{5} \div \frac{5}{9}$
34. $\frac{0}{6} \div \frac{4}{3}$
35. $\frac{4}{9} \div \frac{1}{3}$
36. $\frac{9}{2} \div \frac{5}{8}$
37. $\frac{8}{7} \div 4$
38. $\frac{1}{2} \div \frac{2}{3}$
39. $2 \div \frac{5}{6}$
40. $\frac{7}{2} \div 4$
41. $\frac{3}{4} \div \frac{1}{8}$
42. $\frac{3}{15} \div \frac{5}{3}$
43. $\frac{5}{16} \div \frac{1}{2}$
44. $\frac{3}{2} \div \frac{3}{14}$
45. $\frac{0}{3} \div \frac{5}{8}$
46. $\frac{9}{5} \div \frac{3}{8}$
47. $\frac{1}{5} \div \frac{15}{4}$
48. $\frac{3}{4} \div \frac{3}{4}$

Multiply across and multiply down. Give answers in lowest terms.

49.

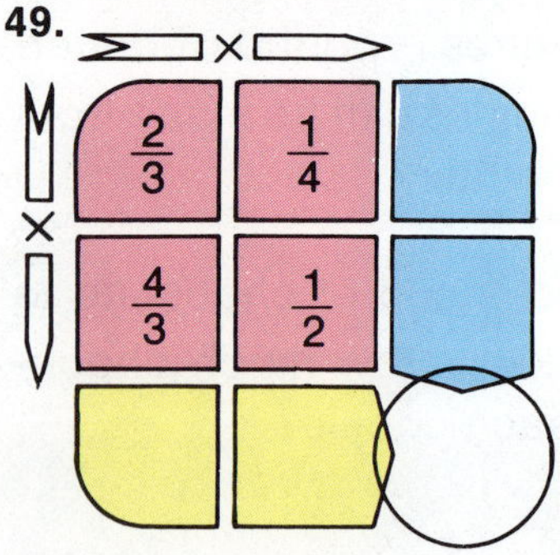

★ 50.

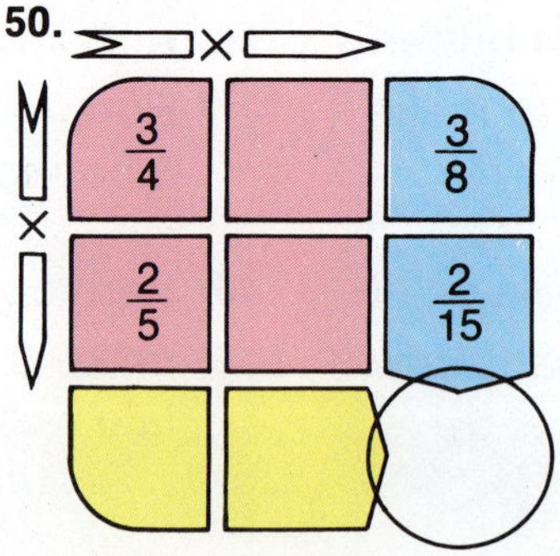

★ 51.

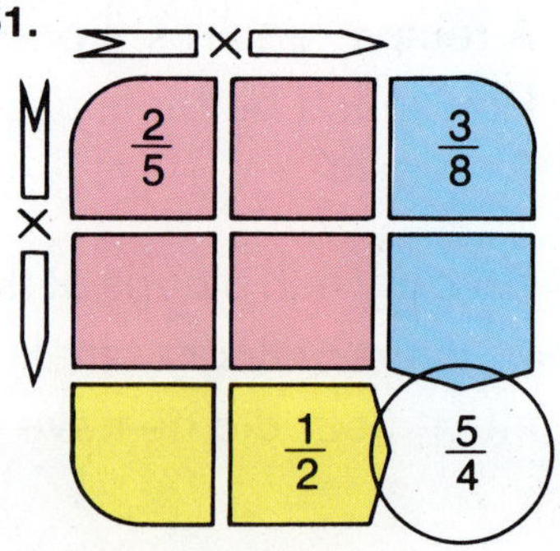

Problem solving

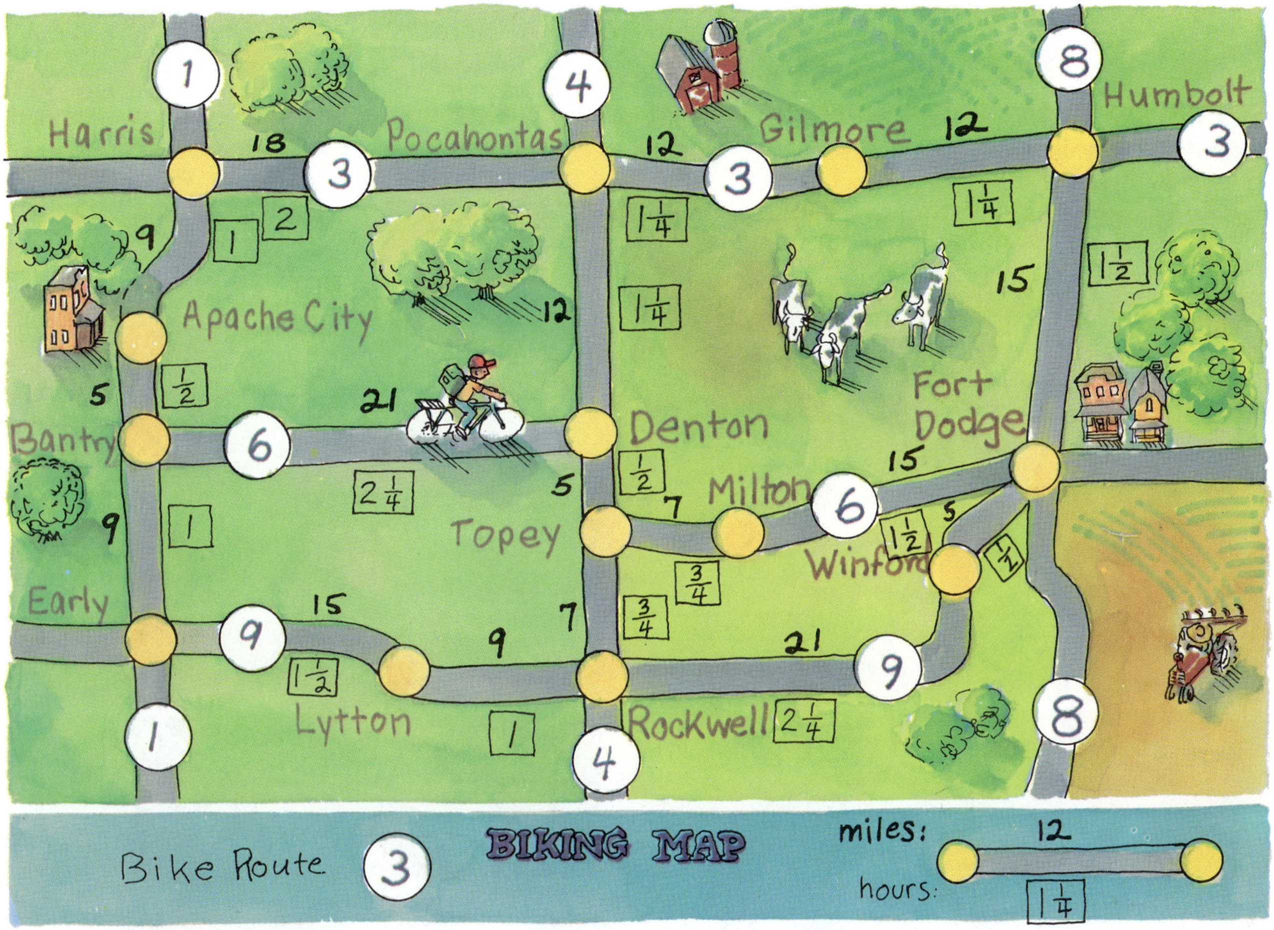

Use the map to solve.

1. a. How many miles is the trip from Fort Dodge to Pocahontas along bike routes ⑥ and ④?
   b. How many hours will this trip take?

2. a. How many miles is the trip from Humbolt to Apache City along bike routes ③ and ①?
   b. How many hours will this trip take?
   c. If a bicyclist travels this route and takes four $\frac{3}{4}$-hour rest stops, how many hours will the trip take?

3. a. How many miles is the shortest bike route from Fort Dodge to Apache City?
   b. How many hours will it take to travel this route?

The cancellation shortcut

We can divide both a numerator and a denominator by a common factor *before* multiplying fractions. This shortcut is called **canceling.**

EXAMPLE.

$$\frac{3}{8} \times \frac{4}{3} = ?$$

Step 1.
Divide by 3.

$$\frac{\overset{1}{\cancel{3}}}{8} \times \frac{4}{\underset{1}{\cancel{3}}} = ?$$

Step 2.
Divide by 4.

$$\frac{\overset{1}{\cancel{3}}}{\underset{2}{\cancel{8}}} \times \frac{\overset{1}{\cancel{4}}}{\underset{1}{\cancel{3}}} = ?$$

Step 3.
Multiply.

$$\frac{\overset{1}{\cancel{3}}}{\underset{2}{\cancel{8}}} \times \frac{\overset{1}{\cancel{4}}}{\underset{1}{\cancel{3}}} = \frac{1}{2}$$

OTHER EXAMPLES.

Divide by 2.
Divide by 2.

$$\frac{\overset{3}{\cancel{6}}}{\underset{\underset{2}{\cancel{4}}}{\cancel{8}}} \times \frac{\overset{1}{\cancel{2}}}{5} = \frac{3}{10}$$

Divide by 2.
Divide by 5.

$$\frac{\overset{3}{\cancel{15}}}{\underset{4}{\cancel{8}}} \times \frac{\overset{3}{\cancel{6}}}{\underset{1}{\cancel{5}}} = \frac{9}{4} = 2\frac{1}{4}$$

EXERCISES
Multiply. Give products in simplest form.

1. $\frac{3}{5} \times \frac{10}{3}$
2. $\frac{4}{7} \times \frac{14}{12}$
3. $\frac{5}{9} \times \frac{3}{15}$
4. $\frac{7}{5} \times \frac{6}{14}$
5. $\frac{6}{4} \times \frac{8}{9}$
6. $\frac{4}{9} \times \frac{18}{12}$
7. $8 \times \frac{1}{12}$
8. $\frac{16}{9} \times \frac{3}{20}$
9. $12 \times \frac{3}{16}$
10. $\frac{15}{21} \times \frac{14}{10}$
11. $\frac{16}{24} \times \frac{18}{14}$
12. $\frac{15}{32} \times \frac{24}{20}$
13. $\frac{35}{8} \times \frac{12}{20}$
14. $\frac{16}{42} \times \frac{36}{24}$
15. $\frac{16}{7} \times 35$
16. $\frac{45}{14} \times \frac{28}{35}$

Divide. Give quotients in simplest form.

EXAMPLE. $\frac{5}{6} \div \frac{2}{3} = ?$

Write as a multiplication problem. Then cancel if you can.

$$\frac{5}{6} \div \frac{2}{3} = \frac{5}{\cancel{6}_{2}} \times \frac{\cancel{3}^{1}}{2}$$

$$= \frac{5}{4} = 1\frac{1}{4}$$

17. $\frac{3}{8} \div \frac{5}{6}$
18. $\frac{4}{7} \div \frac{2}{3}$
19. $\frac{6}{9} \div \frac{1}{3}$
20. $\frac{24}{36} \div 2$
21. $\frac{16}{42} \div \frac{36}{24}$
22. $\frac{45}{13} \div \frac{25}{26}$
23. $\frac{15}{17} \div \frac{15}{17}$
24. $\frac{18}{24} \div \frac{3}{8}$
25. $8 \div \frac{3}{2}$
26. $\frac{4}{5} \div 6$
27. $5 \div \frac{10}{3}$
28. $\frac{3}{4} \div 8$
29. $\frac{4}{5} \div \frac{5}{4}$
30. $\frac{12}{5} \div \frac{4}{10}$
31. $\frac{4}{5} \div \frac{8}{10}$
32. $\frac{5}{9} \div \frac{15}{3}$
33. $\frac{7}{8} \div \frac{21}{4}$
34. $\frac{24}{25} \div \frac{16}{5}$

Solve.

35. There are 32 students in Bill's math class. One day $\frac{1}{8}$ of the students were absent. How many students were present?
36. Meg had $\frac{5}{6}$ of a yard of wire. She used $\frac{1}{2}$ of a yard to hang a picture. What fraction of a yard did she have left?
37. ★ Jack earned \$4.80. He saved $\frac{1}{3}$ of it and spent $\frac{1}{2}$ of the rest going to a movie. How much did the movie cost?
38. ★ Susan went on a bicycle tour at 7:30 A.M. At 9:45 A.M. she had traveled 18 miles. At that rate, how long did she take to travel 24 miles?

KEEPING SKILLS SHARP

Add or subtract. Give answers in lowest terms.

1. $\frac{1}{2} + \frac{3}{8}$
2. $\frac{1}{2} - \frac{3}{8}$
3. $\frac{1}{2} + \frac{1}{3}$
4. $\frac{1}{2} - \frac{1}{3}$
5. $\frac{5}{9} + \frac{1}{6}$
6. $\frac{5}{9} - \frac{1}{6}$
7. $\frac{3}{4} + \frac{1}{12}$
8. $\frac{3}{4} - \frac{1}{12}$
9. $\frac{2}{3} + \frac{1}{5}$
10. $\frac{2}{3} - \frac{1}{5}$
11. $\frac{7}{10} + \frac{1}{4}$
12. $\frac{7}{10} - \frac{1}{4}$
13. $\frac{2}{5} + \frac{3}{10}$
14. $\frac{2}{5} - \frac{3}{10}$
15. $\frac{5}{8} + \frac{1}{6}$
16. $\frac{5}{8} - \frac{1}{6}$

Complex fractions

We can use the fraction bar to indicate the division of whole numbers.

$$\frac{1}{2} = 1 \div 2$$

We can also use the fraction bar to indicate the division of fractions.

$$\frac{\frac{2}{3}}{\frac{5}{6}}$$

This means $\frac{2}{3} \div \frac{5}{6}$.

This fraction is called a **complex fraction.** To simplify a complex fraction, divide the fraction in the numerator by the fraction in the denominator.

$$\frac{\frac{2}{3}}{\frac{5}{6}} = \frac{2}{3} \div \frac{5}{6}$$

$$= \frac{2}{\overset{}{\cancel{3}}_{1}} \times \frac{\overset{2}{\cancel{6}}}{5}$$

$$= \frac{4}{5}$$

EXERCISES

Write each complex fraction in simplest form.

1. $\dfrac{\frac{1}{5}}{\frac{3}{4}}$
2. $\dfrac{\frac{7}{3}}{\frac{14}{9}}$
3. $\dfrac{\frac{5}{9}}{\frac{15}{6}}$
4. $\dfrac{\frac{6}{9}}{\frac{6}{12}}$
5. $\dfrac{\frac{5}{3}}{\frac{10}{6}}$
6. $\dfrac{\frac{8}{5}}{\frac{10}{4}}$
7. $\dfrac{\frac{7}{5}}{\frac{14}{10}}$
8. $\dfrac{\frac{3}{8}}{\frac{9}{4}}$
9. $\dfrac{\frac{5}{6}}{\frac{15}{9}}$
10. $\dfrac{\frac{15}{24}}{\frac{20}{12}}$
11. $\dfrac{\frac{9}{18}}{\frac{6}{15}}$
12. $\dfrac{\frac{7}{15}}{\frac{38}{10}}$

13. $\dfrac{\frac{5}{3}}{\frac{3}{9}}$ 14. $\dfrac{\frac{7}{5}}{\frac{21}{10}}$ 15. $\dfrac{5}{\frac{2}{3}}$ 16. $\dfrac{3}{\frac{3}{4}}$ 17. $\dfrac{\frac{3}{8}}{9}$ 18. $\dfrac{\frac{7}{8}}{\frac{4}{4}}$

19. $\dfrac{\frac{1}{2}}{\frac{3}{4}}$ 20. $\dfrac{\frac{5}{8}}{\frac{5}{4}}$ 21. $\dfrac{\frac{3}{5}}{\frac{6}{3}}$ 22. $\dfrac{7}{\frac{3}{5}}$ 23. $\dfrac{\frac{2}{3}}{\frac{4}{5}}$ 24. $\dfrac{\frac{5}{12}}{\frac{10}{6}}$

25. $\dfrac{\frac{7}{5}}{3}$ 26. $\dfrac{\frac{7}{8}}{\frac{3}{8}}$ 27. $\dfrac{6}{\frac{3}{4}}$ 28. $\dfrac{\frac{8}{9}}{2}$ 29. $\dfrac{\frac{5}{12}}{\frac{1}{6}}$ 30. $\dfrac{\frac{0}{4}}{\frac{1}{2}}$

Solve.

31. Which city has more January precipitation, Fresno or San Diego?

32. How much more January precipitation does Los Angeles have than Fresno?

33. How much less January precipitation does San Diego have than Eureka?

34. Which city has the same amount of January precipitation as two other cities together?

★ 35. What is the average January precipitation for the five cities?

★ 36. How many times as much January precipitation does Los Angeles have than Fresno?

Average January Precipitation in 5 California Cities

City	*Precipitation in inches*
Eureka	$7\frac{2}{5}$
Fresno	$1\frac{4}{5}$
Los Angeles	3
San Diego	$1\frac{9}{10}$
San Francisco	$4\frac{2}{5}$

Fractions, mixed numbers, and decimals

Changing decimals to fractions

If you can *read* a decimal correctly, you can easily change a decimal to a fraction. Then you can write the fraction in simplest form. Study these examples.

2 tenths

$$0.2 = \frac{2}{10} = \frac{1}{5}$$

375 thousandths

$$0.375 = \frac{375}{1000} = \frac{3}{8}$$

3 and 25 hundredths

$$3.25 = 3\frac{25}{100} = 3\frac{1}{4}$$

Changing fractions to decimals

To change a fraction to a decimal, divide the numerator by the denominator. Study these examples.

$5\overline{)2.0}$ = 0.4

$$\frac{2}{5} = 0.4$$

$4\overline{)3.00}$ = 0.75

$$2\frac{3}{4} = 2.75$$

EXERCISES

Change to a fraction or mixed number in simplest form.

1. 0.3 **2.** 0.5 **3.** 0.8 **4.** 0.25 **5.** 0.64 **6.** 0.80

7. 3.375 **8.** 3.1 **9.** 1.08 **10.** 2.16 **11.** 1.875 **12.** 3.48

Change to a decimal.

13. $\frac{1}{2}$ **14.** $\frac{7}{8}$ **15.** $\frac{1}{4}$ **16.** $\frac{1}{8}$ **17.** $\frac{3}{8}$ **18.** $1\frac{1}{2}$

19. $1\frac{1}{4}$ **20.** $\frac{3}{10}$ **21.** $2\frac{1}{8}$ **22.** $3\frac{1}{4}$ **23.** $2\frac{3}{5}$ **24.** $\frac{4}{5}$

<, =, or >?

25. $\frac{1}{8}$ ● 0.12 **26.** $\frac{1}{5}$ ● 0.25 **27.** $\frac{4}{5}$ ● 0.75 **28.** $\frac{3}{8}$ ● 0.375

29. $\frac{3}{4}$ ● 0.8 **30.** $\frac{1}{4}$ ● 0.25 **31.** $\frac{3}{5}$ ● 0.5 **32.** $\frac{5}{8}$ ● 0.63

33. $1\frac{1}{2}$ ● 1.6 **34.** $1\frac{1}{4}$ ● 1.20 **35.** $3\frac{7}{8}$ ● 3.8 **36.** $4\frac{2}{5}$ ● 4.4

Copy and complete these tables.

37.

Fraction or Mixed Number	$\frac{1}{4}$	$\frac{2}{4}$	$\frac{3}{4}$	$\frac{4}{4}$	$1\frac{1}{4}$	$1\frac{2}{4}$
Decimal						

38.

Fraction or Mixed Number	$\frac{1}{5}$	$\frac{2}{5}$	$\frac{3}{5}$	$\frac{4}{5}$	$\frac{5}{5}$	$1\frac{1}{5}$
Decimal						

39.

Fraction or Mixed Number	$\frac{1}{8}$	$\frac{2}{8}$	$\frac{3}{8}$	$\frac{4}{8}$	$\frac{5}{8}$	$\frac{6}{8}$
Decimal						

Many measurements in customary units are made in half-inches, quarter-inches, eighth-inches, sixteenth-inches, and thirty-second-inches. Often these measurements must be changed to decimals. Tables like this make that job easier.

Fraction	$\frac{1}{2}$	$\frac{1}{4}$	$\frac{1}{8}$	$\frac{1}{16}$	$\frac{1}{32}$
Decimal	0.5	0.25	0.125	0.0625	0.03125

Use the table to change to decimals.

40. $\frac{5}{16}$ **41.** $\frac{3}{32}$ **42.** $\frac{5}{8}$ **43.** $\frac{4}{16}$

44. $\frac{3}{8}$ **45.** $\frac{3}{4}$ **46.** $3\frac{7}{8}$ **47.** $6\frac{1}{4}$

48. $2\frac{5}{32}$ **49.** $7\frac{9}{16}$ **50.** $12\frac{5}{8}$ **51.** $16\frac{15}{16}$

KEEPING SKILLS SHARP

Change to a fraction.

1. $2\frac{1}{2}$ **2.** $3\frac{1}{4}$

3. $4\frac{1}{4}$ **4.** $5\frac{1}{3}$

5. $2\frac{2}{3}$ **6.** $2\frac{5}{16}$

7. $3\frac{3}{4}$ **8.** $4\frac{2}{5}$

9. $2\frac{5}{6}$ **10.** $1\frac{7}{8}$

11. $3\frac{5}{9}$ **12.** $2\frac{3}{8}$

13. $4\frac{3}{5}$ **14.** $3\frac{7}{10}$

Change to a mixed number.

15. $\frac{7}{2}$ **16.** $\frac{10}{3}$

17. $\frac{5}{4}$ **18.** $\frac{13}{5}$

19. $\frac{13}{2}$ **20.** $\frac{7}{6}$

21. $\frac{9}{4}$ **22.** $\frac{13}{3}$

23. $\frac{17}{8}$ **24.** $\frac{23}{9}$

25. $\frac{23}{10}$ **26.** $\frac{17}{5}$

27. $\frac{13}{4}$ **28.** $\frac{11}{9}$

Repeating decimals

In these two examples, the division never ends!

$\frac{2}{3} = ?$

$$\begin{array}{r} 0.66 \\ 3\overline{)2.00} \\ -1\,8 \\ \hline 20 \\ -18 \\ \hline 2 \end{array}$$

$\frac{15}{11} = ?$

$$\begin{array}{r} 1.36 \\ 11\overline{)15.00} \\ -11 \\ \hline 4\,0 \\ -33 \\ \hline 70 \\ -66 \\ \hline 4 \end{array}$$

Such decimals are called **repeating decimals.**

There are four ways of writing a repeating decimal.

1 **Use 3 dots.** Write 3 dots after the decimal to show that it repeats.

$\frac{2}{3} = \mathbf{0.66\ldots}$ $\qquad \frac{15}{11} = \mathbf{1.3636\ldots}$

This can be confusing because you can't be sure what part repeats.

2 **Use a bar.** Use a bar over the digits that repeat.

$\frac{2}{3} = \mathbf{0.66\ldots = 0.\overline{6}}$ $\qquad \frac{15}{11} = \mathbf{1.3636\ldots = 1.\overline{36}}$

3 **Write a mixed decimal.** Write the answer as a **mixed decimal.**

mixed decimal

$\frac{2}{3} = \mathbf{0.66\frac{2}{3}}$ Read this as "sixty-six and two-thirds hundredths."

This method is not used often, but it will be useful in your work with percents.

4 **Round.** Round the decimal.

$\frac{2}{3} = \mathbf{0.666\ldots \approx 0.67}$ $\qquad \frac{15}{11} = \mathbf{1.3636\ldots \approx 1.364}$

EXERCISES

Change to decimals. Use three dots to write repeating decimals.

1. $\frac{1}{3}$	**2.** $\frac{2}{3}$	**3.** $\frac{1}{9}$	**4.** $\frac{2}{9}$	**5.** $\frac{3}{9}$	**6.** $\frac{4}{9}$
7. $\frac{5}{9}$	**8.** $\frac{6}{9}$	**9.** $\frac{7}{9}$	**10.** $\frac{8}{9}$	**11.** $\frac{1}{6}$	**12.** $\frac{5}{6}$
13. $\frac{4}{3}$	**14.** $\frac{10}{9}$	**15.** $\frac{7}{6}$	**16.** $\frac{1}{12}$	**17.** $\frac{4}{12}$	**18.** $\frac{5}{12}$

Change to mixed decimals. Give answers in hundredths.

19. $\frac{1}{3}$	**20.** $\frac{1}{8}$	**21.** $\frac{5}{6}$	**22.** $\frac{3}{8}$	**23.** $\frac{1}{9}$	**24.** $\frac{4}{9}$

Change to decimals. Round to the nearest hundredth.

25. $\frac{1}{3}$	**26.** $\frac{1}{6}$	**27.** $\frac{5}{6}$	**28.** $\frac{3}{8}$	**29.** $1\frac{5}{8}$	**30.** $\frac{13}{6}$

$<$, $=$, or $>$?

31. $\frac{1}{3}$ ● 0.3	**32.** $\frac{1}{3}$ ● 0.33	**33.** $\frac{1}{3}$ ● 0.333	**34.** $\frac{1}{3}$ ● 0.3333
35. $\frac{1}{3}$ ● $0.\overline{3}$	**36.** $\frac{2}{3}$ ● 0.6	**37.** $\frac{2}{3}$ ● 0.66	**38.** $\frac{2}{3}$ ● 0.67
39. $\frac{2}{3}$ ● 0.666	**40.** $\frac{2}{3}$ ● 0.6666	**41.** $\frac{2}{3}$ ● $0.\overline{6}$	**42.** $\frac{1}{7}$ ● 0.1

43. Julie, Carlos, and Brenda played a game in which each picked a fraction. The person whose fraction was neither the largest nor the smallest won. Who won?

	Game **1**	Game **2**	Game **3**
Julie:	$\frac{1}{3}$	$\frac{1}{2}$	$\frac{13}{15}$
Carlos:	$\frac{57}{78}$	$\frac{105}{432}$	$\frac{8}{9}$
Brenda:	$\frac{45}{64}$	$\frac{31}{93}$	$\frac{31}{40}$

Multiplying mixed numbers

A custom T-shirt store charges $9\frac{1}{2}$¢ for each letter that is put on a T-shirt. Rudy used 18 letters. How much did he pay for his letters?

Multiply to find how much he paid.

$$18 \times 9\frac{1}{2} = ?$$

Step 1. Change the mixed number to a fraction.

$$18 \times \frac{19}{2} = ?$$

Step 2. Multiply. (Cancel, if possible.)

$$\overset{9}{\cancel{18}} \times \frac{19}{\underset{1}{\cancel{2}}} = 171$$

He paid $1.71.

When multiplying two mixed numbers, you can change them to fractions first. Before you multiply, you can estimate the product by rounding each factor to the nearest whole number.

$$\underset{(1)}{1\frac{1}{4}} \times \underset{(3)}{2\frac{2}{3}} = ?$$

The product is about 3.

$$1\frac{1}{4} \times 2\frac{2}{3} = \frac{5}{\underset{1}{\cancel{4}}} \times \frac{\overset{2}{\cancel{8}}}{3}$$

$$= \frac{10}{3}$$

$$= 3\frac{1}{3}$$

The answer is close to the estimate.

EXERCISES

First estimate. Then multiply. Give answers in simplest form.

1. $3\frac{3}{4} \times 1\frac{2}{3}$	2. $5\frac{3}{4} \times 2\frac{3}{4}$	3. $3 \times 4\frac{2}{3}$	4. $1\frac{5}{8} \times 3\frac{1}{3}$
5. $5\frac{1}{6} \times 8\frac{1}{5}$	6. $2\frac{1}{2} \times 2\frac{1}{2}$	7. $6\frac{2}{7} \times 4$	8. $9 \times 2\frac{1}{6}$
9. $7\frac{3}{4} \times 9$	10. $3\frac{3}{4} \times 5\frac{1}{6}$	11. $5\frac{3}{4} \times 2\frac{2}{3}$	12. $6\frac{1}{3} \times 4\frac{2}{5}$
13. $4\frac{3}{8} \times 2$	14. $5 \times 2\frac{5}{8}$	15. $3 \times 4\frac{2}{5}$	16. $6\frac{2}{3} \times 8$
17. $1\frac{2}{5} \times 2\frac{3}{8}$	18. $4 \times 4\frac{5}{6}$	19. $3\frac{2}{3} \times 5\frac{1}{2}$	20. $5\frac{3}{8} \times 7\frac{3}{4}$

Compute each total cost. Remember that the letters cost $9\frac{1}{2}$¢ each. Round each cost up to the nearest cent.

21. SKATE BOARD WIZARD

22. PLAN AHEAD

23. ROADRUNNER

24. THIS SPACE AVAILABLE FOR ADVERTISING

When one factor is a whole number do this:

$$\begin{array}{r} 2\frac{1}{3} \\ \times 8 \\ \hline \end{array}$$

Step 1. Multiply the fraction. → $2\frac{2}{3}$ $\left(\frac{8}{3}\right)$

Step 2. Multiply the whole number. → 16

Step 3. Add. → $18\frac{2}{3}$

★ **Use the method shown above to multiply. Give answers in simplest form.**

25. $9\frac{1}{2} \times 8$	26. $2\frac{3}{4} \times 4$	27. $3\frac{1}{3} \times 6$	28. $5 \times 4\frac{1}{3}$

Dividing mixed numbers

Two girls hiked from Base Camp to Clear Lake Camp in $2\frac{3}{4}$ hours. You can divide $8\frac{1}{2}$ miles by $2\frac{3}{4}$ hours to find how many miles they averaged per hour.

To divide mixed numbers, change each mixed number to an equivalent fraction and divide.

$$8\frac{1}{2} \div 2\frac{3}{4} = \frac{17}{2} \div \frac{11}{4}$$

$$= \frac{17}{{}_1\cancel{2}} \times \frac{\cancel{4}^2}{11}$$

$$= \frac{34}{11} = 3\frac{1}{11}$$

They averaged $3\frac{1}{11}$ miles per hour.

We can estimate the quotient by rounding each mixed number to the nearest whole number and dividing.

$$\overset{9}{8\frac{1}{2}} \div \overset{3}{2\frac{3}{4}} = ?$$

The quotient should be about 3. The answer is close to the estimate.

EXERCISES

First estimate the quotient. Then give the quotient in simplest form.

1. $5\frac{1}{2} \div 1\frac{1}{3}$
2. $3\frac{1}{4} \div 2$
3. $4\frac{2}{3} \div 1\frac{1}{4}$
4. $6\frac{2}{3} \div 4\frac{1}{5}$
5. $5 \div 2\frac{1}{3}$
6. $5\frac{3}{8} \div 4\frac{4}{5}$
7. $8\frac{1}{5} \div 3\frac{3}{4}$
8. $4\frac{1}{2} \div 2$
9. $6\frac{1}{2} \div 2\frac{2}{3}$
10. $5\frac{1}{2} \div 2\frac{1}{4}$
11. $1\frac{1}{3} \div 5$
12. $5\frac{3}{8} \div 2\frac{1}{3}$
13. $1\frac{3}{4} \div 4\frac{1}{2}$
14. $3\frac{1}{2} \div 1\frac{1}{8}$
15. $6\frac{3}{4} \div 3$
16. $5\frac{2}{3} \div 1\frac{1}{4}$
17. $2\frac{3}{8} \div 1\frac{1}{3}$
18. $3\frac{1}{2} \div 2\frac{3}{4}$
19. $4 \div 1\frac{1}{9}$
20. $5 \div 3\frac{5}{8}$
21. $8\frac{3}{4} \div 3$
22. $6 \div 2\frac{3}{4}$
23. $4\frac{1}{3} \div 2\frac{1}{2}$
24. $2 \div 1\frac{1}{2}$
25. $2\frac{1}{9} \div 2\frac{1}{4}$
26. $8 \div 4\frac{2}{3}$
27. $9\frac{2}{3} \div 3\frac{1}{2}$

Solve.

28. Laura packed $4\frac{3}{4}$ pounds of food. If she rationed herself to $1\frac{1}{2}$ pounds per day, how many days' worth of food did she pack?
29. Leslie packed $4\frac{1}{2}$ pounds of food. The first day she ate $1\frac{3}{4}$ pounds. How many pounds did she have left?
30. Laura and Leslie rented a canoe at Clear Lake for $3\frac{1}{2}$ hours. How much did it cost if the canoe rented for $1.20 an hour?

★ 31. They canoed an average of $2\frac{1}{2}$ miles an hour during the $3\frac{1}{2}$-hour trip. They canoed $5\frac{1}{2}$ miles during the first 2 hours. How far did they canoe during the last $1\frac{1}{2}$ hours?

KEEPING SKILLS SHARP

Add. Write answers in simplest form.

1. $\frac{4}{9} + \frac{1}{9}$
2. $\frac{3}{5} + \frac{2}{5}$
3. $\frac{1}{2} + \frac{1}{4}$
4. $\frac{3}{8} + \frac{1}{4}$
5. $\frac{1}{2} + \frac{1}{3}$
6. $\frac{2}{3} + \frac{3}{4}$
7. $\frac{5}{8} + \frac{2}{3}$
8. $\frac{5}{9} + \frac{5}{6}$
9. $\frac{0}{8} + \frac{3}{2}$
10. $\frac{3}{2} + \frac{2}{3}$
11. $\frac{4}{9} + \frac{1}{2}$
12. $\frac{1}{6} + \frac{3}{4}$
13. $\frac{2}{3} + \frac{3}{5}$
14. $\frac{1}{2} + \frac{4}{5}$
15. $\frac{3}{5} + \frac{1}{4}$
16. $\frac{8}{9} + \frac{2}{3}$

SKILL GAME

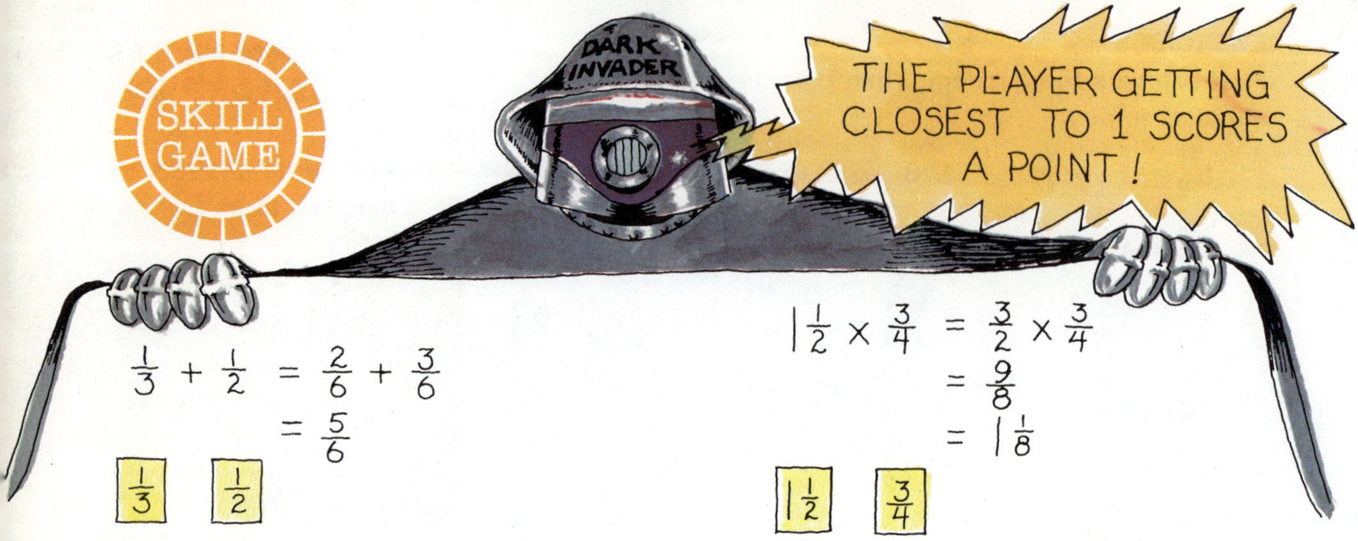

Which is closer to 1, a or b?

1. a. $1\frac{3}{4} \div 1\frac{1}{4}$
   b. $1 - \frac{1}{6}$
2. a. $\frac{3}{8} + \frac{3}{4}$
   b. $\frac{3}{4} \times 1\frac{1}{3}$
3. a. $1\frac{5}{6} - \frac{7}{8}$
   b. $2\frac{1}{4} \div 1\frac{5}{6}$
4. a. $\frac{5}{8} + \frac{1}{3}$
   b. $1\frac{7}{8} - \frac{3}{4}$
5. a. $1\frac{5}{6} - \frac{5}{8}$
   b. $2 \div 1\frac{7}{8}$
6. a. $2\frac{1}{4} - 1$
   b. $1 \div \frac{7}{8}$

Play the game.

1. Make this set of cards.

$\frac{1}{2}$	$\frac{1}{3}$	$\frac{1}{4}$	$\frac{3}{4}$	$\frac{3}{8}$	$\frac{5}{8}$	$\frac{7}{8}$	$\frac{1}{6}$	$\frac{5}{6}$	1	$1\frac{1}{4}$	$1\frac{1}{3}$
$1\frac{1}{2}$	$1\frac{3}{4}$	$1\frac{3}{8}$	$1\frac{5}{8}$	$1\frac{7}{8}$	$1\frac{1}{6}$	$1\frac{5}{6}$	2	$2\frac{1}{4}$	$2\frac{1}{3}$	$2\frac{1}{2}$	$2\frac{3}{8}$

2. Divide the class into two teams.
3. Mix up the cards. A player from each team picks two cards.
4. Each player may either add, subtract, multiply, or divide the two numbers he/she picks. The player who gets closer to 1 earns a point for his/her team.
5. After all players have played, the team with the greater total wins the game.

Mathematics in careers

Workers in a greenhouse must often use mathematics in their work. They follow recipes for mixing soil, apply fertilizers according to formulas, and compute costs, prices, and taxes.

Here is a recipe for potting soil:

2 parts loam

$1\frac{1}{2}$ parts peat moss

1 part coarse sand

1 lb of bone meal for each bushel of mixture

Solve.

1. A worker mixing potting soil started with 4 bushels of loam. How much peat moss did he need?
2. How much peat moss would be used with 3 bushels of sand?
3. How much loam would be used with 6 bushels of peat moss?
4. How much bone meal would be used in a mixture that started with 8 bushels of loam?

★ 5. A fertilizer bag marked 5-10-5 is $\frac{5}{100}$ nitrogen, $\frac{10}{100}$ phosphorus, and $\frac{5}{100}$ potash. How much nitrogen is in a 50-lb bag of the fertilizer?

★ 6. How much phosphorus is in a 25-lb bag of fertilizer marked 10-15-15?

Problem solving

A class had a bake sale to raise money for a field trip. They used the two recipes shown.

Nutty Carrot Cake
Makes 15 servings

8 ounces grated carrots

$\frac{1}{4}$ cup soy flour

$1\frac{1}{4}$ cups whole wheat flour

1 cup unbleached white flour

$1\frac{1}{2}$ teaspoons baking powder

$\frac{3}{4}$ cup safflower or vegetable oil

1 cup honey

4 eggs

$\frac{1}{4}$ cup orange juice

1 teaspoon vanilla

$1\frac{1}{2}$ teaspoons cinnamon

$1\frac{1}{2}$ teaspoons nutmeg

$\frac{1}{2}$ teaspoon cloves

$\frac{1}{2}$ teaspoon salt

1 cup chopped sunflower seeds or chopped nuts

In a large bowl, cream oil and honey together until fluffy. Beat in the eggs, orange juice, and vanilla. Sift together the dry ingredients and add to creamed mixture. Stir in the carrots and nuts. Pour into a greased 9″ by 13″ cake pan. Bake at 350° for 35 to 50 minutes. Cake is done when golden brown on top and toothpick inserted in center comes out clean. Cool completely before removing from pan. Cut into pieces. Serve with applesauce.

Judy's Whole Wheat Rounds
Makes 3 dozen

1 teaspoon baking powder

$\frac{1}{2}$ teaspoon baking soda

$\frac{1}{2}$ teaspoon nutmeg

1 tablespoon milk powder

2 cups whole wheat flour

$\frac{1}{2}$ cup butter

$\frac{1}{2}$ cup honey

1 tablespoon grated orange peel

1 teaspoon vanilla

1 egg

$\frac{1}{2}$ teaspoon cinnamon

Combine all the dry ingredients in a large bowl. Cut the butter into the flour mixture. Beat together the honey, orange peel, vanilla, and egg and add to the flour mixture. Mix well. Shape the dough into 1-inch balls. Place them on a baking sheet 2 inches apart. Flatten each lightly with a fork. Sprinkle cinnamon over rounds. Bake at 375° for 8 to 10 minutes.

A group of students made 10 dozen whole wheat rounds.

Solve.

1. How many recipes did they have to make to get 10 dozen rounds?
2. How much flour did they use?
3. How much baking soda did they use?
4. How much of each other ingredient did they use?
5. If they had 6 cups of butter before baking the rounds, how much should they have had after baking the rounds?
6. How many rounds can be made with $1\frac{1}{2}$ cups of whole wheat flour?

Another group of students made 100 servings of carrot cake.

7. How many recipes did they have to make to get 100 servings of cake?
8. How much soy flour did they use?
9. How much oil did they use?
10. How much of each other ingredient did they use?
11. How much flour did they use altogether?
12. How many servings of carrot cake can be made with 5 cups of whole wheat flour?

CHAPTER CHECKUP

Give each product in simplest form. [pages 202–203, 212, 214, 228–229]

1. $\frac{3}{4} \times \frac{1}{2}$
2. $\frac{2}{3} \times 2$
3. $\frac{7}{8} \times \frac{2}{3}$
4. $\frac{5}{9} \times \frac{3}{4}$
5. $\frac{5}{8} \times \frac{8}{5}$
6. $\frac{3}{2} \times \frac{5}{6}$
7. $9 \times \frac{4}{3}$
8. $\frac{8}{3} \times \frac{3}{4}$

Solve. [pages 204–205]

9. $\frac{1}{2}$ of $16 = ?$
10. $\frac{2}{3}$ of $81 = ?$
11. $\frac{4}{3}$ of $132 = ?$
12. $\frac{1}{3}$ of $24 = ?$
13. $\frac{5}{8}$ of $48 = ?$
14. $\frac{3}{2}$ of $36 = ?$
15. A record that regularly sells for \$5.76 is on sale for $\frac{2}{3}$ of its regular price. What is the sale price?
16. A girls' club wants to raise \$120 to buy books for the library. It has raised $\frac{4}{5}$ of the money. How much has it raised?

Give each quotient in simplest form. [pages 208–212, 215]

17. $\frac{3}{4} \div \frac{1}{4}$
18. $3 \div \frac{2}{3}$
19. $\frac{4}{3} \div \frac{2}{5}$
20. $\frac{7}{8} \div \frac{3}{4}$
21. $\frac{5}{6} \div \frac{2}{3}$
22. $\frac{7}{8} \div \frac{3}{2}$
23. $\frac{5}{6} \div 4$
24. $\frac{9}{4} \div \frac{3}{8}$

Copy and complete this table. Give fractions in lowest terms. [pages 218–221]

	25.	26.	27.	28.	29.	30.	31.	32.
Fraction	$\frac{6}{10}$		$\frac{1}{4}$			$\frac{3}{2}$	$\frac{5}{4}$	$\frac{2}{3}$
Decimal		0.27		0.3	0.68			

Multiply. Give the product in simplest form. [pages 222–223]

33. $2\frac{1}{2} \times 3$
34. $4 \times 2\frac{3}{4}$
35. $1\frac{1}{2} \times 3\frac{3}{4}$
36. $2\frac{5}{8} \times 3\frac{2}{3}$

Divide. Give the quotient in simplest form. [pages 224–225]

37. $8\frac{1}{2} \div 1\frac{3}{4}$
38. $6\frac{2}{3} \div 2$
39. $9\frac{1}{4} \div 2\frac{3}{8}$
40. $10\frac{5}{6} \div 3\frac{2}{3}$

CHAPTER PROJECT

1. For a typical school day, estimate the fraction of a whole day (24 hours) spent on each activity.
   - **a.** sleeping
   - **b.** eating
   - **c.** going to and from school
   - **d.** doing classwork
   - **e.** doing homework
   - **f.** socializing
   - **g.** watching television
   - **h.** other

2. Show how you spend your time by making a graph.

   EXAMPLE.

   Suppose $\frac{5}{12}$ of the day is spent sleeping.

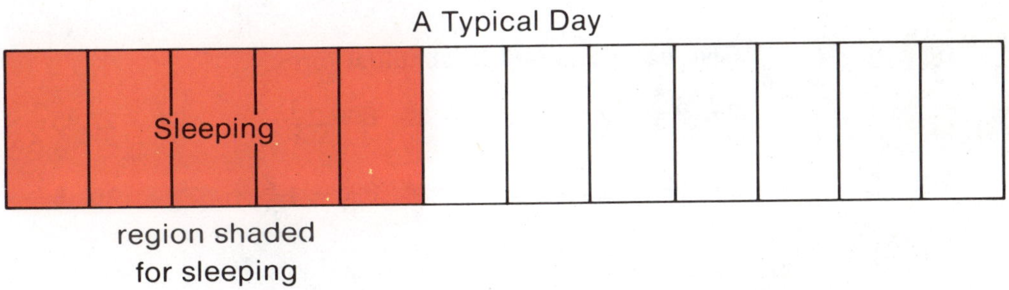

CHAPTER REVIEW

$\frac{3}{5} \times \frac{3}{2} = \frac{9}{10}$

Give each product in simplest form.

1. $3 \times \frac{1}{2}$
2. $\frac{7}{8} \times \frac{2}{5}$
3. $\frac{2}{3} \times 4$
4. $\frac{3}{4} \times \frac{2}{3}$
5. $\frac{5}{6} \times \frac{6}{5}$
6. $\frac{5}{9} \times \frac{3}{8}$

To find a fraction of a quantity, divide by the denominator and multiply by the numerator.

$\frac{2}{3}$ of 36 = 24

Complete.

7. $\frac{1}{4}$ of 20 = ?
8. $\frac{1}{3}$ of 24 = ?
9. $\frac{3}{4}$ of 36 = ?
10. $\frac{3}{2}$ of 18 = ?
11. $\frac{5}{8}$ of 40 = ?
12. $\frac{7}{3}$ of 42 = ?

To divide, multiply by the reciprocal.

$$\frac{4}{3} \div \frac{2}{5} = \frac{4}{3} \times \frac{5}{2} = \frac{20}{6} = \frac{10}{3} = 3\frac{1}{3}$$

Give each quotient in simplest form.

13. $\frac{5}{6} \div \frac{1}{2}$
14. $\frac{2}{3} \div 3$
15. $\frac{7}{8} \div \frac{3}{4}$
16. $\frac{5}{9} \div \frac{5}{3}$
17. $4 \div \frac{7}{8}$
18. $\frac{6}{5} \div \frac{5}{9}$

$$2\frac{1}{2} \times 3\frac{3}{4} = \frac{5}{2} \times \frac{15}{4} = \frac{75}{8} = 9\frac{3}{8}$$

Give each product in simplest form.

19. $2\frac{1}{2} \times 6$
20. $2\frac{1}{5} \times 3\frac{1}{4}$
21. $8 \times 3\frac{2}{3}$
22. $1\frac{1}{5} \times 5\frac{1}{2}$
23. $9\frac{3}{4} \times 3\frac{1}{3}$
24. $4\frac{3}{5} \times 2\frac{3}{8}$

$$8\frac{2}{3} \div 1\frac{3}{4} = \frac{26}{3} \div \frac{7}{4} = \frac{26}{3} \times \frac{4}{7} = \frac{104}{21} = 4\frac{20}{21}$$

Give each quotient in simplest form.

25. $9\frac{1}{2} \div 3$
26. $8 \div 2\frac{1}{2}$
27. $5\frac{1}{2} \div 1\frac{1}{4}$
28. $6\frac{3}{5} \div 2\frac{1}{3}$
29. $8\frac{3}{8} \div 1\frac{2}{3}$
30. $6\frac{2}{3} \div 1\frac{2}{5}$

CHAPTER CHALLENGE

Use the clues to tell how much money is in each bank.

1\.

Clues

- There are only dimes and quarters in this bank.
- There are fewer than 10 coins.
- $\frac{3}{5}$ of the coins are dimes.

2\.

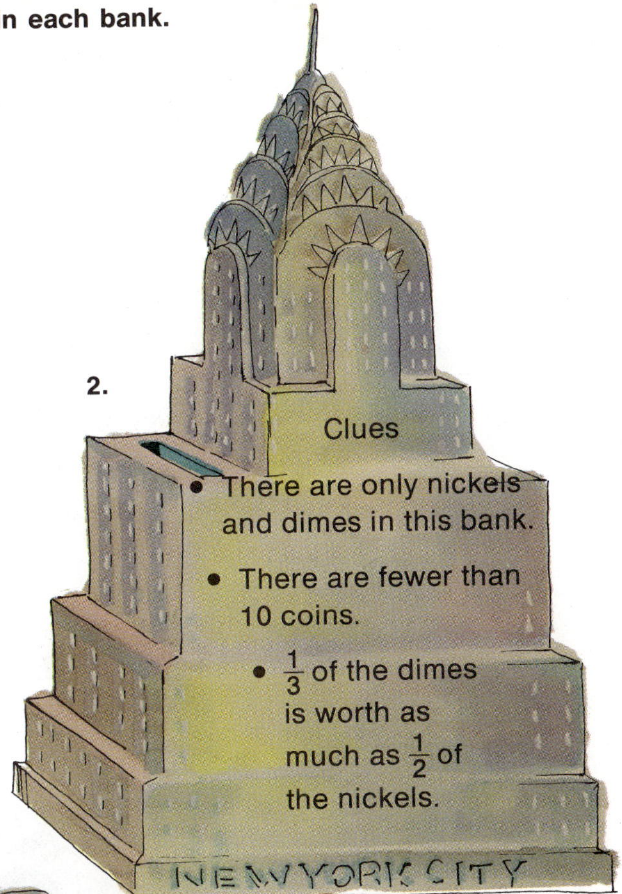

Clues

- There are only nickels and dimes in this bank.
- There are fewer than 10 coins.
- $\frac{1}{3}$ of the dimes is worth as much as $\frac{1}{2}$ of the nickels.

3\.

Clues

- There are only pennies in this bank.
- There are fewer than 20 pennies.
- $\frac{1}{4}$ of the pennies is 5¢ less than $\frac{2}{3}$ of the pennies.

MAJOR CHECKUP

STANDARDIZED FORMAT

Choose the correct letter.

1. Round 5.3692 to the nearest thousandth.
   a. 5.370
   b. 5.37
   c. 5.369
   d. none of these

2. Find the sum.
   82.5 + 6.98
   a. 89.48
   b. 15.23
   c. 152.3
   d. none of these

3. Find the difference.
   38.2 − 3.45
   a. 34.75
   b. 2.55
   c. 347.5
   d. none of these

4. Multiply.
   5.42×31.4
   a. 170.188
   b. 160.088
   c. 4.336
   d. none of these

5. Divide.
   $0.23\overline{)2.944}$
   a. 0.128
   b. 12.8
   c. 1.28
   d. none of these

6. 5.23 m = ___?___ cm
   a. 52.3
   b. 0.523
   c. 523
   d. none of these

7. Which number is composite?
   a. 2
   b. 11
   c. 29
   d. none of these

8. Add.
   $5\frac{3}{4} + 2\frac{1}{3}$
   a. $7\frac{1}{12}$
   b. $8\frac{1}{12}$
   c. $7\frac{5}{12}$
   d. none of these

9. Subtract.
   $8\frac{1}{4} - 2\frac{5}{8}$
   a. $6\frac{5}{8}$
   b. $6\frac{3}{8}$
   c. $5\frac{3}{8}$
   d. none of these

10. Multiply.
    $\frac{2}{3} \times 6$
    a. 4
    b. 9
    c. $\frac{1}{9}$
    d. none of these

11. Eggs cost $.96 a dozen. How much do 6 eggs cost?
    a. $5.76
    b. $.48
    c. $.16
    d. none of these

12. Dean hiked 19 miles in 4 hours. How many miles did he average per hour?
    a. $4\frac{3}{4}$
    b. 4
    c. 76
    d. none of these

Perimeter, Area, and Volume

8

READY OR NOT?

Complete.

1. 1 m = ? mm
2. 1 km = ? m
3. 50 mm = ? cm
4. 1.8 m = ? cm
5. 750 m = ? km
6. 750 cm = ? m
7. 1 ft = ? in.
8. 3 yd = ? ft
9. 2 ft 3 in. = ? in.
10. 2 yd 1 ft = ? ft

Add or subtract.

11. 2 ft 6 in. + 1 ft 8 in.

12. 5 yd 1 ft + 3 yd 2 ft

13. 6 ft 4 in. − 1 ft 3 in.

14. 8 yd 1 ft − 3 yd 2 ft

Perimeter

A square pen measures 8 meters on a side. How many meters of fence were used to build it?

To answer the question, we find the distance around the pen. We find the **perimeter.**

A drawing might help you solve the problem. Add the lengths of the sides to find the perimeter.

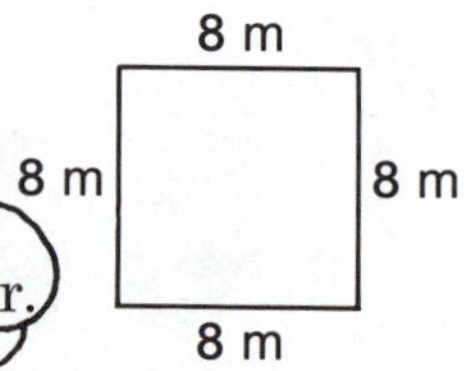

$$\begin{array}{r} 8\text{ m} \\ 8\text{ m} \\ 8\text{ m} \\ +8\text{ m} \\ \hline 32\text{ m} \end{array}$$

Formula P (perimeter) = $4s$ (side length)

$P = 4 \times 8$ m

$P = 32$ m

4s is a short way of writing 4×s.

It took 32 meters of fence to build the pen.

Here is the formula for the perimeter of a rectangle.

Formula $P = 2l + 2w$

(P: perimeter, l: length, w: width)

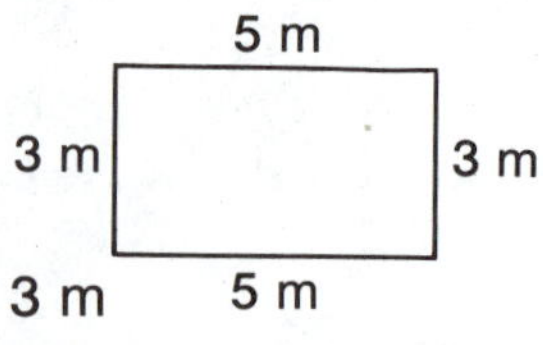

$P = 2 \times 5$ m $+ 2 \times 3$ m

$P = 10$ m $+ 6$ m

$P = 16$ m

EXERCISES

Give the perimeter of each figure.

1.

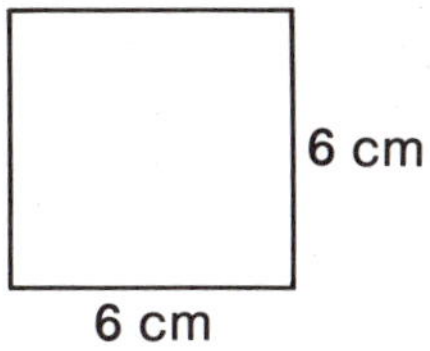

2.

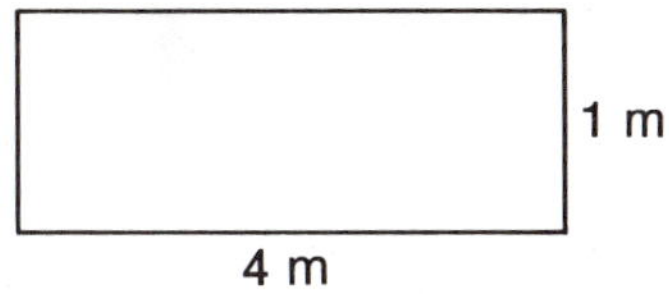

3.

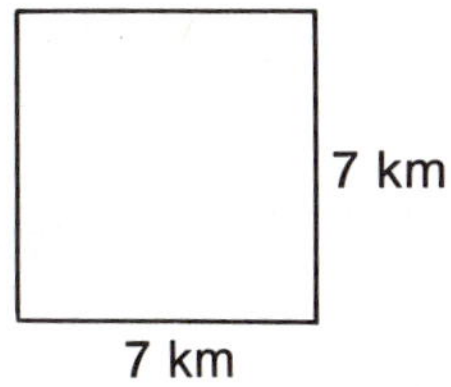

4.

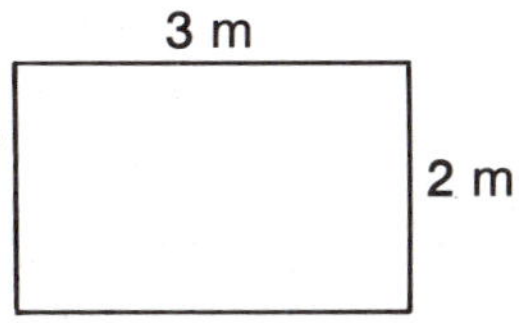

5.

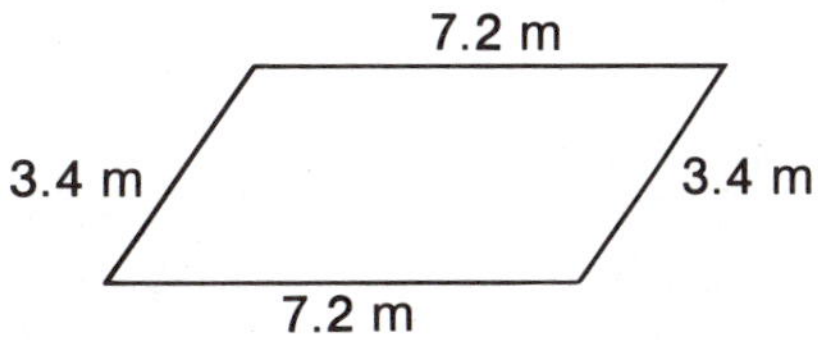

6.

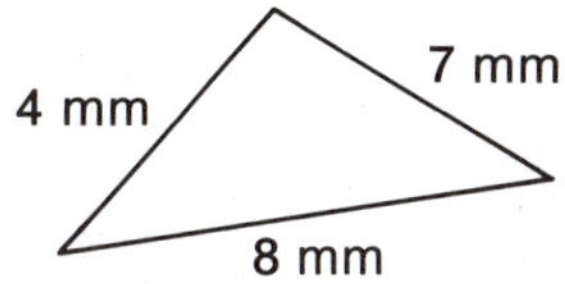

7.

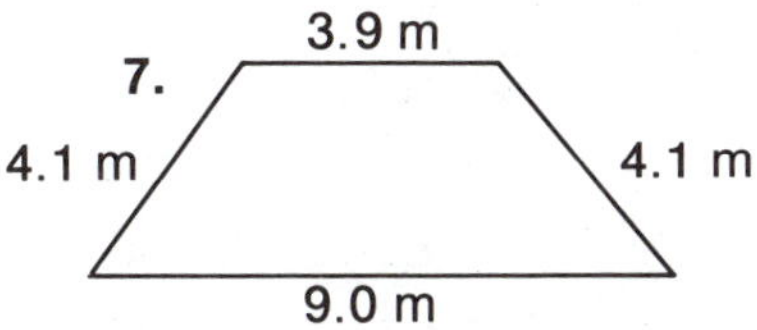

8.

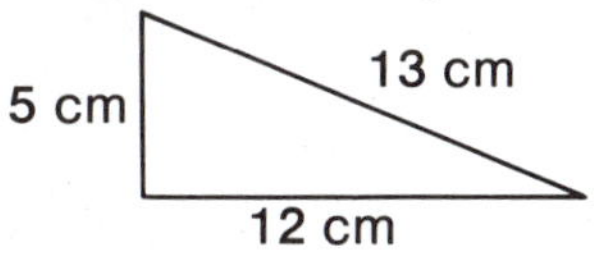

9.

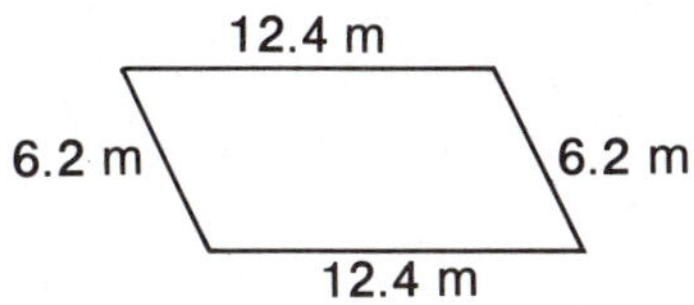

Help the carpenter set up the fairground. How much fence is needed for each project?

10. **11.**

12. Compute the cost of each fence in exercises 10 and 11, using the following information:

Fencing costs $15 per meter.
There is one post every 1.5 meters.
Posts cost $3 each.
There are no gates.

(*Hint:* Draw a picture to place the posts.)

Circumference

The distance around a circle is called the **circumference.** The circumference of a circle (C) is equal to π (pi—a number slightly greater than 3) times the diameter (d). We'll use 3.14 as a decimal approximation for π.

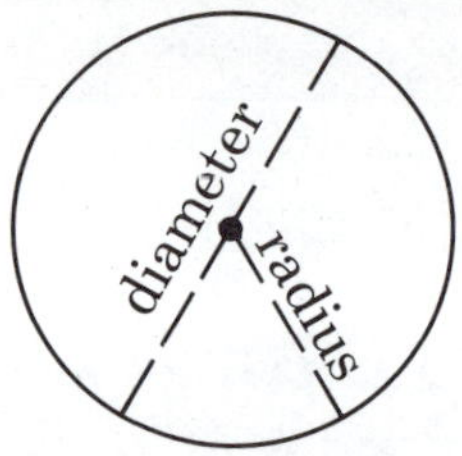

Notice that the **diameter** is twice the **radius.**

EXAMPLE.

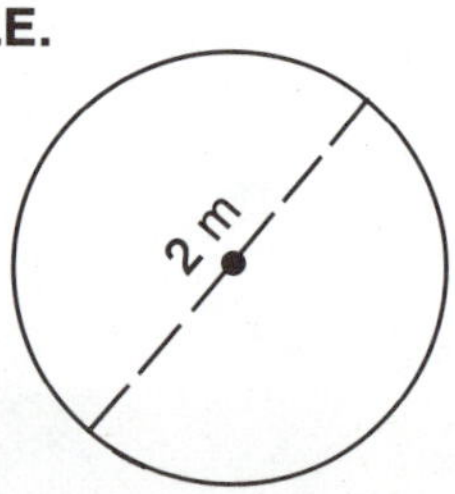

$C = \pi d$ *Formula*

$C \approx 3.14 \times 2$ m

$C \approx 6.28$ m

$\approx$ is read "is approximately equal to."

EXERCISES

Give the circumference of each circle. Use 3.14 as an approximation for π.

1. 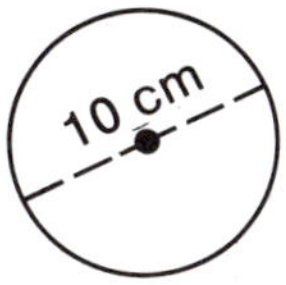

2. 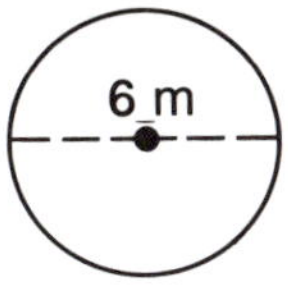

3. 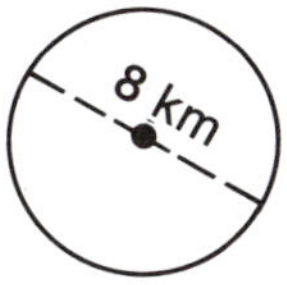

4. 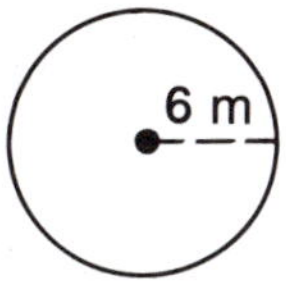

5. 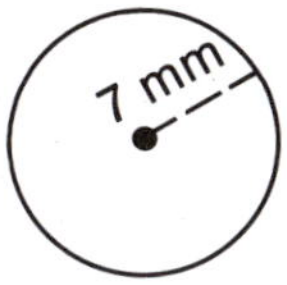

6. 

7. 

8. 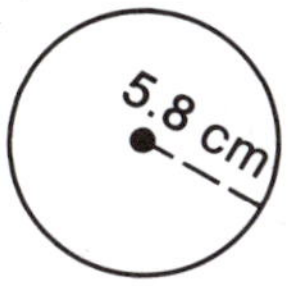

9. 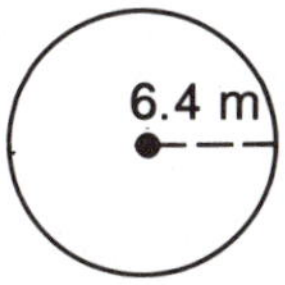

10. 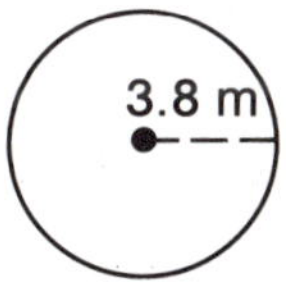

11. 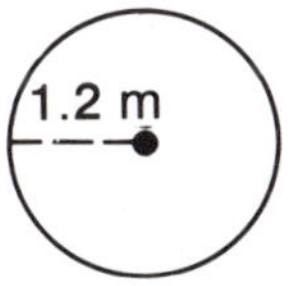

12. 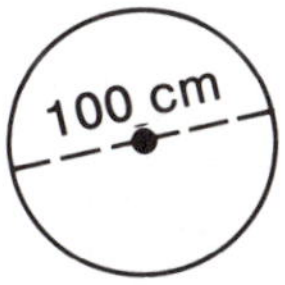

13. The diameter of a certain bicycle wheel is 65 centimeters.
    a. How far does the bicycle travel when the wheel turns 1 revolution?
    ★ b. How many revolutions does the wheel make in a kilometer?

★ 14. Mary's bicycle wheel has a circumference of 226.08 cm. What is the radius?

KEEPING SKILLS SHARP

Add or subtract. Give answers in simplest form.

1. $4\frac{1}{3} + 2\frac{1}{3}$
2. $3\frac{1}{2} + 1\frac{1}{4}$
3. $5\frac{1}{2} + 2$
4. $2\frac{5}{8} + 2\frac{1}{2}$
5. $4\frac{2}{3} + 3\frac{5}{6}$
6. $2\frac{1}{6} + 1\frac{5}{6}$
7. $7\frac{2}{9} + 4\frac{1}{6}$
8. $5\frac{3}{8} - 4\frac{2}{3}$
9. $4\frac{5}{9} - 2\frac{1}{9}$
10. $3\frac{5}{8} - 1\frac{1}{4}$
11. $7\frac{1}{2} - 4\frac{1}{3}$
12. $6 - 1\frac{1}{2}$
13. $2\frac{3}{4} - 1\frac{5}{6}$
14. $7\frac{5}{9} - 4\frac{1}{6}$

Area

The **area** of a region is the number of square units that it takes to cover the region.

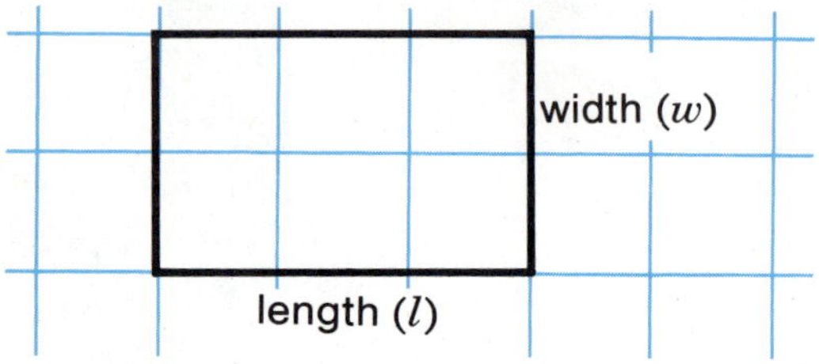

Formula

RECTANGLE

$A = lw$

$A = 3\text{ cm} \times 2\text{ cm}$

$A = 6\text{ cm}^2$

cm^2 is a short way to write square centimeter.

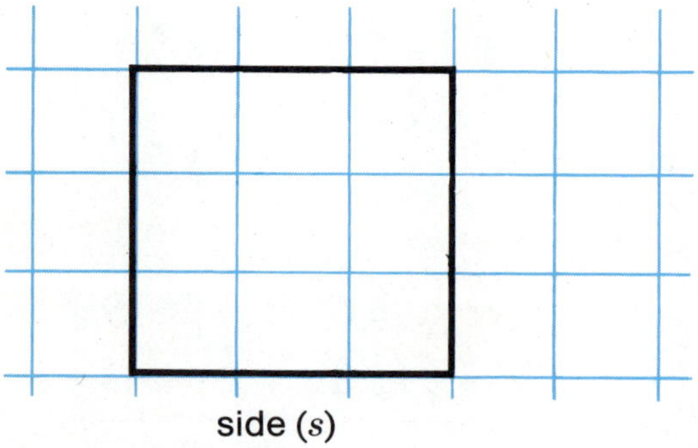

A square is a rectangle with the length equal to the width.

Formula

SQUARE

$A = s^2$

$A = 3\text{ cm} \times 3\text{ cm}$

$A = 9\text{ cm}^2$

s^2 means $s \times s$. It is read "s squared."

EXERCISES

Give each area.

1.

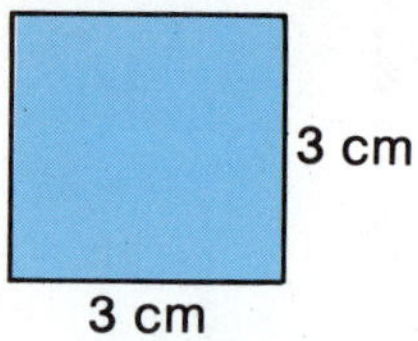

2.

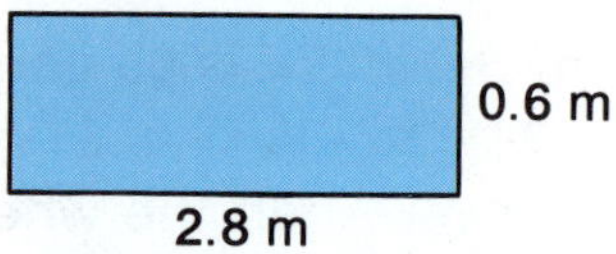

3.

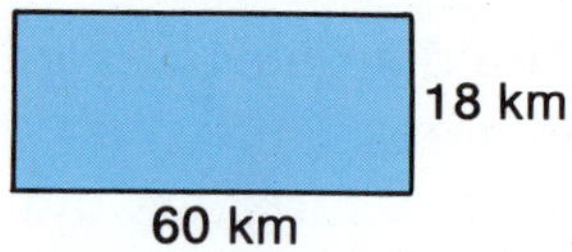

4.

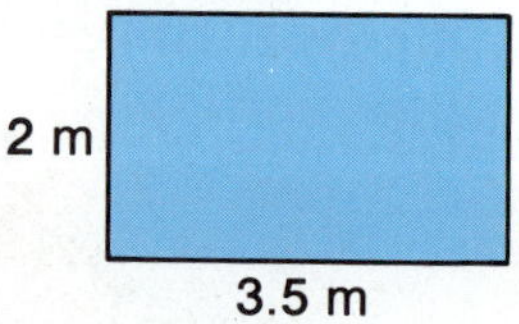

5.

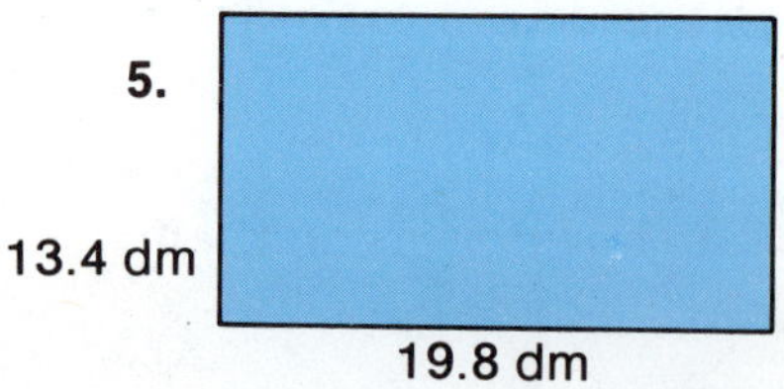

6.

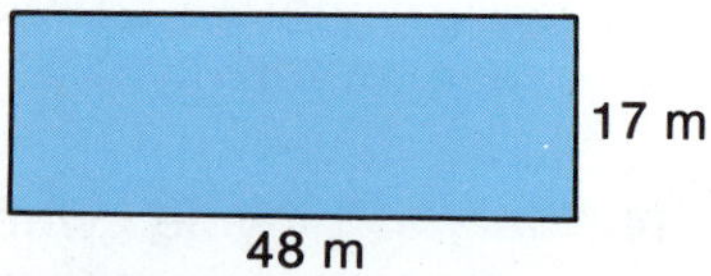

First measure the dimensions to the nearest mm. Then compute the area.

7. 

8. 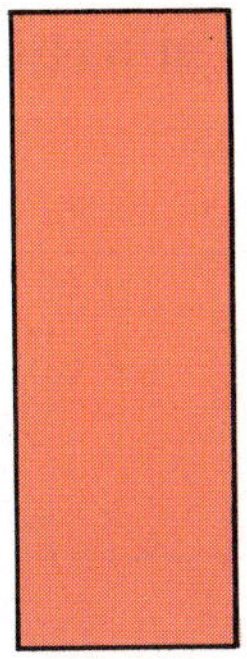

9. 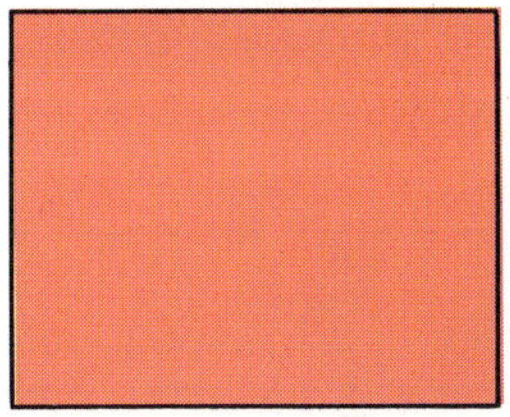

Solve.

10. Some carpet costs $11.50 per square meter. How much would the carpet cost for a rectangular room 4.1 meters by 6.8 meters?

11. A floor tile 20 cm by 20 cm costs 38¢. How much would it cost to tile the floor of a 9-meter by 5-meter recreation room?

12. A square lot measuring 25 meters on a side is to be seeded with grass seed. One kilogram of seed will cover 200 square meters. If grass seed costs $3.52 per kilogram, how much will it cost to seed the lot?

13. A carpet layer is to carpet two rectangular rooms, one 5 meters by 8 meters and another 3.1 meters by 6 meters. If the charge is $1.80 per square meter, how much will the charge be for the two rooms?

★ 14. **a.** What is the area?

b. On graph paper draw five different shapes that have this area.

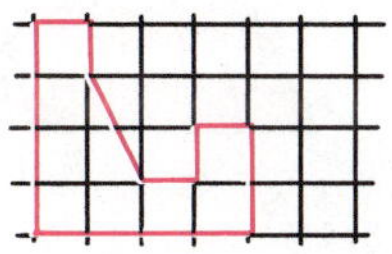

Area of a parallelogram

In each of these parallelograms, the base (b) is 3 cm and the height (h) is 2 cm.

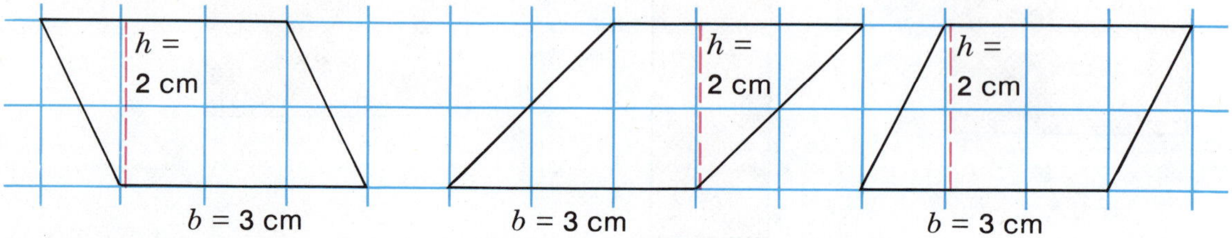

Count the squares to find the area of each parallelogram.

The area of each parallelogram is 6 cm^2.

cm^2 is read as "square centimeters."

There is a formula that you can use to find the area of a parallelogram.

Formula $A = bh$

area, base, height

EXAMPLE. What is the area of this parallelogram?

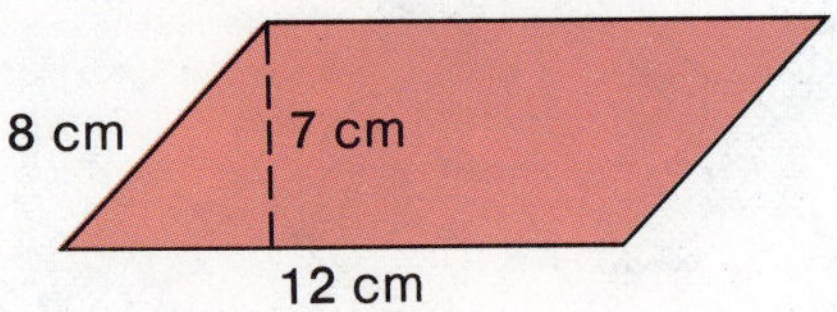

$A = bh$

$A = 12 \text{ cm} \times 7 \text{ cm}$

$A = 84 \text{ cm}^2$

EXERCISES

Find each area.

1.

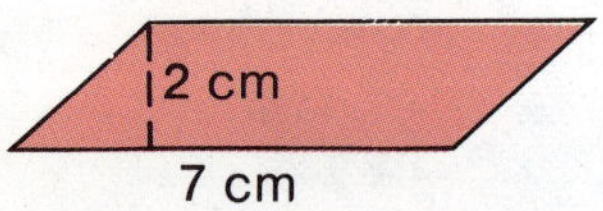

2.

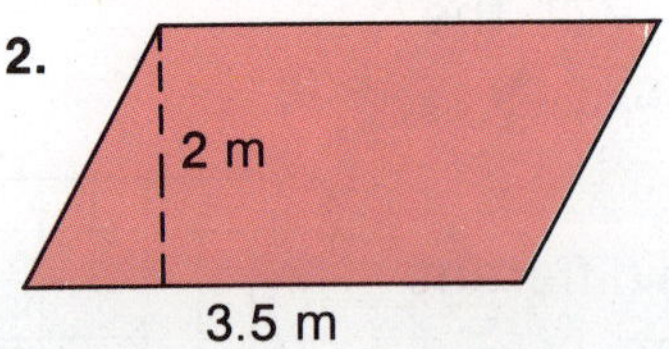

3.

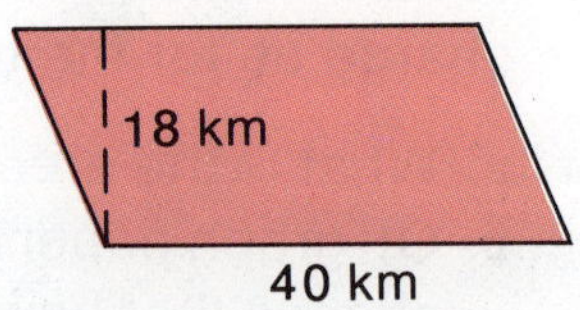

4.

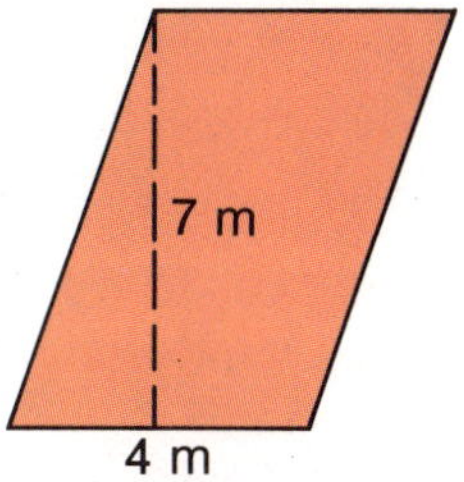

5.

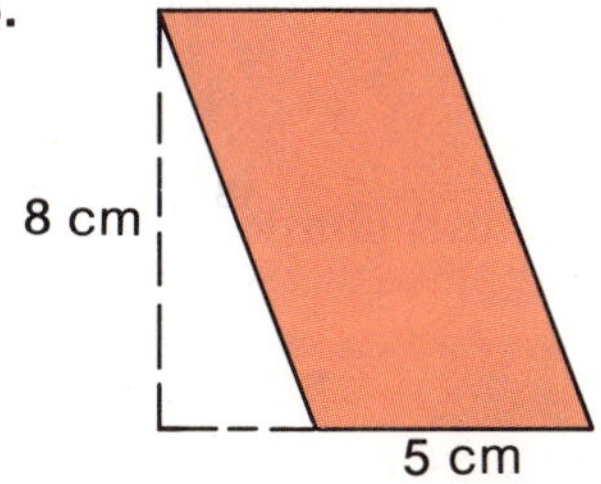

6.

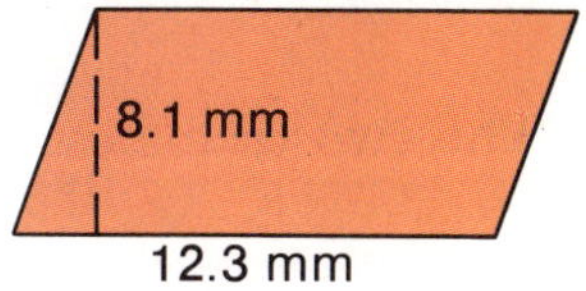

7.

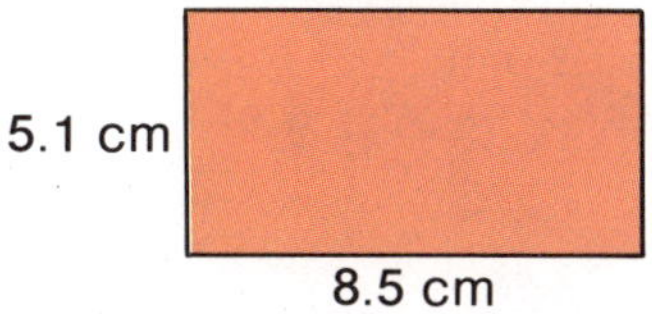

8.

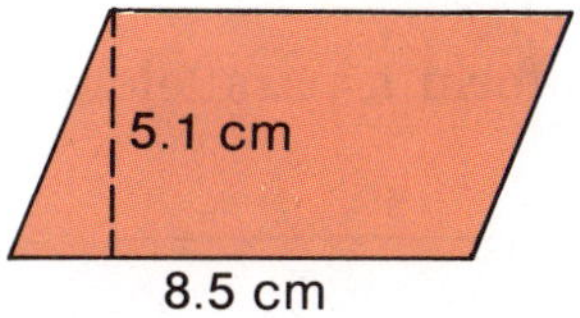

9.

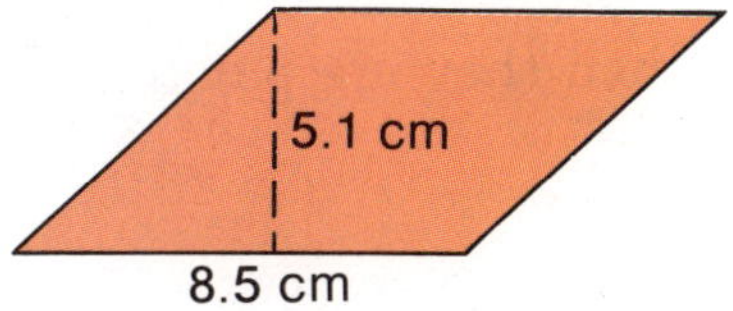

Measure the base and height to the nearest millimeter. Then compute the area.

10.

11.

12. **a.** What is the base of this parallelogram?
b. What is the height of the parallelogram?
c. What is the area of the parallelogram?

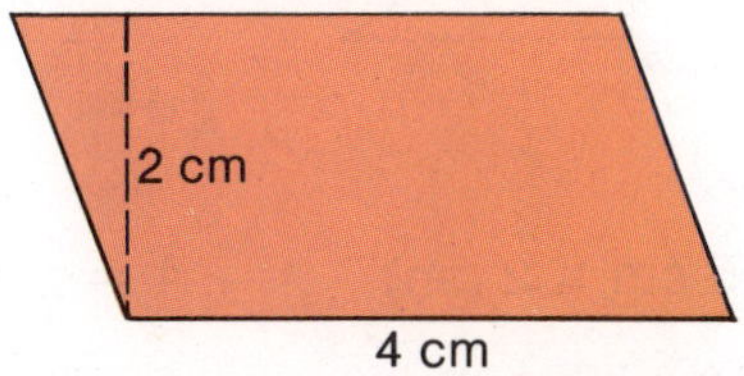

13. **a.** The left end of the parallelogram in exercise 12 was moved to the right end of it. What is the shape of the new region?
b. What is its base?
c. What is its height?
d. What is its area?

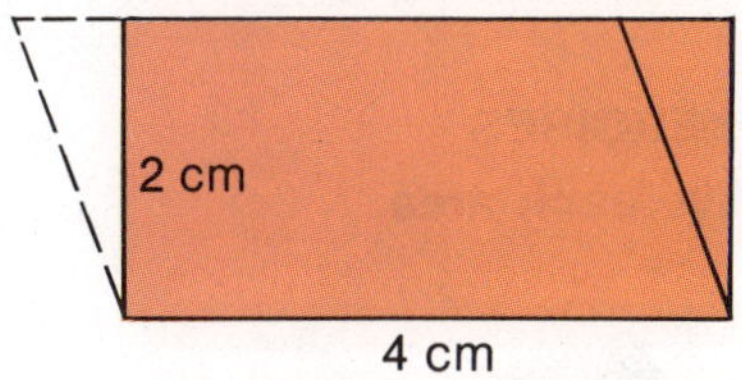

★ **14.** Is it possible to change any parallelogram into a rectangle with the same base, height, and area?

Area of a triangle

The two triangles are the same size and shape.

When they are put together, they form a parallelogram.

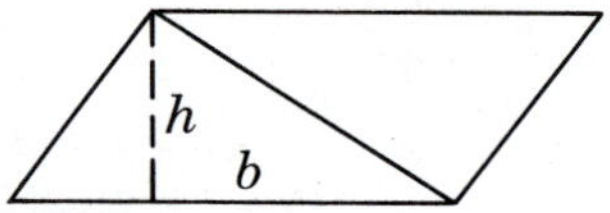

The area of each triangle is half the area of the parallelogram.

Formula $A = \frac{1}{2}bh$

EXAMPLE 1.

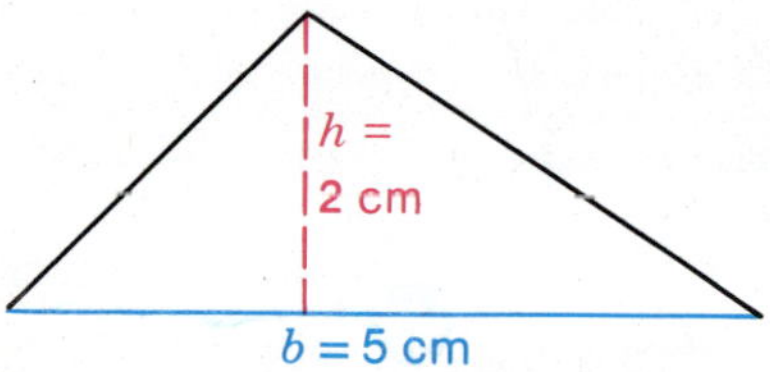

$A = \frac{1}{2} \times (5\text{ cm} \times 2\text{ cm})$

$A = \frac{1}{2} \times 10\text{ cm}^2$

$A = 5\text{ cm}^2$

EXAMPLE 2.

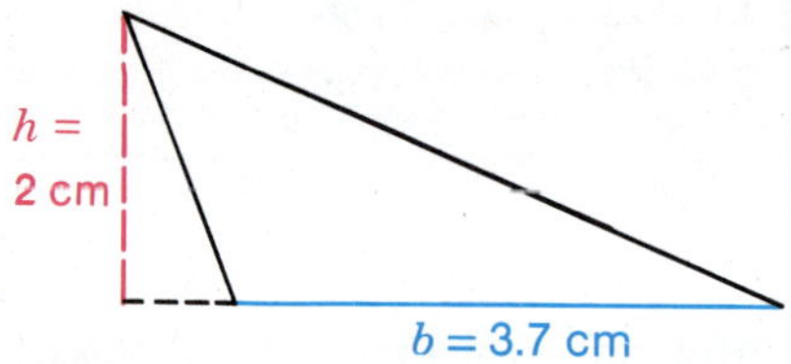

$A = \frac{1}{2} \times (3.7\text{ cm} \times 2\text{ cm})$

$A = \frac{1}{2} \times 7.4\text{ cm}^2$

$A = 3.7\text{ cm}^2$

EXAMPLE 3.

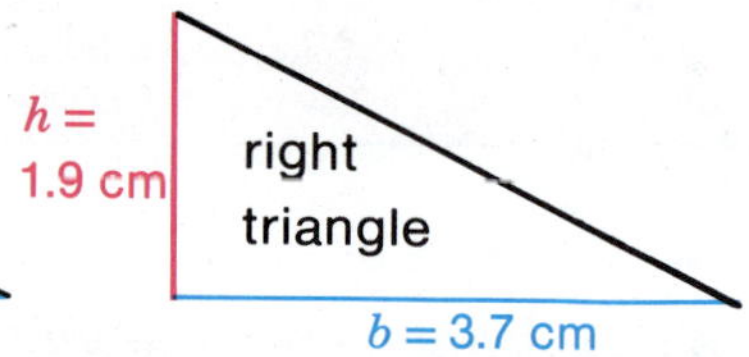

$A = \frac{1}{2} \times (3.7\text{ cm} \times 1.9\text{ cm})$

$A = \frac{1}{2} \times 7.03\text{ cm}^2$

$A = 3.515\text{ cm}^2$

EXERCISES

Give each area.

1.

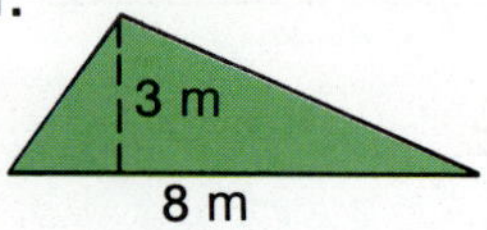

2.

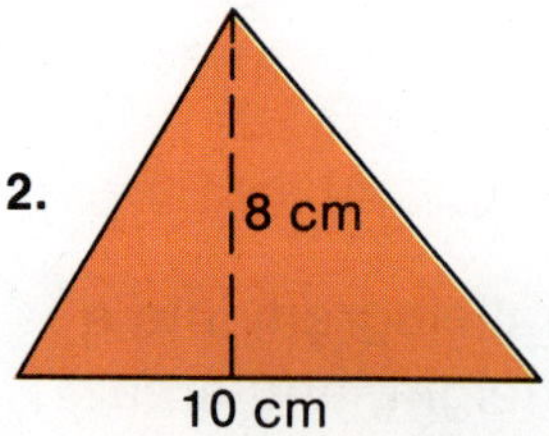

3.

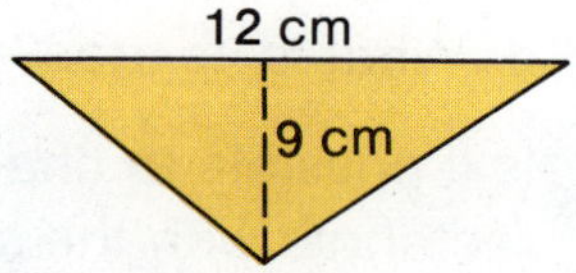

4.

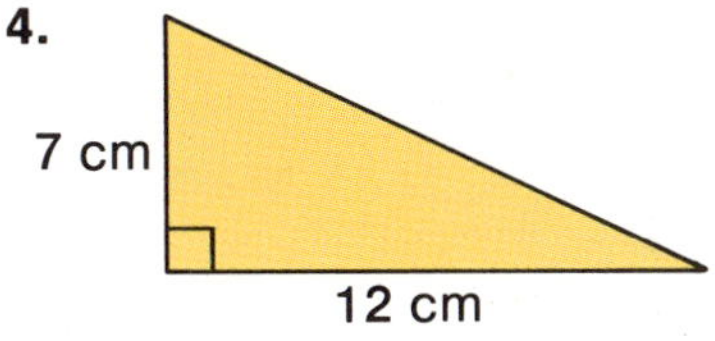

5.

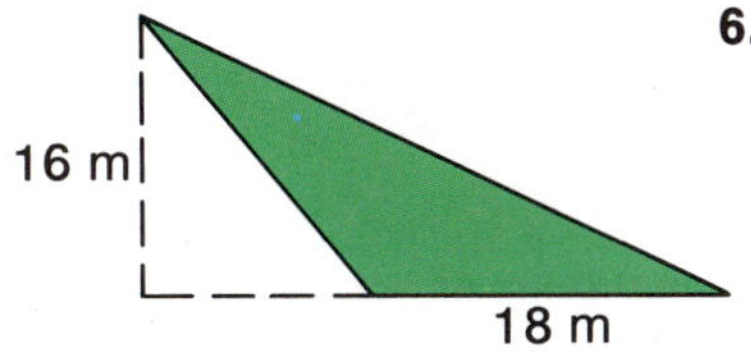

6.

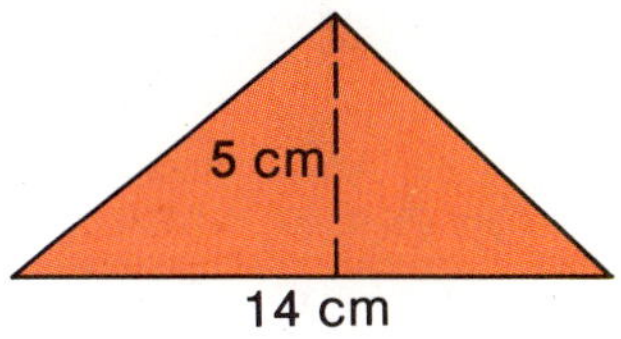

Measure the base and height to the nearest millimeter. Then compute the area.

7.

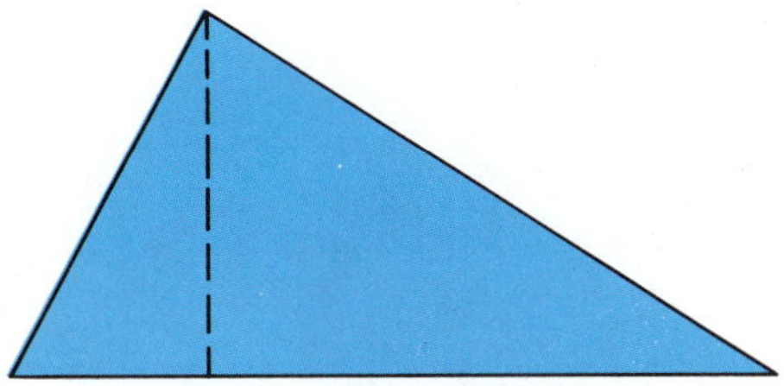

8.

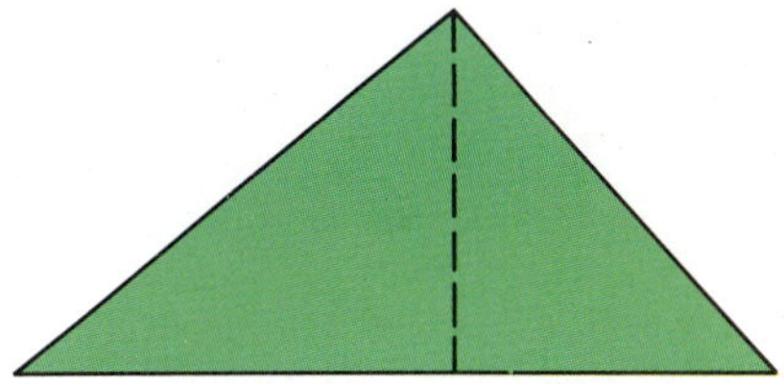

Is the question about perimeter or area?

9. How much picture framing is needed to frame a picture?
10. How much ceramic tile is needed to cover a table top?
11. How much fencing is needed to fence in a baseball field?
12. How much sod is needed to cover a baseball field?
13. How much molding is needed to go around the ceiling?
14. How much tile is needed to cover your classroom floor?
15. How much construction paper is needed to cover the classroom bulletin board?

The figures below are called **trapezoids.**

Find the area of each. (*Hint:* Use what you know about triangles.)

★ **16.**

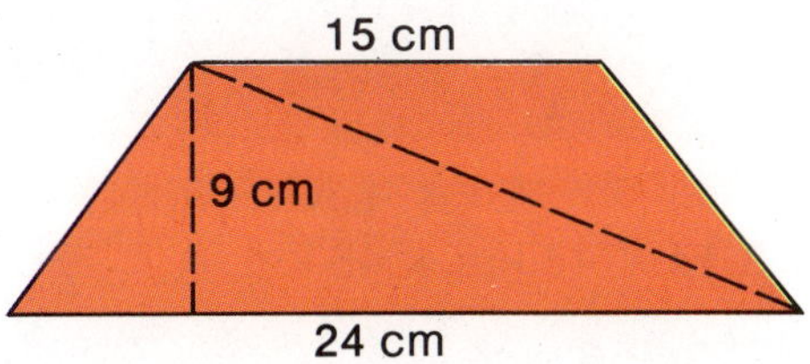

★ **17.**

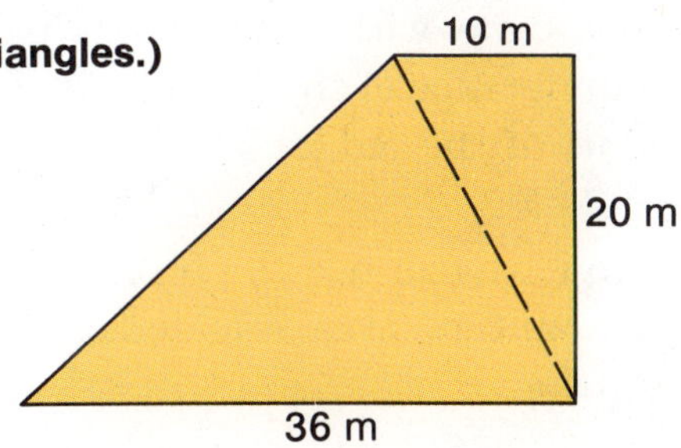

Area of a circle

On Oct. 2, 1969, an early morning earthquake shook Sonoma County, California. Damage was extensive, and strong tremors could be felt as much as 48 kilometers away. How large an area was affected?

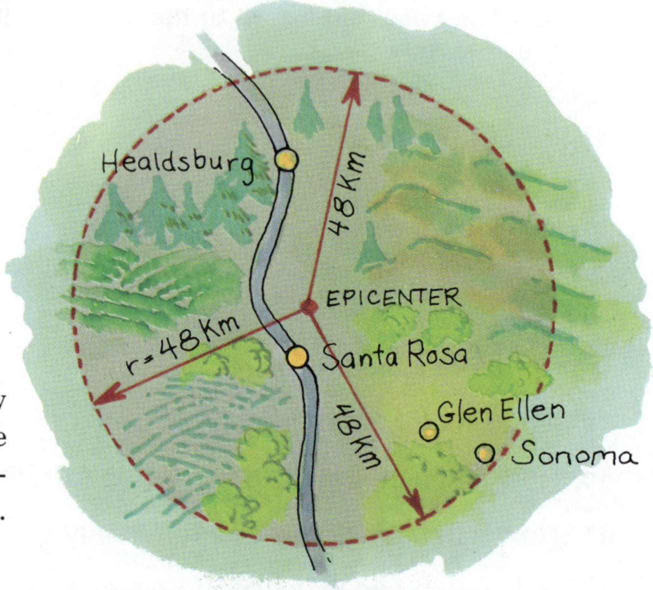

To answer the question, we need to know how to find the area of a circle. These pictures will help you to see that the formula for the area of a circle is reasonable.

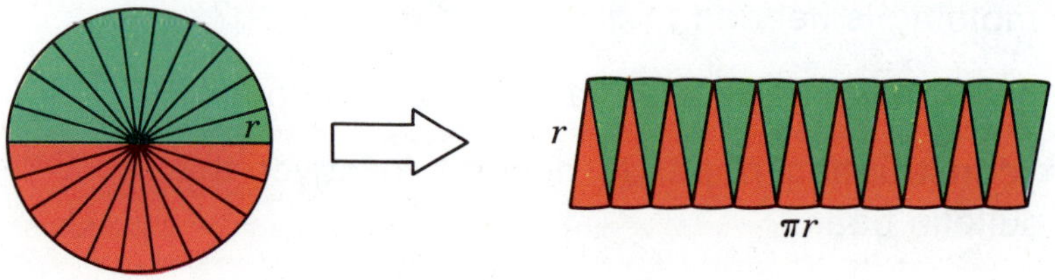

The figure made from the circle is like a rectangle with length πr (half the circumference of the circle) and width r (radius of the circle).

Formula $A = \pi r^2$

Now we can use the formula to find the amount of area affected by the earthquake.

$$A = \pi r^2$$
$$\approx 3.14 \times 48 \text{ km} \times 48 \text{ km}$$
$$\approx 7234.56 \text{ km}^2$$

EXERCISES

Find the area of each circle. Use 3.14 for π.

1. 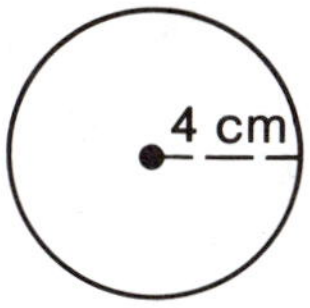

2. 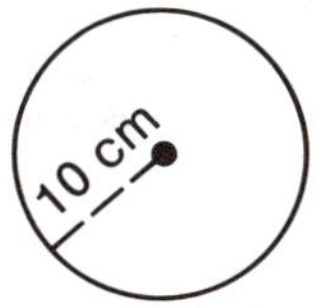

3. 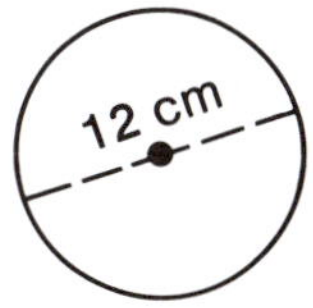

4. 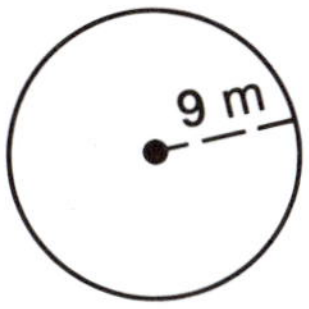

5. 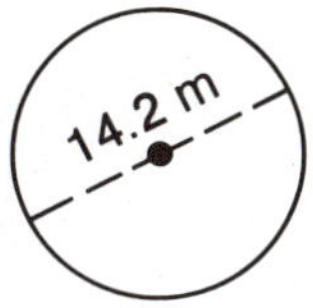

6. 

Find the area of each region. Note that you can cut regions and add areas.

7. 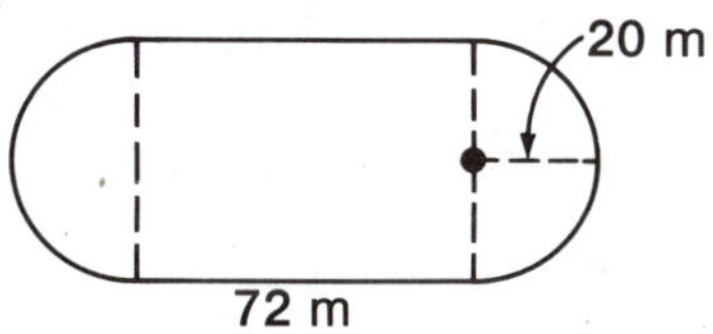

8. 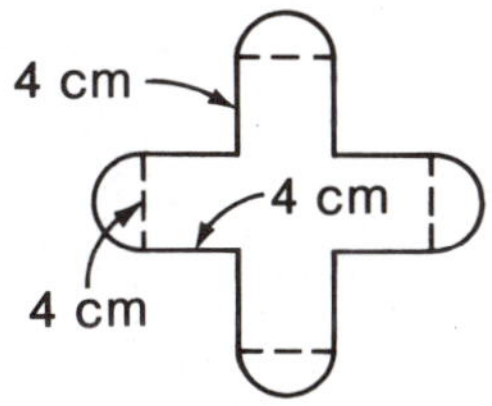

9. 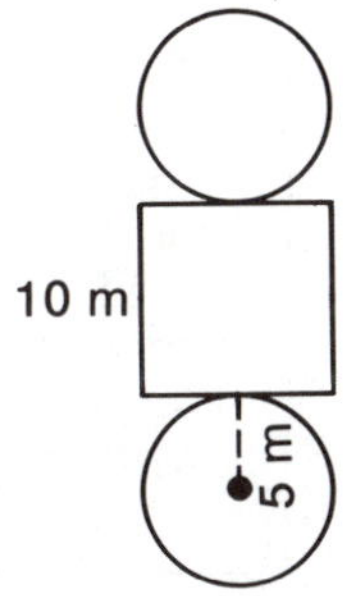

Find the area of each colored region. Note that you can subtract areas.

10. 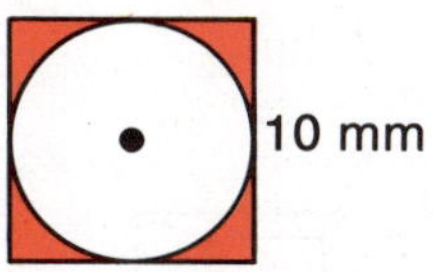

11. 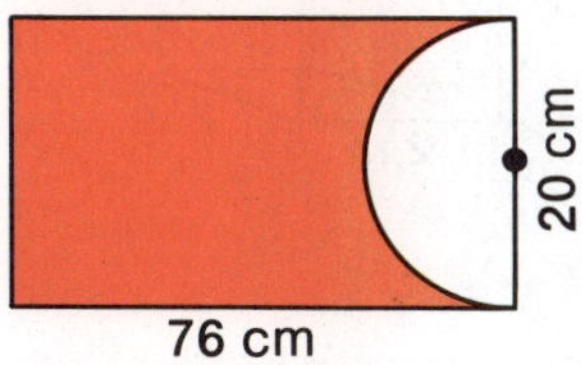

12. 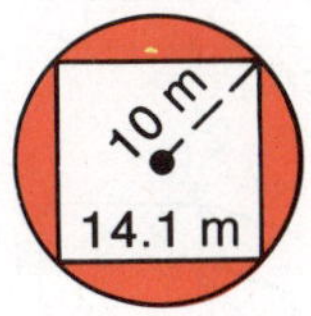

Surface area

The Band Club planned a chili supper to raise money for band uniforms.

Judy and Jeremiah want to cover the box that will hold the raffle tickets. How much paper will they need?

The box is called a **rectangular prism.** It has 6 **faces.** To answer the question, we find the areas of the six faces. Then we find their sum. This sum is the **surface area** of the rectangular prism.

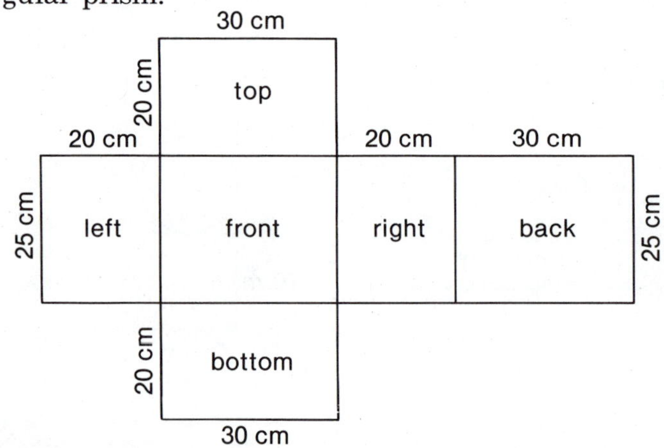

Area of

front . . . . 750 cm^2

back . . . . 750 cm^2

top . . . . 600 cm^2

bottom . . . . 600 cm^2

left side . . . . 500 cm^2

right side . . . . 500 cm^2

Surface Area = 3700 cm^2

EXERCISES

Find the surface area of each rectangular prism.

1.

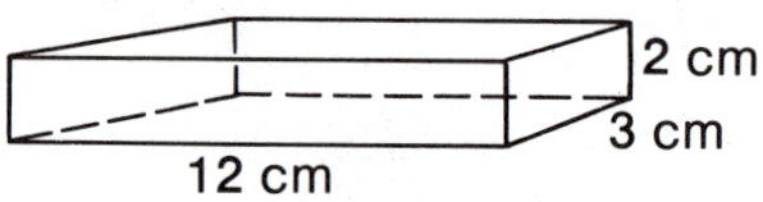

2.

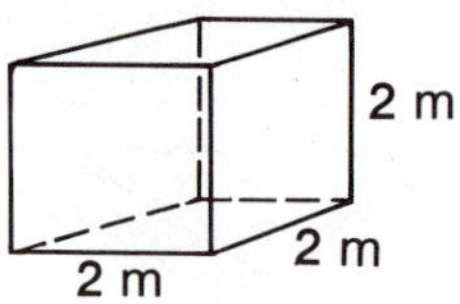

3.

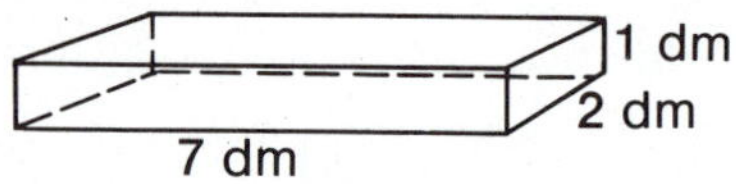

4.

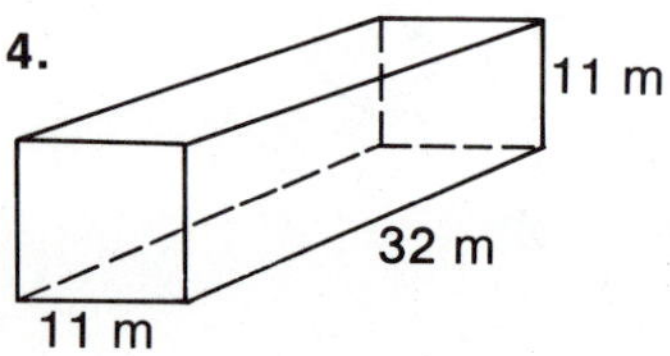

5.

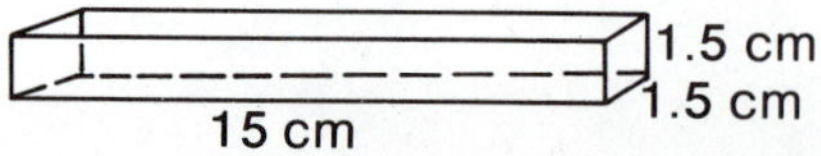

6.

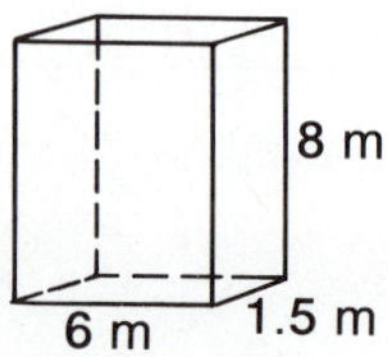

Problem solving—drawing a picture

1. Study and understand.
2. Plan and do.
3. Answer and check.

These steps show how to draw a rectangular prism:

Step 1

Step 2

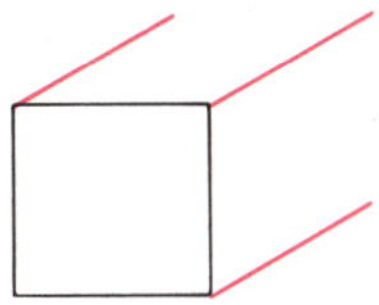

Step 3

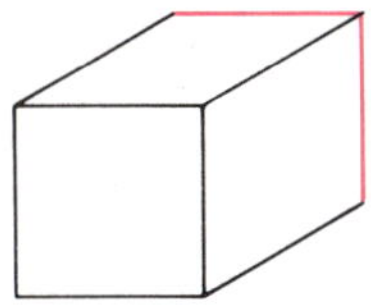

EXERCISES

Draw and label rectangular prisms from the given information.

1. length: 20 cm
width: 15 cm
height: 10 cm

2. length: 24 cm
width: 16 cm
height: 12 cm

3. Julia is the production chief for the Super Crax Cereal Co. New box designs require new calculations. How much cardboard would be needed for these box designs?

a. Honey Super Crax: 8 cm by 24 cm by 28 cm

b. Puffo Crax: 6 cm by 20 cm by 24 cm

★ **4.** The sales department is planning a special offer. They want to wrap two Honey Super Crax boxes back to back. How much cellophane is needed?

★ **5.** What is the surface area of this can? Remember to add all areas. (*Hint:* First draw what you would get if you cut off the top and bottom and cut straight down one side of the can.)

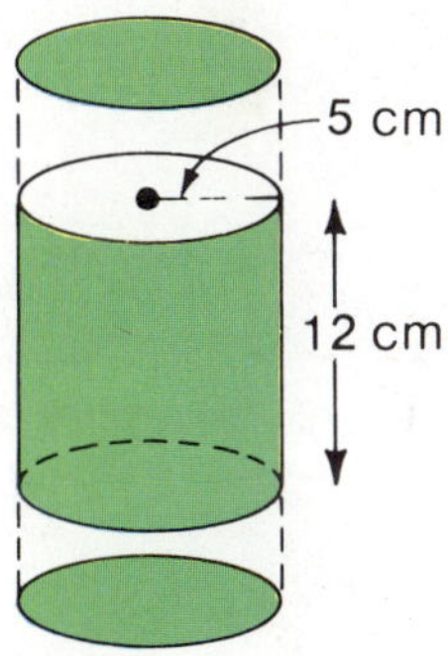

Volume of a prism

1 cm^3

To find the **volume** of this rectangular prism, we use a cubic centimeter (cm^3) as the unit. Then we see how many of the units are needed to fill the prism. There are 12 cm^3 in each layer and there are 5 layers. So the volume is 60 cm^3.

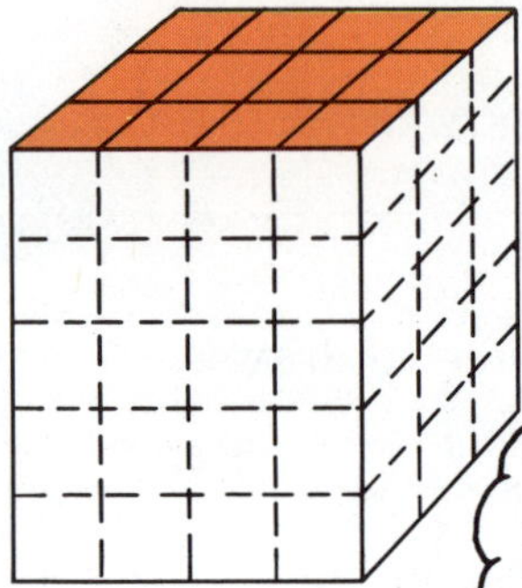

The base is a rectangle that has an area of 12 cm^2.

We can also find the volume by this formula:

$V = lwh$ ← *Formula*

$V = 4\text{ cm} \times 3\text{ cm} \times 5\text{ cm}$

$V = 60\text{ cm}^3$

Notice that the volume of a rectangular prism is equal to the area of a base times the height. So the volume can also be found by using the formula $V = Bh$. (B stands for the area of a base.)

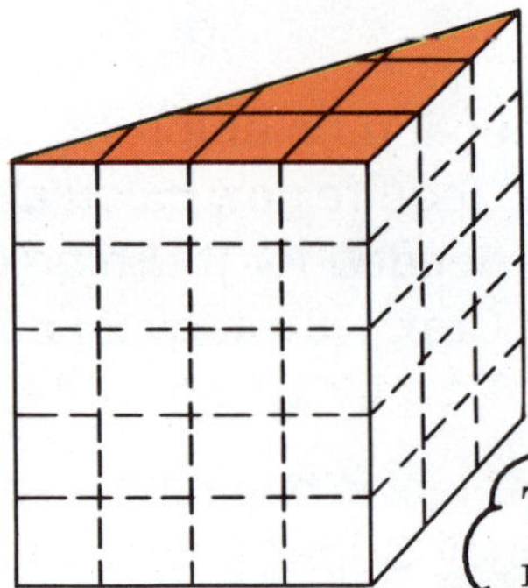

The base is a right triangle that has an area of 6 cm^2.

This prism has 5 faces. It is called a **triangular prism.** To find the volume, multiply the area of a base (B) by the height (h).

$V = Bh$ ← *Formula*

$V = \frac{1}{2} \times (4\text{ cm} \times 3\text{ cm}) \times 5\text{ cm}$

(6 square cm) $V = 6\text{ cm}^2 \times 5\text{ cm}$

(30 cubic cm) $V = 30\text{ cm}^3$

EXERCISES

Give each volume.

1.
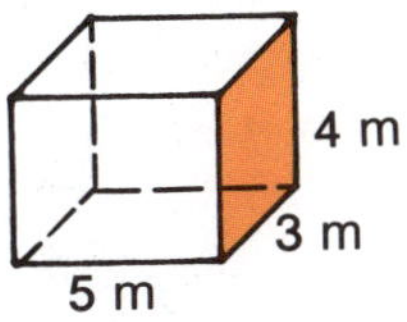

2.
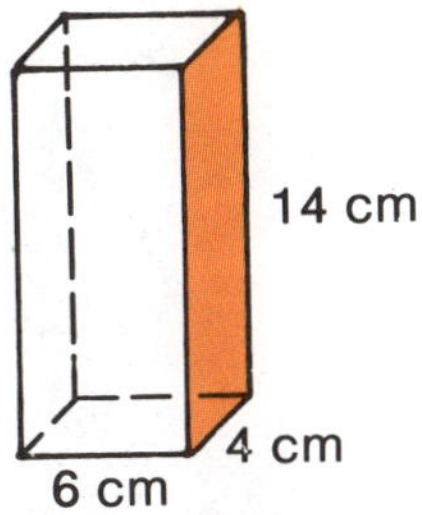

3.
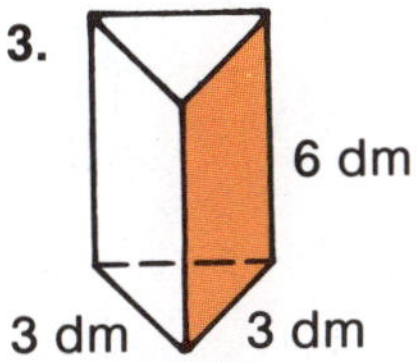

Bases are right triangles.

4.
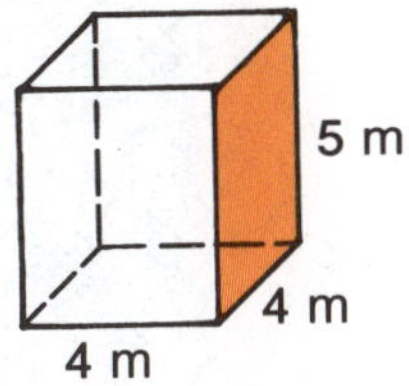

5.
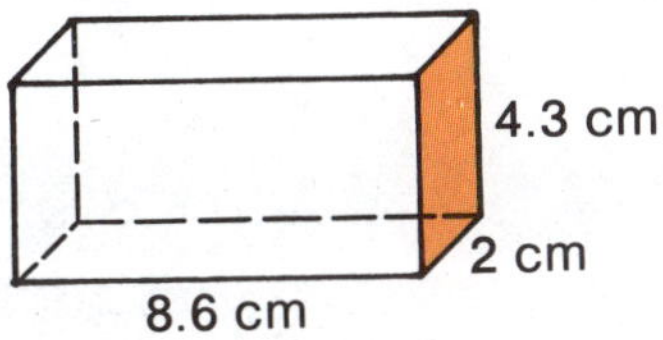

6.
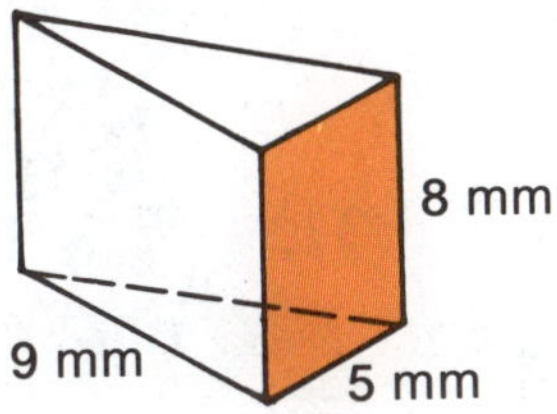

Bases are right triangles.

Draw and label rectangular prisms from the given information. Then find each volume.

7. length: 8 cm
   width: 4 cm
   height: 6 cm

8. length: 4 m
   width: 3 m
   height: 2.5 m

Solve.

9. A rectangular form that measures 120 decimeters long, 50 decimeters wide, and 24 decimeters high is to be filled with concrete. If concrete costs $36 a cubic meter, what will be the total cost of the concrete?

10. Find the volume of your classroom in cubic meters. Make your measurements to the nearest 0.1 meter.

KEEPING SKILLS SHARP

Multiply or divide. Give answers in simplest form.

1. $\frac{1}{2} \times \frac{3}{8}$
2. $\frac{1}{2} \div \frac{3}{8}$
3. $\frac{1}{2} \times \frac{1}{3}$
4. $\frac{1}{2} \div \frac{1}{3}$
5. $\frac{4}{9} \times 6$
6. $\frac{4}{9} \div 6$
7. $\frac{3}{4} \times \frac{1}{12}$
8. $\frac{3}{4} \div \frac{1}{12}$
9. $10 \times \frac{1}{2}$
10. $10 \div \frac{1}{2}$
11. $\frac{2}{5} \times \frac{5}{8}$
12. $\frac{2}{5} \div \frac{5}{8}$
13. $\frac{5}{6} \times \frac{2}{3}$
14. $\frac{5}{6} \div \frac{2}{3}$
15. $\frac{1}{6} \times \frac{3}{10}$
16. $\frac{1}{6} \div \frac{3}{10}$

Perimeter and circumference—customary units

Remember that the perimeter of a figure is the distance around the figure. Recall these formulas:

Square

$P = 4s$

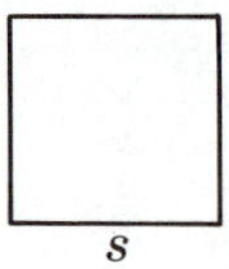

Rectangle

$P = 2l + 2w$

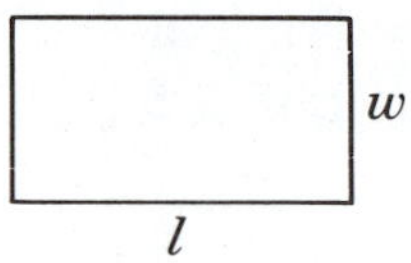

Circle

$C = \pi d$

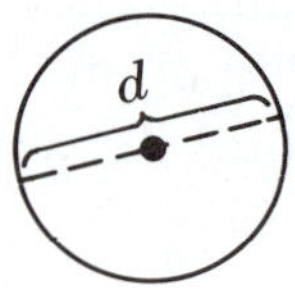

EXAMPLE.

George jogs around a circular track that has a diameter of 40 yards 2 feet. How far is it around the track?

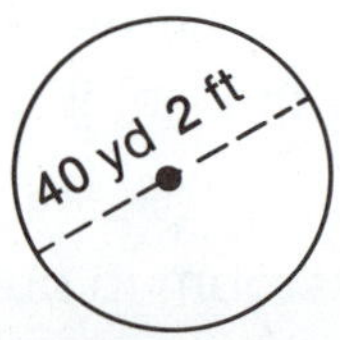

40 yd
×3
120 ft
+2 ft
122 ft

$C = \pi d$
$\approx 3.14 \times 122$ ft
≈ 383.08 ft

It is 383.08 feet around the track.

EXERCISES

Give the perimeter or circumference of each figure. Use 3.14 for π.

1.
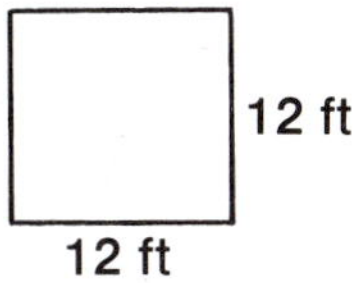

2.
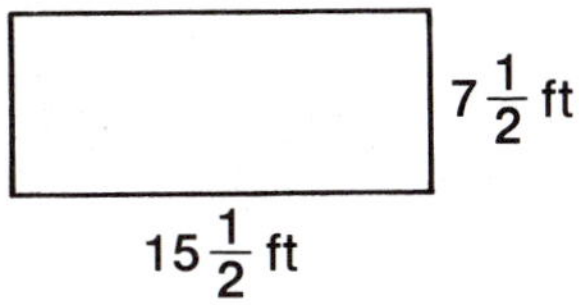

3.
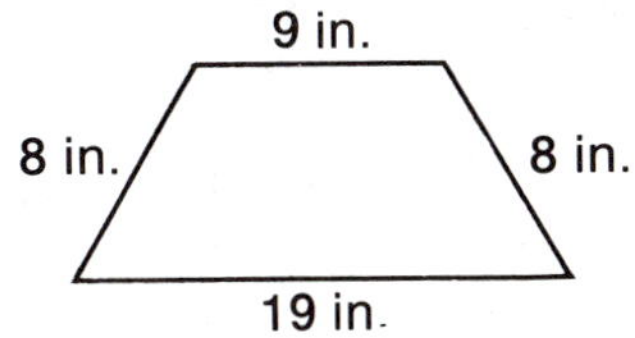

4.
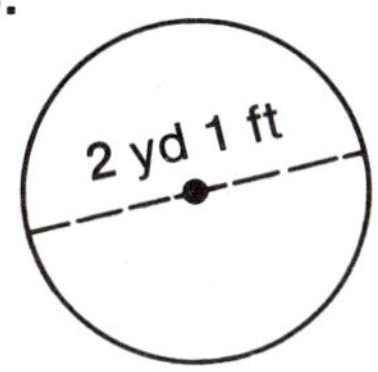

5.
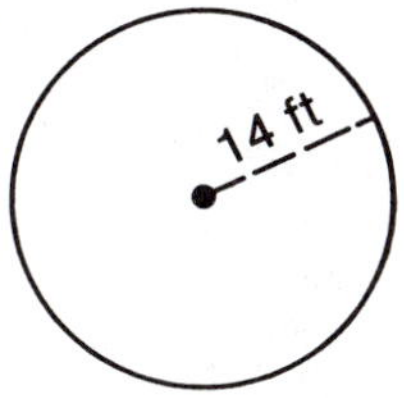

★ 6.
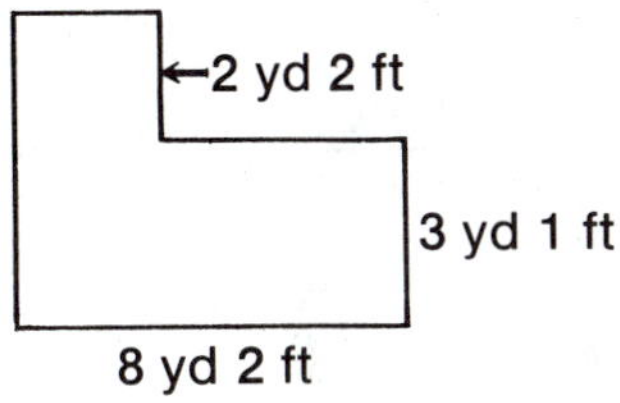

Solve.

7. a. Steve jogs around this track. How far is it around the track?

   b. How many miles does he jog if he goes around the track 10 times?

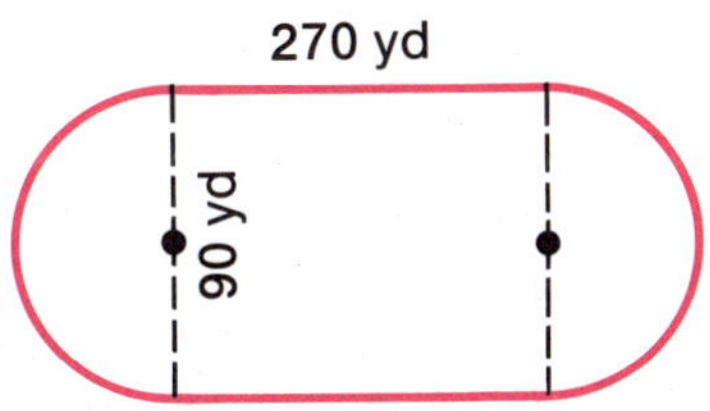

8. a. Each block in Jill's neighborhood is 250 yards long and 175 yards wide. Each street is 10 yards wide. Jill's jogging path is shown in red. How long is it?

   b. Find a route in Jill's neighborhood that is exactly 1 mile. Be sure to start and end at Jill's house.

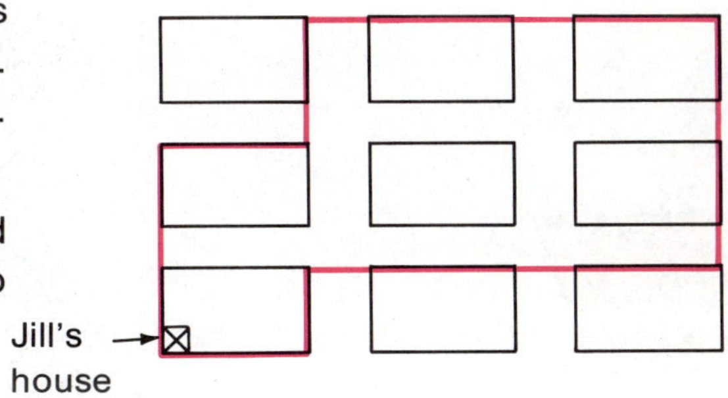

Area—customary units

Earlier you learned these formulas for area:

Square

$A = s^2$

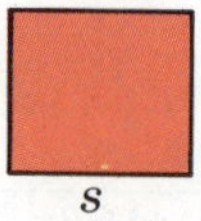

Rectangle

$A = lw$

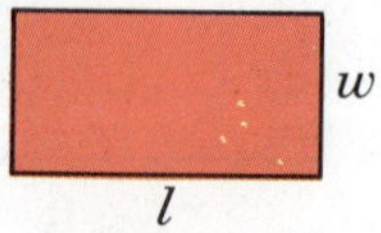

Circle

$A = \pi r^2$

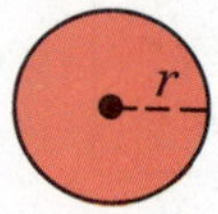

Parallelogram

$A = bh$

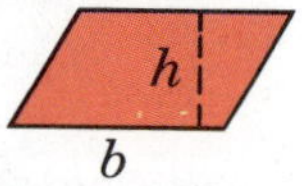

Triangle

$A = \frac{1}{2}bh$

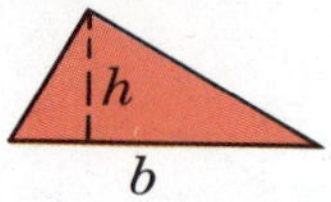

Now you can use the formulas. The measures are in customary units.

EXERCISES

Give each area. Use 3.14 as an approximation for π.

1.

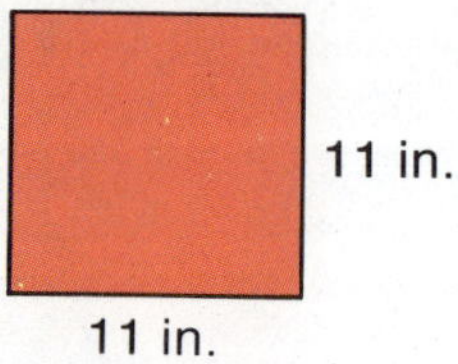

2.

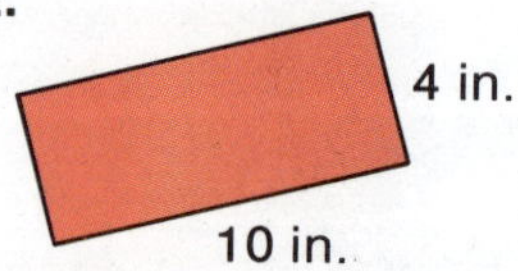

3.

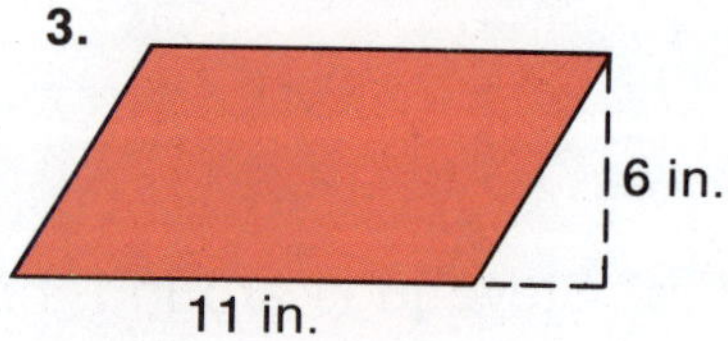

4.

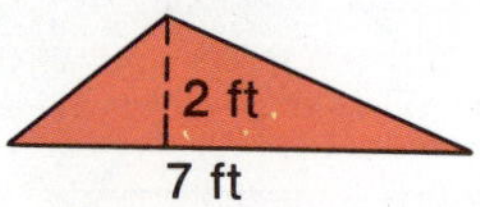

5.

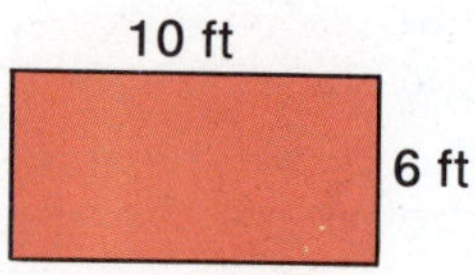

6.

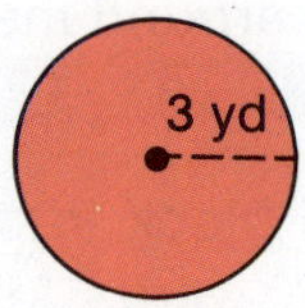

7.

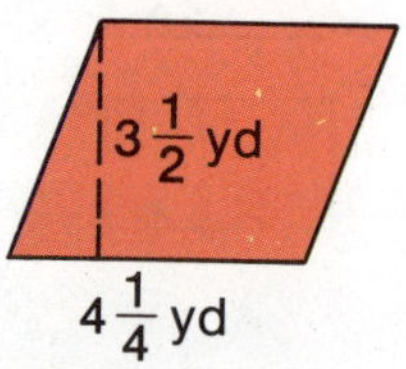

8.

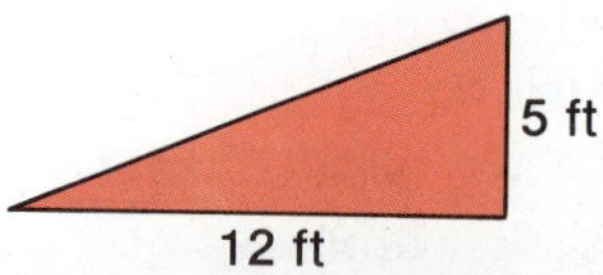

9.

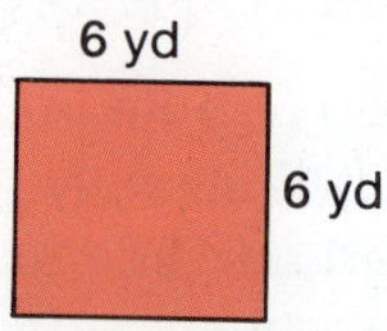

Hint: **First change to a single unit.**

10.

11.

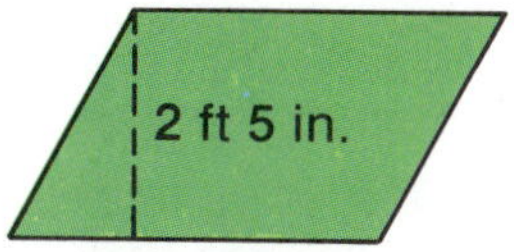

12.

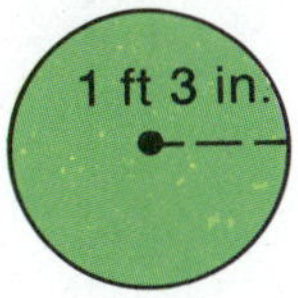

13.

Solve. ***Hint:*** **First draw a picture.**

14. How many square inches are in one square foot?
15. How many square feet are in one square yard?
16. How many square inches are in one square yard?
17. What is the area in square feet of a circular garden with a diameter of 10 yards?

18. THE GRAZING GOAT

A goat is tied to the corner of the barn. The rope is 15 yd long. On how much area can the goat graze?

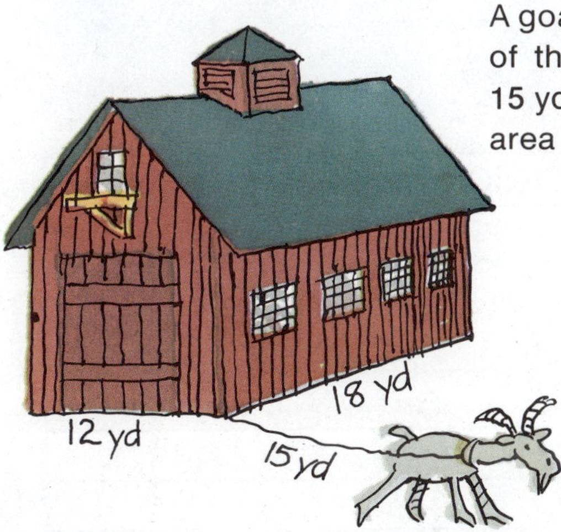

KEEPING SKILLS SHARP

Multiply or divide. Give answers in simplest form.

1. $1\frac{1}{2} \times \frac{3}{4}$
2. $1\frac{1}{2} \div \frac{3}{4}$
3. $\frac{5}{6} \times 3\frac{1}{3}$
4. $\frac{5}{6} \div 3\frac{1}{3}$
5. $6 \times 2\frac{2}{3}$
6. $6 \div 2\frac{2}{3}$
7. $1\frac{4}{5} \times 2$
8. $1\frac{4}{5} \div 2$
9. $1\frac{3}{4} \times 1\frac{1}{4}$
10. $1\frac{3}{4} \div 1\frac{1}{4}$
11. $1\frac{2}{3} \times 1\frac{1}{5}$
12. $1\frac{2}{3} \div 1\frac{1}{5}$
13. $2\frac{1}{4} \times 1\frac{1}{8}$
14. $2\frac{1}{4} \div 1\frac{1}{8}$
15. $1\frac{1}{6} \times 3\frac{1}{2}$
16. $1\frac{1}{6} \div 3\frac{1}{2}$

Mathematics in careers

Carpet layers generally are paid for each square yard of carpet they lay. They often charge extra for carpeting a small room such as a bath or for carpeting stairs.

EXERCISES

Copy and complete these bills.

1.

JACKSON'S CARPET SERVICE		
Living room	15 yd² @ $2.25/yd²	
Dining room	12 yd² @ $2.25/yd²	
Hall	4 yd² @ $2.25/yd²	
Stairs	10 steps @ $1.25/step	
Total		

2.

JACKSON'S CARPET SERVICE		
Living room	14 yd² @ $2.25/yd²	
Bedroom	10 yd² @ $2.25/yd²	
Bath	3 yd² @ $3.75/yd²	
Hall	5 yd² @ $2.25/yd²	
Stairs	11 steps @ $1.25/step	
Total		

Carpet layers have to compute area to determine the amount of carpet laid. (Since they charge by the square yard, each computed area must be in *square yards*.) The area of each "shaded room" shown below was computed and rounded up to the nearest square yard.

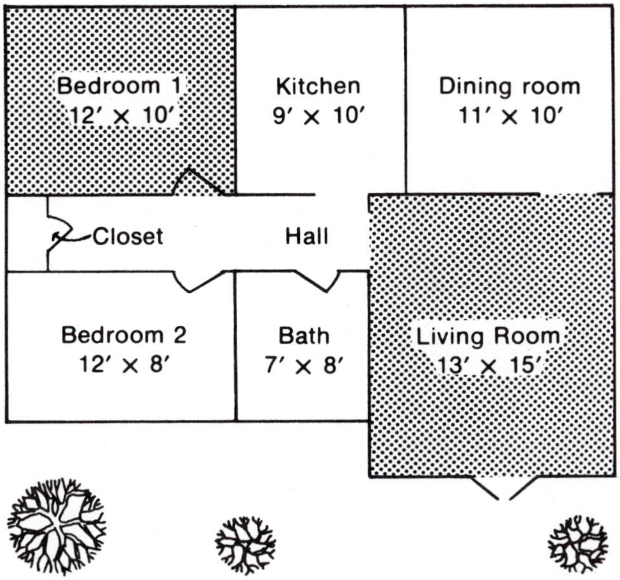

$A = 12' \times 10'$

$= 120 \text{ ft}^2$

Divide by 9 to change to square yards. Round up to 14 yd^2.

$$\begin{array}{r} 13 \\ 9\overline{)120} \\ -9 \\ \hline 30 \\ -27 \\ \hline 3 \end{array}$$

$A = 13' \times 15'$

$= 195 \text{ ft}^2$

Change to square yards. Round up to 22 yd^2.

$$\begin{array}{r} 21 \\ 9\overline{)195} \\ -18 \\ \hline 15 \\ -9 \\ \hline 6 \end{array}$$

Using the rate of \$2.25 a square yard, compute the charge for laying the carpet in each of the following rooms. Round each area up to the nearest square yard.

3. 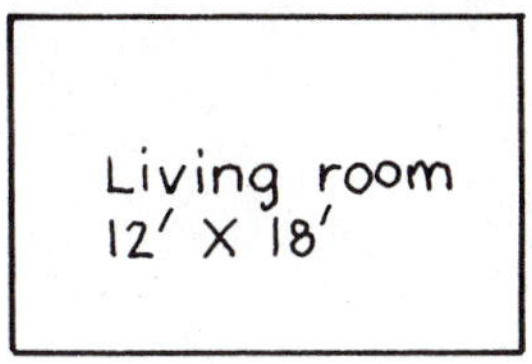

4. 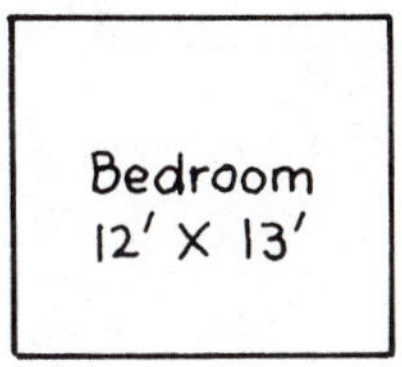

5. 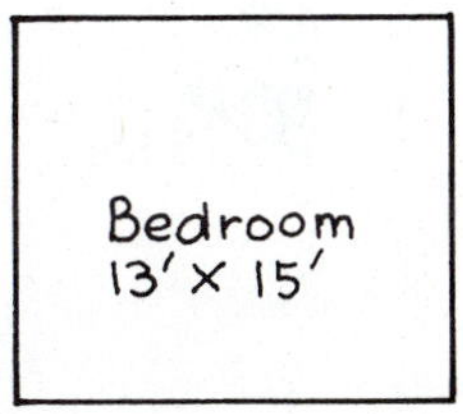

6. 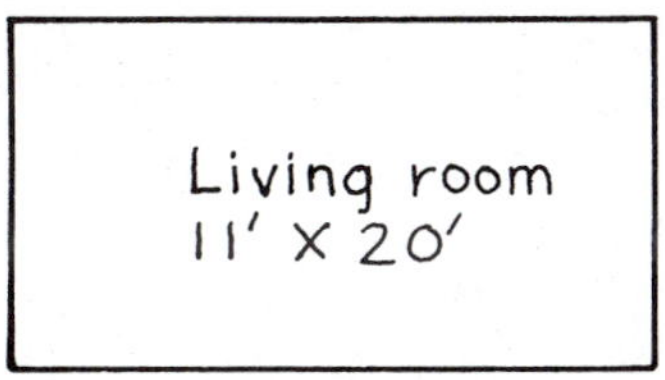

7. 

★ **8.** 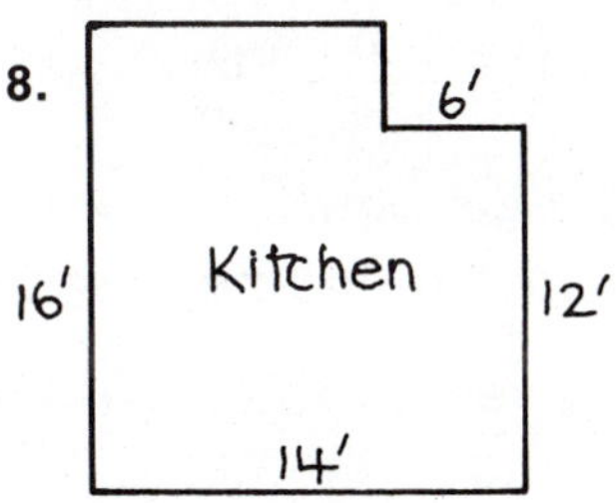

Volume—customary units

Study this example.

Important:

When measurements are given in more than one unit, change to a single unit before multiplying.

EXAMPLE.

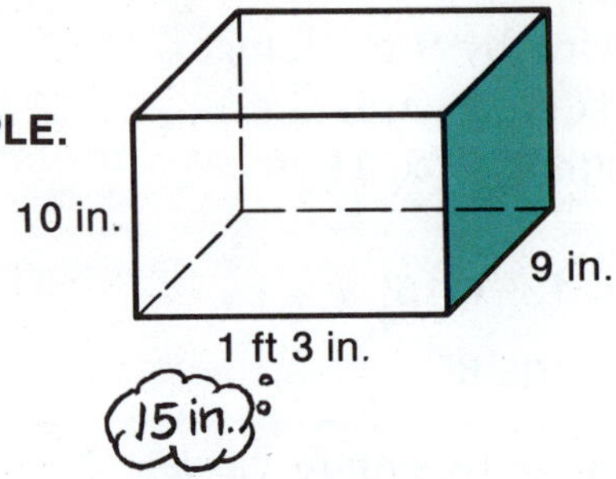

$V = lwh$

$V = (15 \text{ in.} \times 9 \text{ in.}) \times 10 \text{ in.}$

$V = 135 \text{ in.}^2 \times 10 \text{ in.}$

$V = 1350 \text{ in.}^3$

EXERCISES

Give each volume.

1.

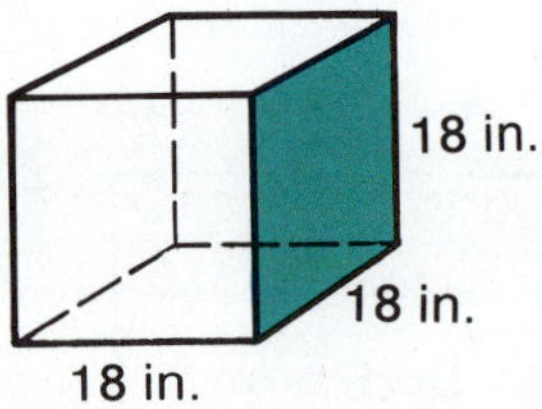

2.

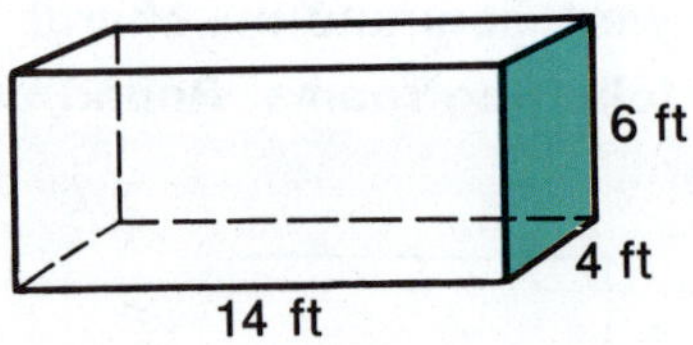

3.

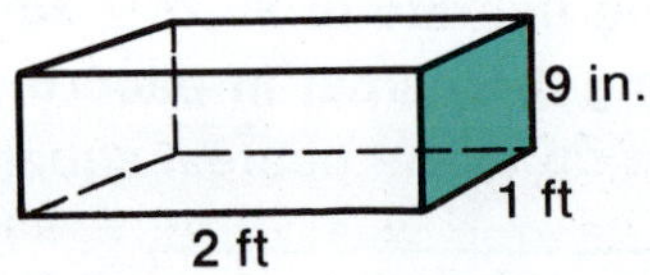

Complete.

4. $1 \text{ yd}^3 = \underline{\quad ? \quad} \text{ ft}^3$

5. $1 \text{ ft}^3 = \underline{\quad ? \quad} \text{ in.}^3$

6. $1 \text{ yd}^3 = \underline{\quad ? \quad} \text{ in.}^3$

7. Convert your answer for exercise 2 to yd^3. Round to the nearest 0.1 yd^3.

8. Convert your answer for exercise 3 to ft^3. Round to the nearest 0.1 ft^3.

9. A crepe-paper border is to be placed around some rectangular tables that measure 1 yard by 2 yards. How many yards of crepe paper will be needed for 7 tables?

10. A fence is to be built around a rectangular lot that is 22 yd long and 13 yd wide. If fencing costs $1.25 a foot, how much will it cost to fence the lot?

11. A triangular sail has a base of 5 yards and a height of 8 yards. How many square yards of sail is that?

12. A dog is leashed to a stake. If the leash is 10 feet long, how many square feet does the dog have to roam?

13. Concrete is sold by the cubic yard. How many cubic yards of concrete are needed for a basement wall that is 3 yd high, 10 yd long, and 0.25 yd thick?

14. A revolving lawn sprinkler sprays water a distance of 28 feet. How many square feet of lawn will it sprinkle without being moved?

15. A concrete garage floor is 23 ft by 23 ft and 3 in. thick. How many cubic yards of concrete were used? Round your answer to the nearest 0.1 yd^3.

16. A pizza house sells large pizzas (15 inches in diameter) for $4.80 and medium pizzas ($7\frac{1}{2}$ inches in diameter) for $3.20. Which is the better buy?

17. **CUBE COUNT**

The large block was made by stacking 27 white blocks together and then painting the outside green.

How many small blocks have

0 green faces? 1 green face? 2 green faces?
3 green faces? 4 green faces?

CHAPTER CHECKUP

Find the perimeter or circumference. [pages 236–239]
Use 3.14 for π.

1. 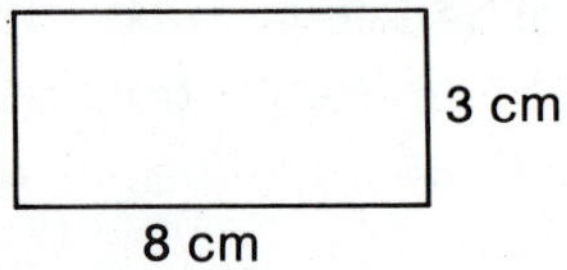

2. 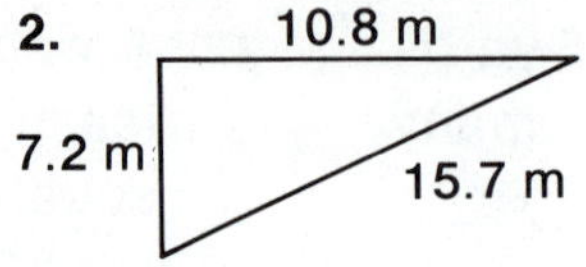

3. 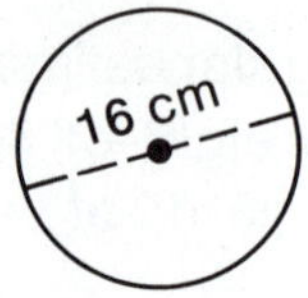

Find each area. [pages 240–247]
Use 3.14 for π.

4. 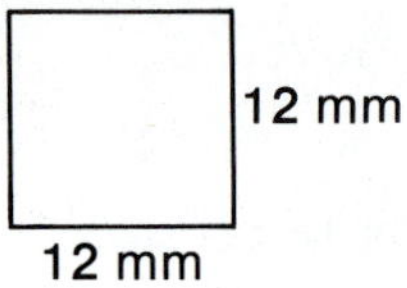

5. 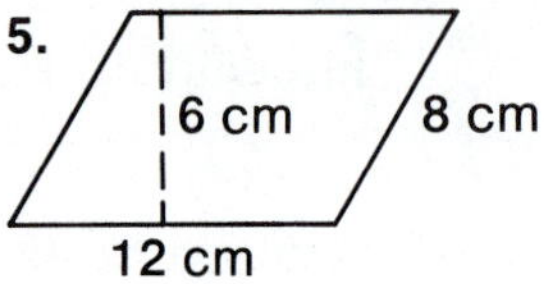

6. 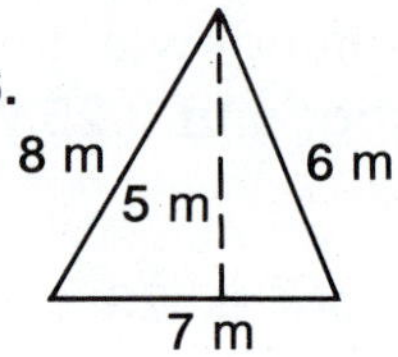

7. 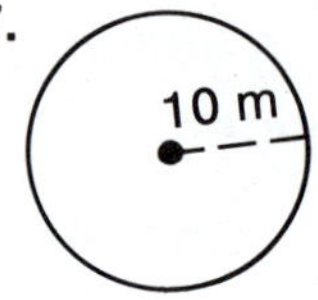

8. Find the surface area. [pages 248–249]
9. Find the volume. [pages 250–251]

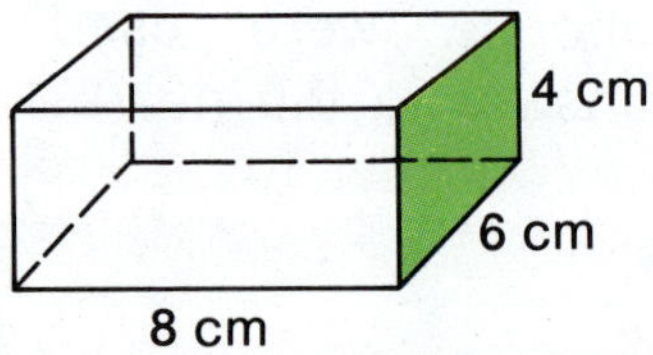

Find the perimeter or circumference. [pages 252–253]
Use 3.14 for π.

10. 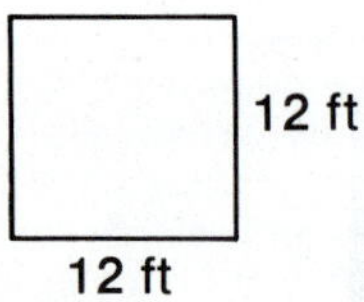

11. 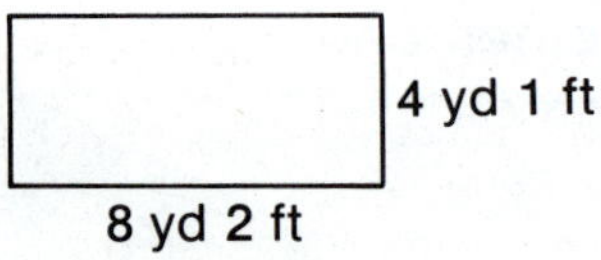

12. 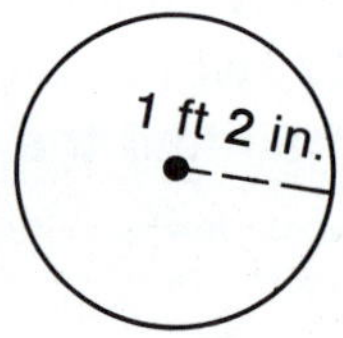

Find the area. [pages 254–257]
Use 3.14 for π.

13. 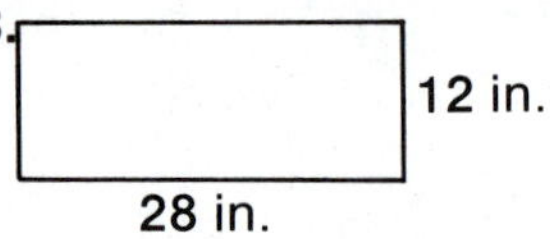

14. 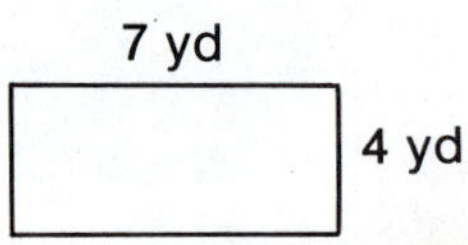

15. 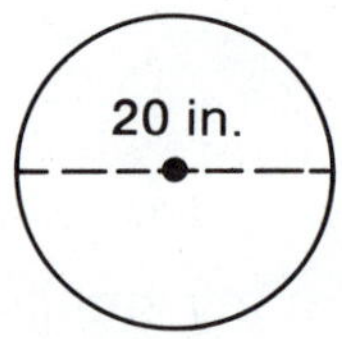

16. Find the volume. [pages 258–259]

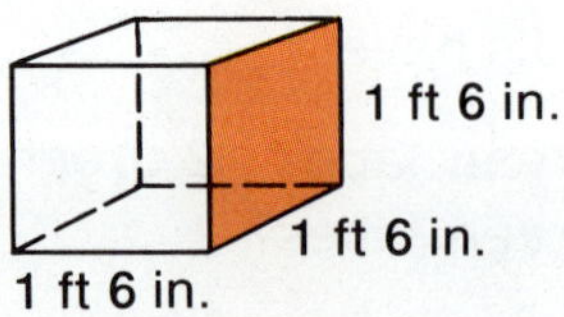

CHAPTER PROJECT

A metric unit for measuring land area is the **hectare.** A square that is 100 meters on a side has an area of 1 hectare. The hectare is used to measure areas of such regions as parks and farms. Working in a group, lay out a hectare on your school grounds. Estimate the total area of your school grounds in hectares.

Answer these questions.

1. How many square meters are in a hectare?
2. How many hectares are in a square kilometer?

CHAPTER REVIEW

Perimeter or circumference is the distance around.
Square: $P = 4s$
Rectangle: $P = 2l + 2w$
Circle: $C = \pi d$

Give the perimeter or circumference. Use 3.14 for π.

1.

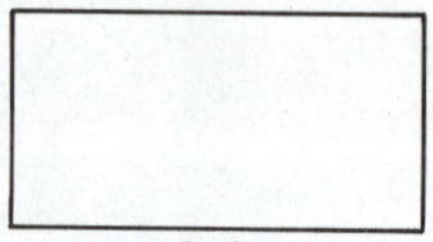

5.2 m

10.6 m

2.

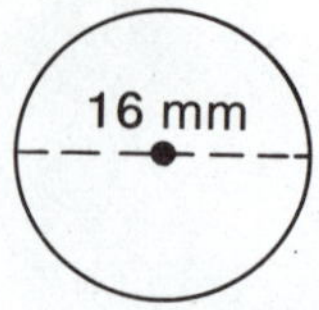

Square: $A = s^2$
Rectangle: $A = lw$
Parallelogram: $A = bh$
Triangle: $A = \frac{1}{2}bh$
Circle: $A = \pi r^2$

Give the area. Use 3.14 for π.

3.

7 cm

7 cm

4.

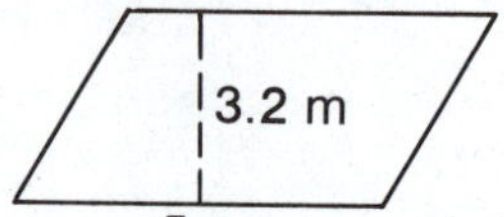

5 m

5.

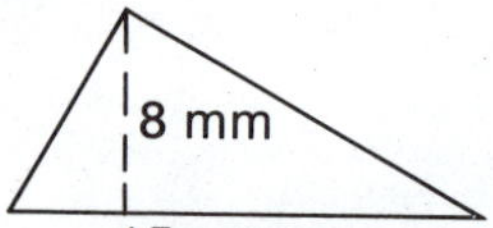

15 mm

6.

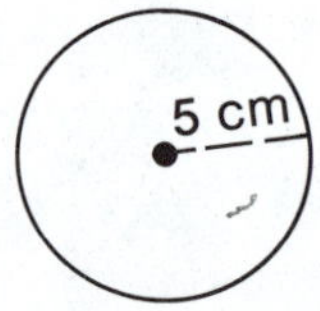

The surface area of a rectangular prism is the sum of the areas of the faces.

Give the surface area.

7.

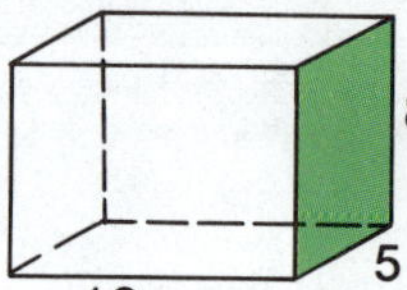

8 cm

5 cm

12 cm

The volume of a rectangular prism is given by $V = lwh$.

Give the volume.

8.

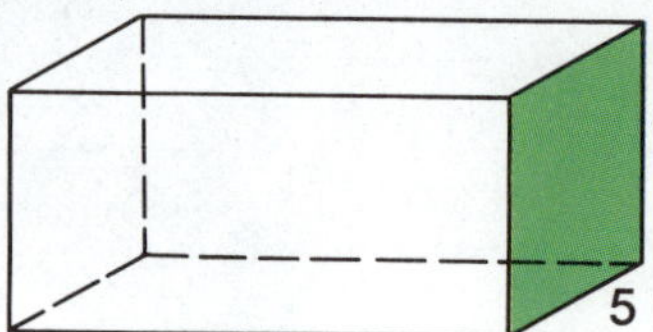

7 m

5 m

10 m

Remember to change all measures to the same single unit.

Give the perimeter.

9.

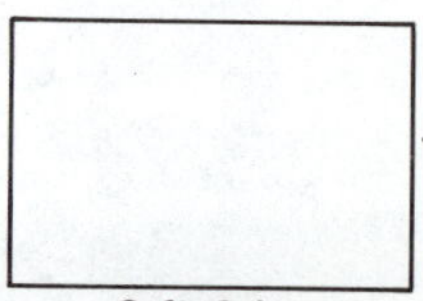

1 ft 9 in.

3 ft 6 in.

Give the area.

10.

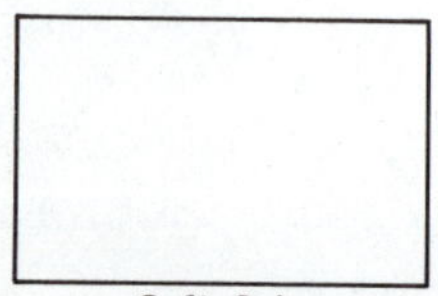

1 ft 3 in.

2 ft 9 in.

CHAPTER CHALLENGE

Remember that you can find the volumes of prisms like these by multiplying the area of a base (B) by the height (h).

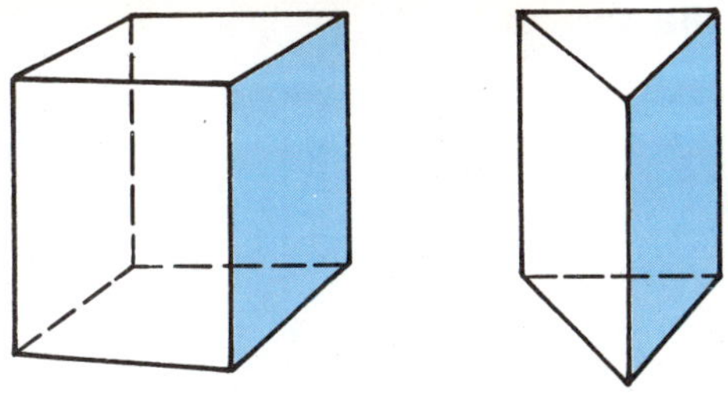

The volume of a cylinder can be found in the same way. Use 3.14 as an approximation for π.

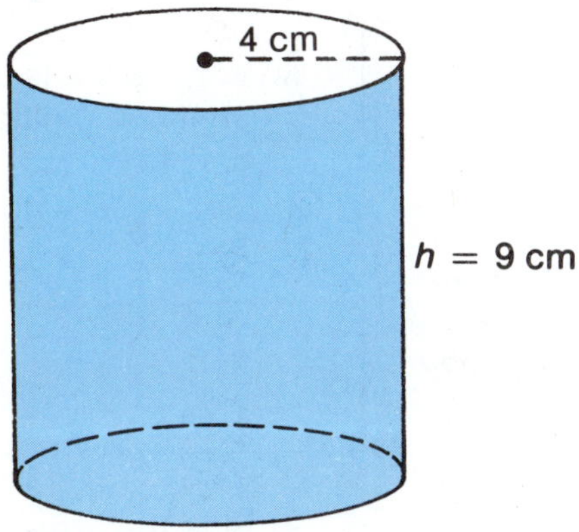

$V = Bh$

$V = \pi r^2 h$

$V = (\pi \times 16\text{ cm}^2) \times 9\text{ cm}$

$V \approx (3.14 \times 16\text{ cm}^2) \times 9\text{ cm}$

$V \approx 50.24\text{ cm}^2 \times 9\text{ cm}$

$V \approx 452.16\text{ cm}^3$

EXERCISES

Give the volume of each cylinder. Use 3.14 as an approximation for π.

1.

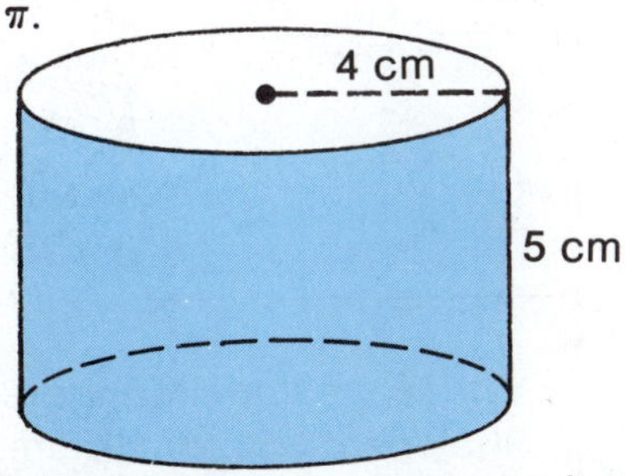

2.

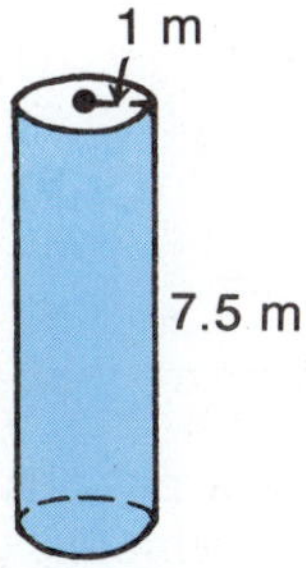

3.

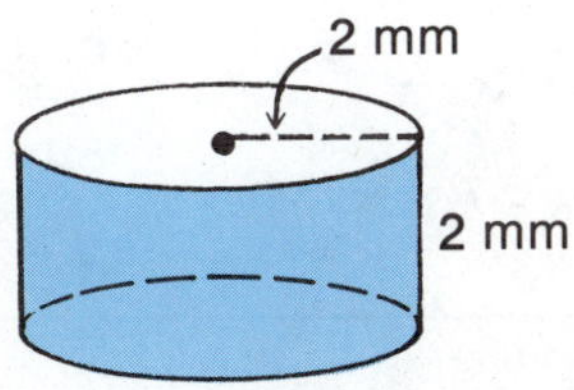

★ **4.** Which solid has the greatest volume?

a.

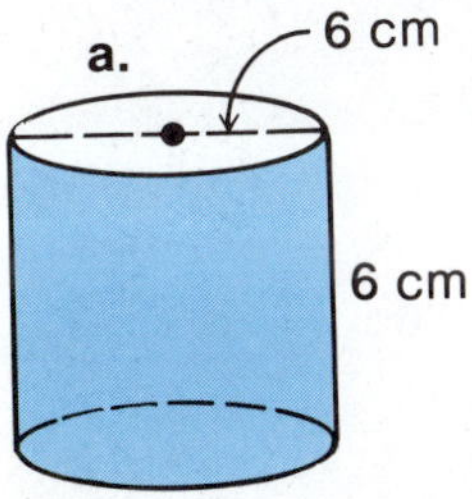

b.

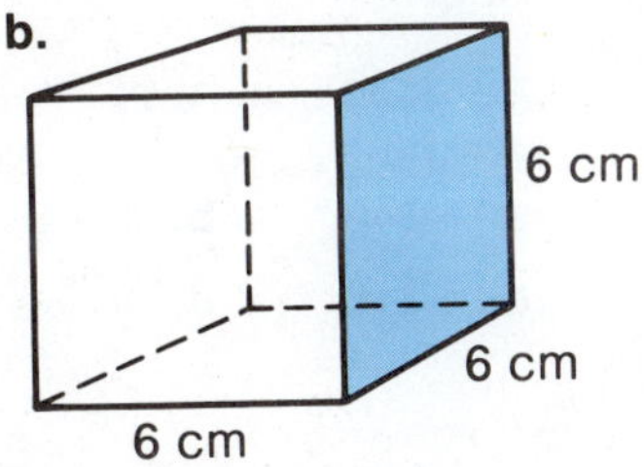

c.

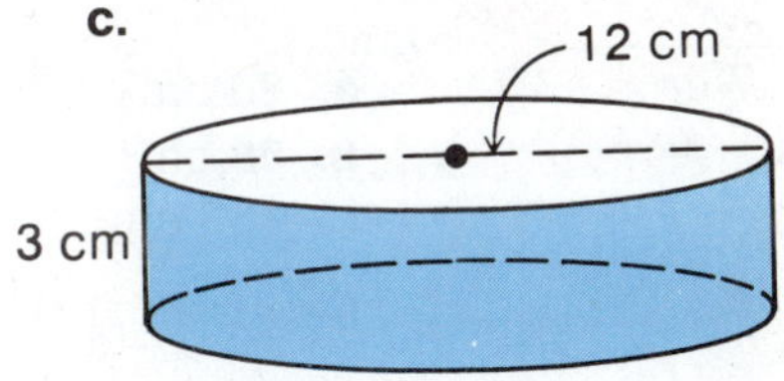

MAJOR CHECKUP
STANDARDIZED FORMAT

Choose the correct letter.

1. Subtract.

$$\begin{array}{r} 6025 \\ -1773 \\ \hline \end{array}$$

a. 4252
b. 5252
c. 5752
d. none of these

2. Multiply.

$$\begin{array}{r} 248 \\ \times 72 \\ \hline \end{array}$$

a. 18,836
b. 17,836
c. 18,856
d. none of these

3. Give the sum.

6.7 + 0.488

a. 0.555
b. 7.188
c. 55.5
d. none of these

4. Give the quotient to the nearest hundredth.

$2.7\overline{)41.33}$

a. 1.53
b. 1.54
c. 1.5
d. none of these

5. 43.6 m = ? cm

a. 0.436
b. 0.0436
c. 4360
d. none of these

6. What is the least common multiple of 9 and 12?

a. 3
b. 18
c. 36
d. none of these

7. Subtract.

$$\begin{array}{r} 7\frac{3}{4} \\ -5\frac{1}{6} \\ \hline \end{array}$$

a. $2\frac{7}{12}$
b. $1\frac{7}{12}$
c. $1\frac{5}{12}$
d. none of these

8. Divide.

$\frac{7}{8} \div \frac{1}{4}$

a. $\frac{7}{32}$
b. $3\frac{1}{2}$
c. $\frac{2}{7}$
d. none of these

9. Find the area.

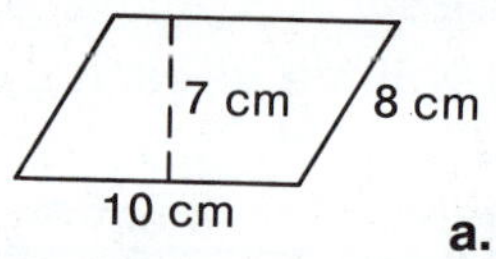

a. 80 cm^2
b. 70 cm^2
c. 35 cm^2
d. none of these

10. Find the volume.

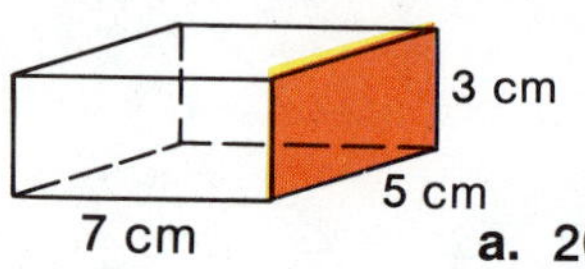

a. 20 cm^3
b. 63 cm^3
c. 180 cm^3
d. none of these

11. Carpet costs $12 a square meter. How much does a 4-meter by 5-meter carpet cost?

a. $240 **b.** $108
c. $216 **d.** none of these

12. Tanya had 4 quarters. She spent 38¢ for 2 plums. How much money did she have left?

a. 24¢ **b.** 62¢
c. 76¢ **d.** none of these

Ratios and Percent

9

READY OR NOT?

What fraction of the marbles are red?

1. 

2. 

3. 

4. 

Copy and complete.

5. $\frac{1}{2} = \frac{?}{4}$

6. $\frac{2}{3} = \frac{?}{6}$

7. $\frac{3}{4} = \frac{?}{12}$

8. $\frac{4}{3} = \frac{?}{12}$

9. $\frac{3}{2} = \frac{?}{18}$

10. $\frac{3}{8} = \frac{12}{?}$

11. $\frac{9}{4} = \frac{36}{?}$

12. $\frac{5}{6} = \frac{30}{?}$

Ratios

Monica Eastman makes designs out of stained glass. One of her most popular designs, Sunburst, uses 3 red and 4 yellow pieces of glass.

The **ratio** of red pieces to yellow pieces is 3 to 4.

We write ratios in three different ways:

$$3 \text{ to } 4 \qquad \frac{3}{4} \qquad 3:4$$

We read each ratio as "3 to 4."

You can find **equal ratios** by finding equivalent fractions.

	1 sunburst		2 sunbursts		3 sunbursts
red → / yellow →	$\frac{3}{4}$	=	$\frac{6}{8}$	=	$\frac{9}{12}$

EXERCISES

Here are some of Monica's designs. Write a ratio for each color pair.

1. red to blue

2. green to red

3. yellow to red

Give each ratio as a fraction in lowest terms.

4. 5:10
5. 6 to 4
6. 12 to 8
7. 9:24
8. 8:6
9. 16 to 36

Copy and complete.

10. $\frac{1}{3} = \frac{12}{?}$
11. $\frac{7}{2} = \frac{?}{8}$
12. $\frac{8}{5} = \frac{40}{?}$
13. $\frac{4}{3} = \frac{12}{?}$
14. $\frac{5}{8} = \frac{15}{?}$
15. $\frac{7}{8} = \frac{42}{?}$
16. $\frac{4}{8} = \frac{?}{48}$
17. $\frac{6}{5} = \frac{?}{30}$
18. $\frac{6}{5} = \frac{30}{?}$
19. $\frac{5}{8} = \frac{?}{16}$
20. $\frac{3}{4} = \frac{24}{?}$
21. $\frac{2}{3} = \frac{18}{?}$

For the pieces of glass pictured, give each ratio as a fraction in lowest terms.

22. triangles to squares
23. circles to triangles
24. squares to circles
25. yellow shapes to blue shapes
26. blue triangles to yellow triangles
27. red circles to blue triangles

Solve.

★ 28. Monica orders pieces of glass in the ratio of 2 yellow to 3 red. How many red pieces should she order if she wants 240 yellow pieces?

Proportions

An equation stating that two ratios are equal is called a **proportion.** Each proportion has a related multiplication equation.

Proportion	Multiplication equation
$\frac{3}{4} = \frac{9}{12}$	$3 \times 12 = 4 \times 9$

The two products 3×12 and 4×9 are often called **cross products.** The cross products and the ratios have a special relationship:

If the ratios are equal, so are the cross products
and
if the cross products are equal, so are the ratios.

Since the cross products are equal, the ratios are equal.

$$\frac{3}{5} = \frac{6}{10}$$

Since the cross products are not equal, the ratios are not equal.

$$\frac{5}{9} \neq \frac{1}{2}$$

EXAMPLE. Solve: $\frac{3}{7} = \frac{12}{n}$

Step 1. Write the related multiplication equation.

3n is a short way of writing 3×n.

$$3n = 7 \times 12$$

Step 2. Solve the multiplication equation.

$$3n = 7 \times 12$$

$$3n = 84$$

$$n = 28$$

$$3\overline{)84}\text{ } = 28$$

Check: $\frac{3}{7} = \frac{12}{28}$

EXERCISES

$=$ or $\neq$?

1. $\frac{3}{4}$ ● $\frac{6}{8}$
2. $\frac{2}{1}$ ● $\frac{3}{2}$
3. $\frac{9}{8}$ ● $\frac{6}{5}$
4. $\frac{6}{9}$ ● $\frac{4}{6}$
5. $\frac{3}{6}$ ● $\frac{4}{8}$
6. $\frac{9}{6}$ ● $\frac{6}{4}$
7. $\frac{6}{5}$ ● $\frac{7}{6}$
8. $\frac{7}{7}$ ● $\frac{8}{8}$
9. $\frac{9}{12}$ ● $\frac{6}{8}$

Copy and complete.

10. $\frac{5}{2} = \frac{n}{3}$

    $2n = 5 \times 3$

    $n = $ __?__

11. $\frac{5}{6} = \frac{4}{n}$

    $5n = 6 \times 4$

    $n = $ __?__

Hint: Divide 15 by 2.

12. $\frac{n}{8} = \frac{3}{5}$

    $5n = $ __?__

    $n = $ __?__

13. $\frac{4}{n} = \frac{6}{9}$

    $6n = $ __?__

    $n = $ __?__

Solve each proportion.

14. $\frac{n}{3} = \frac{5}{9}$
15. $\frac{7}{n} = \frac{4}{6}$
16. $\frac{4}{8} = \frac{2}{n}$
17. $\frac{6}{n} = \frac{4}{3}$
18. $\frac{9}{n} = \frac{7}{4}$
19. $\frac{8}{n} = \frac{5}{6}$
20. $\frac{n}{8} = \frac{6}{5}$
21. $\frac{3}{3} = \frac{n}{3}$
22. $\frac{n}{8} = \frac{3}{4}$
23. $\frac{5}{16} = \frac{n}{11}$
24. $\frac{17}{23} = \frac{n}{15}$
25. $\frac{8}{5} = \frac{6}{n}$
26. $\frac{13}{38} = \frac{24}{n}$
27. $\frac{9}{100} = \frac{n}{50}$
28. $\frac{n}{100} = \frac{5}{6}$

KEEPING SKILLS SHARP

Give each sum.

1. 3.8 + 9.6
2. 10.3 + 7.8
3. 15.8 + 28.3
4. 29.9 + 37.4
5. 16 + 3.8
6. 12.8 + 19
7. 2.9 + 4.05
8. 3.68 + 29.8
9. 15.8 + 32.9
10. 34.78 + 25.09

Give each difference.

11. 8 − 2.1
12. 15.7 − 2.3
13. 18 − 6.5
14. 15.7 − 3.8
15. 23.4 − 7.8
16. 52.6 − 3.74
17. 83 − 2.97
18. 75 − 3.75
19. 43.8 − 22.91
20. 86.5 − 3.75

Proportions in problem solving

The Wright brothers made the world's first flight on December 17, 1903, at Kitty Hawk, North Carolina.

1903 Wright Brothers

Wingspan: 12.29 meters
Length: 6.43 meters

First airplane to make a successful flight

A model of the Wright brothers' plane is to be $\frac{1}{20}$ the size of the real airplane. Find its wingspan.

model airplane → 1; real airplane → 20; model airplane → n; real airplane → 12.29

$$\frac{1}{20} = \frac{n}{12.29}$$

$$20n = 12.29$$

$$n = 0.6145$$

The length of the model should be 0.6145 meter.

1924 Douglas *Chicago*

Wingspan: 15.24 meters
Length: 10.82 meters

First airplane to be flown around the world

Ford Trimotor

Wingspan: 22.60 meters
Length: 15.19 meters

In 1929 Commander Richard E. Byrd and Bernt Balchen, using a Ford Trimotor, became the first pilots to fly over the South Pole.

Lockheed Vega

Wingspan: 12.50 meters
Length: 8.38 meters

In 1932 Amelia Earhart, using a Lockheed Vega, became the first woman to fly across the Atlantic solo and nonstop.

EXERCISES

Solve.

1. What are the wingspan and length of a model Douglas Chicago if it is $\frac{1}{25}$ the actual size?
2. Compute the wingspan and length of a Ford Trimotor model if it is $\frac{1}{80}$ the actual size.
3. What is the ratio of wingspan to length of the Lockheed Vega? What is the length of a model that has a wingspan of 3.0 meters?
4. To get a special shade of green for painting the Douglas Chicago, 2 bottles of midnight blue are mixed with 3 bottles of canary yellow. How much canary yellow should be mixed with 1 bottle of midnight blue to get the same color?
5. A model of the Wright brothers' airplane is to be 0.5 meter long. What should the wingspan be?
6. A model of Ford Trimotor is to have a wingspan of 0.5 meter. What should the length be?

Rates

You can think of a **rate** as a ratio of two different quantities. Here is a problem involving rates. We can solve the problem by using a proportion.

In 1932 Hubert Opperman set a 24-hour distance record for a bicycle rider when he rode 1380 kilometers. At that rate, how far did he ride in 7 hours?

0 km — n km — 1380 km
0 h — 7 h — 24 h

kilometers

$$\frac{1380}{24} = \frac{n}{7}$$

hours

$$24n = 9660$$

$$n = 402.5$$

He rode 402.5 kilometers in 7 hours.

EXERCISES

Study each picture. Then use a proportion to solve the problem.

1. 0 kilometers — 123 km — n km
   0 hours — 2 h — 3 h

2. 0 kilometers — 96 km — 240 km
   0 hours — n h — 5 h

3. 0 kilometers — 17 km — 25 km
   0 hours — 5 h — n h

4. 0 kilometers — n km — 45 km
   0 hours — $1\frac{3}{4}$ h — 3 h

Use the given rate to solve the problem. Round answers to the nearest tenth. (*Hint:* You may want to draw a picture.)

5. 30 km in 7 hours
   How far in 3 hours?

6. 62 km in 2.5 hours
   How far in 4 hours?

7. 96 km in 1.6 hours
   How long to go 60 km?

8. 168 km in 3.25 hours
   How long to go 240 km?

Solve. Round answers to the nearest tenth or cent.

9. Ms. Lowry paid \$6.76 for 15.4 L of gasoline. At that rate, how much would 20 L cost?

10. Ms. Lowry drove her car 48 km on 3.8 L of gasoline. At that rate, how much gasoline would be needed to drive 125 km?

11. A car traveled 268 km in 4.5 hours. At that rate, how long would it take to travel 380 km?

12. A toll road costs \$1.80 to travel 120 km. At that rate, how much would it cost to travel 200 km?

★ 13. The gasoline tank of a car was full when the odometer read 5938.2 km. At 6104.1 km it took 13.8 L to fill the tank. At that rate, how much gasoline would be needed to travel the next 200 km?

★ 14. It took the Andersons $2\frac{1}{4}$ hours to travel the first $\frac{1}{3}$ of their trip. At that rate, how long would it take to finish the trip?

15. LETTER PERFECT

What letters are missing?
Hint: Think of the positions of the letters in the alphabet.

$$\frac{B}{E} = \frac{H}{?} \qquad \frac{F}{I} = \frac{?}{R} \qquad \frac{P}{?} = \frac{D}{C}$$

Problem solving—scale drawing

A map is an example of a scale drawing. One centimeter on the map stands for 24 meters on the island.

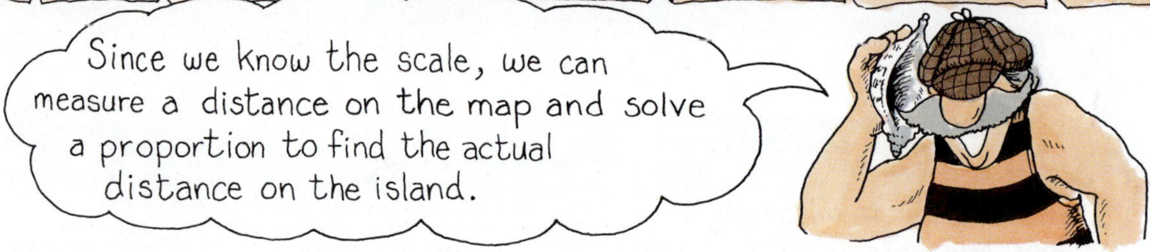

EXAMPLE.

The distance from Hook's Landing to the Swashbuckler Hotel on the map is 4.5 centimeters. What is the actual distance from the landing to the hotel?

cm on map → 1 and 4.5 ← cm on map

m on island → 24 and n ← m on island

$$\frac{1}{24} = \frac{4.5}{n}$$

$$n = 4.5 \times 24$$

$$n = 108$$

The actual distance from Hook's Landing to the hotel is 108 meters.

Solve.

1. On the map, measure the straight-line distance from Barado Point to the Swashbuckler Hotel. How many meters is it from the hotel to Barado Point?
2. What is the actual distance from Yorke Cove to the Swashbuckler Hotel?
3. What is the actual distance from Barado Point to Hook's Landing?
4. What is the actual distance from Yorke Cove to Hook's Landing?
5. What is the actual distance from tree A to tree I?
6. This treasure note was found in a bottle near Hook's Landing.

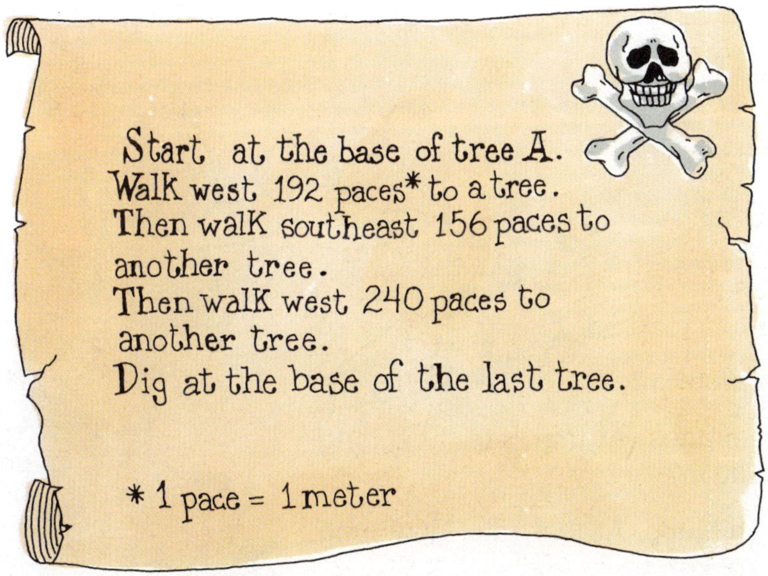

At the base of which tree would you dig to find the treasure?

KEEPING SKILLS SHARP

Estimate to find the incorrect answer.

1. a. $0.95 \times 36 = 34.2$

b. $8.6 \times 4.5 = 3.87$

Multiply.

2. 4.5×2.1

3. 0.58×5.2

4. 5.3×2.9

5. 6.7×0.83

6. 48×3.4

7. 2.86×8.1

8. 1.78×8.3

9. 8.96×1.42

Practice exercises

Solve.

1. $\frac{5}{2} = \frac{n}{4}$
2. $\frac{6}{n} = \frac{3}{5}$
3. $\frac{28}{12} = \frac{n}{3}$
4. $\frac{8}{n} = \frac{24}{18}$
5. $\frac{n}{6} = \frac{15}{30}$
6. $\frac{3}{4} = \frac{15}{n}$
7. $\frac{13}{8} = \frac{39}{n}$
8. $\frac{12}{42} = \frac{n}{7}$
9. $\frac{3}{5} = \frac{n}{6}$
10. $\frac{7}{9} = \frac{4}{n}$
11. $\frac{5}{2} = \frac{8}{n}$
12. $\frac{7}{9} = \frac{3}{n}$
13. $\frac{6}{4} = \frac{5}{n}$
14. $\frac{n}{12} = \frac{9}{5}$
15. $\frac{3}{8} = \frac{n}{2}$
16. $\frac{7}{16} = \frac{4}{n}$
17. $\frac{n}{8} = \frac{3}{7}$
18. $\frac{5}{2} = \frac{n}{9}$
19. $\frac{n}{6} = \frac{1}{8}$
20. $\frac{8}{5} = \frac{n}{9}$
21. $\frac{n}{\frac{1}{2}} = \frac{3}{4}$
22. $\frac{\frac{2}{3}}{4} = \frac{5}{n}$
23. $\frac{2\frac{1}{2}}{3} = \frac{n}{6}$
24. $\frac{4}{n} = \frac{1\frac{1}{2}}{2}$
25. $\frac{3\frac{2}{3}}{4} = \frac{7\frac{1}{3}}{n}$
26. $\frac{n}{2\frac{3}{4}} = \frac{4}{5}$
27. $\frac{1\frac{1}{4}}{2} = \frac{\frac{3}{4}}{n}$
28. $\frac{n}{\frac{5}{8}} = \frac{\frac{1}{5}}{3}$

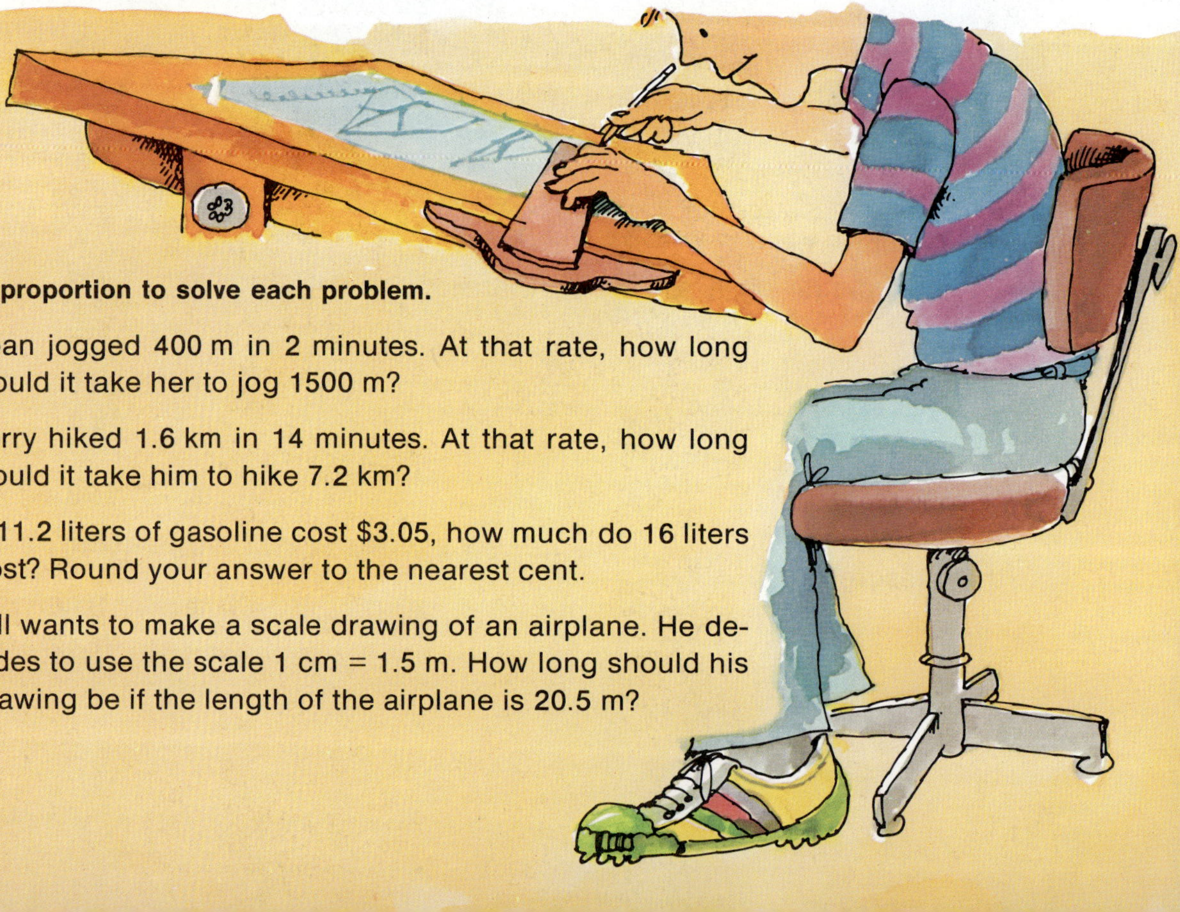

Use a proportion to solve each problem.

29. Joan jogged 400 m in 2 minutes. At that rate, how long would it take her to jog 1500 m?

30. Jerry hiked 1.6 km in 14 minutes. At that rate, how long would it take him to hike 7.2 km?

31. If 11.2 liters of gasoline cost \$3.05, how much do 16 liters cost? Round your answer to the nearest cent.

32. Bill wants to make a scale drawing of an airplane. He decides to use the scale 1 cm = 1.5 m. How long should his drawing be if the length of the airplane is 20.5 m?

Mathematics in careers

Concrete contractors construct such things as foundations, walls, walks, patios, driveways, and roads. Concrete is made from portland cement, sand, gravel, and water. Portland cement comes in a 94-pound bag, which holds 1 cubic foot of cement. The ratio of the ingredients depends on the use of the concrete.

EXERCISES

Use your completed table in exercises 2 and 3.

1. Copy and complete the following table.

Makes 1 Cubic Yard of Concrete			
Mix ratio cement : sand : gravel	Portland cement (ft^3)	Sand (ft^3)	$\frac{3}{4}$-inch gravel (ft^3)
$1:1:1\frac{3}{4}$	10	10	$17\frac{1}{2}$
$1:3:5$	$4\frac{1}{2}$		
$1:2\frac{1}{2}:3\frac{1}{2}$	6		

2. It is recommended that the $1:3:5$ mix be used for footings (the bases on which foundation walls rest). How many cubic feet of each material are needed for a footing having a volume of 3 cubic yards?

3. The $1:2\frac{1}{2}:3\frac{1}{2}$ mix can be used for driveways. How much of each material is needed for a driveway having a volume of 5 cubic yards?

★ 4. Here is a scale drawing of a patio:

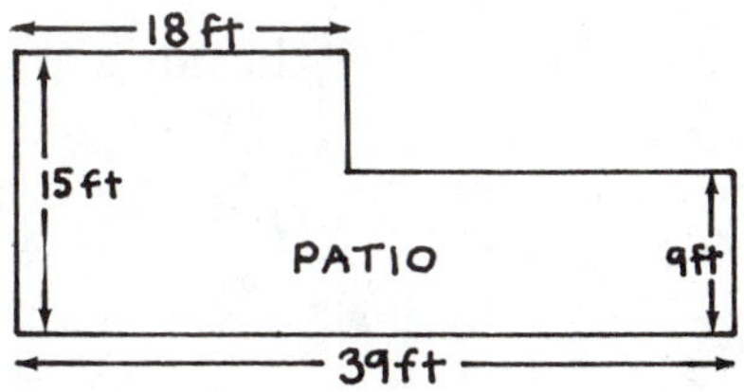

If concrete costs \$38 a cubic yard, how much will a 4-inch slab for this patio cost? Round the volume up to the nearest $\frac{1}{2}$ cubic yard.

Percent

Marvin Thompson's hobby is electronics. He invented Marvin's Marvelous Message Board. The board has 100 lights. He used 64 lights to say "Hi."

We can write:

He used 64 hundredths of the lights.

He used $\frac{64}{100}$ of the lights.

He used 0.64 of the lights.

We can also write:

He used 64% (64 percent) of the lights.

"Percent" means "hundredths."

EXERCISES

What percent of the lights on Marvin's board are used for each letter?

1.

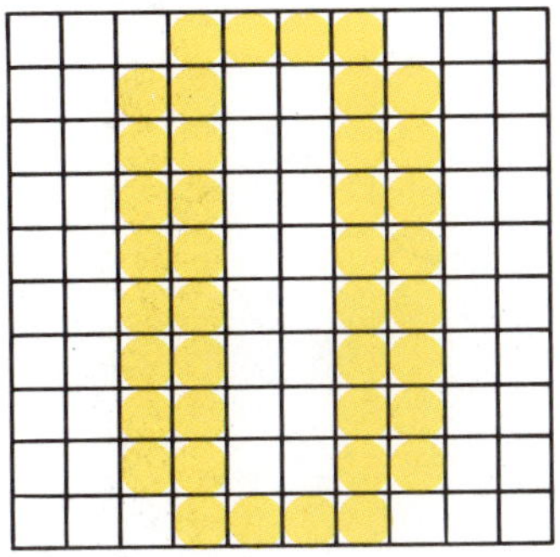

2.

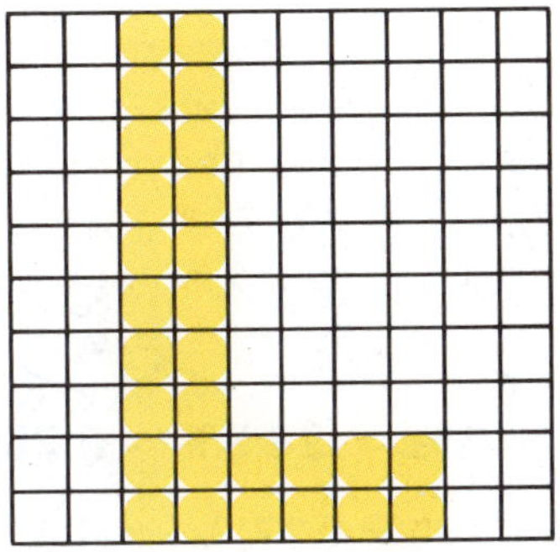

3.

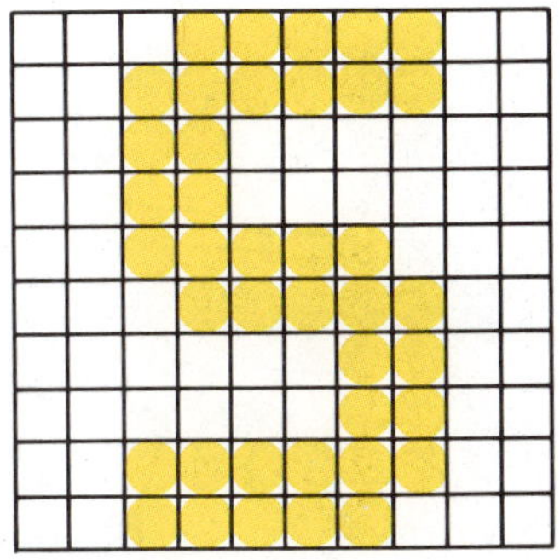

Copy and complete.

	4.	5.	6.	7.	8.	9.
Decimal	0.48	0.25	0.30	0.18	0.06	0.73
Fraction						
Percent						

Write each fraction as a percent. ***Hint:*** **First change to an equivalent fraction having a denominator of 100.**

10. $\frac{1}{2}$ ($\frac{50}{100}$) **11.** $\frac{1}{4}$ ($\frac{?}{100}$) **12.** $\frac{3}{4}$ ($\frac{?}{100}$) **13.** $\frac{2}{5}$ ($\frac{?}{100}$)

14. $\frac{1}{5}$ **15.** $\frac{3}{10}$ **16.** $\frac{4}{4}$ **17.** $\frac{3}{5}$ **18.** $\frac{9}{10}$ **19.** $\frac{4}{5}$

20. $\frac{3}{2}$ **21.** $\frac{5}{4}$ **22.** $\frac{5}{2}$ **23.** $\frac{7}{4}$ **24.** $\frac{1}{10}$ **25.** $\frac{1}{20}$

Write each percent as a fraction in lowest terms.

26. 10% ($\frac{10}{100}$) **27.** 20% **28.** 50% **29.** 60% **30.** 75% **31.** 80%

32. 5% **33.** 32% **34.** 48% **35.** 64% **36.** 25% **37.** 65%

More about percents

When you have to compute with percents, you must change them to fractions or decimals.

28.3% = .28.3% = 0.283

37% = .37.% = 0.37

100% = 1.00.% = 1.00

4% = 0.04.% = 0.04

0.52 = 0.52.% = 52%

0.314 = 0.31.4% = 31.4%

2 = 2.00.% = 200%

0.07 = 0.07.% = 7%

EXERCISES

Change to decimals.

1. 58%	**2.** 77%	**3.** 95%	**4.** 16%	**5.** 3%
6. 9%	**7.** 1%	**8.** 4.6%	**9.** 13.7%	**10.** 46.8%
11. 150%	**12.** 395%	**13.** 1000%	**14.** 186.3%	**15.** 0.4%
16. 0.6%	**17.** 3.8%	**18.** 75%	**19.** 0.32%	**20.** 0.51%

Change to percents.

21. 0.82 **22.** 0.75 **23.** 0.50 **24.** 0.4 **25.** 0.7

26. 3 **27.** 2 **28.** 3.6 **29.** 0.02 **30.** 0.07

31. 0.251 **32.** 0.384 **33.** 0.462 **34.** 0.8 **35.** 4

36. 0.001 **37.** 0.01 **38.** 0.1 **39.** 1 **40.** 0.173

Change to percents. ***Hint:*** **First change to decimals.**

EXAMPLE.

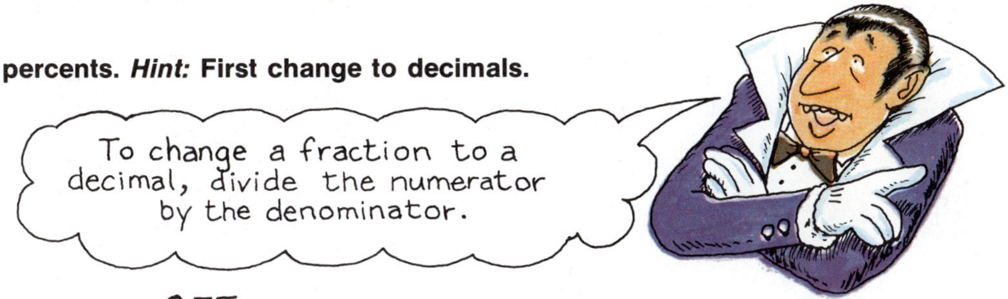

$\frac{3}{4} = ?$

$$\begin{array}{r} 0.75 \\ 4\overline{)3.00} \\ -2\,8 \\ \hline 20 \\ -20 \\ \hline 0 \end{array}$$

$\frac{3}{4} = 75\%$

41. $\frac{1}{2}$ **42.** $\frac{3}{2}$ **43.** $\frac{1}{4}$ **44.** $\frac{7}{8}$ **45.** $\frac{5}{4}$

46. $\frac{7}{4}$ **47.** $\frac{1}{5}$ **48.** $\frac{2}{5}$ **49.** $\frac{3}{5}$ **50.** $\frac{4}{5}$

51. $\frac{7}{5}$ **52.** $\frac{1}{8}$ **53.** $\frac{3}{8}$ **54.** $\frac{5}{8}$ **55.** $\frac{7}{8}$

Change each number to a percent.

56. More than $\frac{1}{2}$ of the babies born are boys.

57. About $\frac{3}{4}$ of the earth is covered by water.

58. About 0.55 of the students in one school are girls.

59. Jim scored 0.35 of the team's points.

60. Karen answered $\frac{3}{5}$ of the questions correctly.

61. Marcia is $\frac{4}{5}$ as tall as Mary.

62. Ken earned 0.7 as much as Carl.

63. Bill spent 0.65 of the money he earned.

64. During a sale the price was reduced by $\frac{1}{4}$.

65. There is a sales tax of $\frac{1}{20}$ of the price of a football.

Fractions and percents

The Valleyview JHS basketball team won 18 of its 24 games. The Mountainview JHS basketball team won 13 of its 24 games. What percent of its games did each team win?

Remember that you can change a fraction to a percent by first changing it to a decimal.

$\frac{18}{24} = ?\%$

$$\begin{array}{r} 0.75 \\ 24\overline{)18.00} \\ -16\,8 \\ \hline 1\,20 \\ -1\,20 \\ \hline 0 \end{array}$$

$\frac{18}{24} = 75\%$

The Valleyview team won 75% of its games.

$\frac{13}{24} = ?\%$

$$\begin{array}{r} 0.54\frac{1}{6} \\ 24\overline{)13.00} \\ -12\,0 \\ \hline 1\,00 \\ -96 \\ \hline 4 \end{array}$$

$\frac{13}{24} = 54\frac{1}{6}\%$

The Mountainview team won $54\frac{1}{6}\%$ of its games.

EXERCISES

Change each fraction to a percent.

1. $\frac{1}{3}$	**2.** $\frac{1}{8}$	**3.** $\frac{1}{6}$	**4.** $\frac{2}{3}$	**5.** $\frac{1}{9}$	**6.** $\frac{5}{8}$
7. $\frac{4}{9}$	**8.** $\frac{1}{16}$	**9.** $\frac{7}{8}$	**10.** $\frac{5}{16}$	**11.** $\frac{5}{3}$	**12.** $\frac{2}{9}$
13. $\frac{3}{8}$	**14.** $\frac{4}{3}$	**15.** $\frac{15}{8}$	**16.** $\frac{5}{6}$	**17.** $\frac{7}{6}$	**18.** $\frac{9}{8}$

Solve.

19. 20 = __?__ % of 40	**20.** 7 = __?__ % of 28	**21.** 9 = __?__ % of 27
22. 11 = __?__ % of 53	**23.** 12 = __?__ % of 8	**24.** 50 = __?__ % of 86

Percent Equivalents		
Fraction	Decimal	Percent
* $\frac{1}{2}$	0.50	50%
* $\frac{1}{3}$	$0.33\frac{1}{3}$	$33\frac{1}{3}$%
* $\frac{1}{4}$	0.25	25%
* $\frac{1}{5}$	0.20	20%
$\frac{1}{6}$	$0.16\frac{2}{3}$	$16\frac{2}{3}$%
$\frac{1}{7}$	$0.14\frac{2}{7}$	$14\frac{2}{7}$%
* $\frac{1}{8}$	$0.12\frac{1}{2}$	$12\frac{1}{2}$%
$\frac{1}{9}$	$0.11\frac{1}{9}$	$11\frac{1}{9}$%
* $\frac{1}{10}$	0.10	10%

Use the chart to write these percents.

25. $\frac{3}{4}$ = ___?___ %

$\frac{1}{4} = 25\%$, so $\frac{3}{4} = 75\%$

26. $\frac{3}{5}$ = ___?___ %

27. $\frac{2}{7}$ = ___?___ %

28. $\frac{3}{8}$ = ___?___ %

29. $\frac{4}{5}$ = ___?___ %

30. $\frac{5}{6}$ = ___?___ %

31. $\frac{2}{3}$ = ___?___ %

32. $\frac{2}{9}$ = ___?___ %

33. $\frac{5}{8}$ = ___?___ %

Solve.

34.

WEAVER JUNIOR HIGH SCHOOL

Starting Line-up	Shots Made	Shots Attempted
Jones	40	96
Casey	36	80
Allen	36	72
Davis	50	112
Garcia	34	64

a. What percent of his shots did Jones make?

b. What percent of his shots did Garcia miss?

c. How many players made less than 50% of their shots?

d. Who made the higher percent of shots, Jones or Davis?

Finding a percent of a number

Most of the percent problems that you will have to solve involve finding a percent of a number. Here is an example.

What is the sale price of the bicycle?

Equation: $75\% \text{ of } 160 = n$

Since you already know how to find a fraction of a number, you can solve the percent equation by solving a fraction equation.

$\frac{75}{100}$

$$75\% \text{ of } 160 = \frac{3}{4} \text{ of } 160$$
$$= \frac{3}{4} \times 160$$
$$= 120$$

So, the sale price is $120.

Here are some other examples of finding a percent of a number.

$\frac{50}{100}$

$$50\% \text{ of } 24 = \frac{1}{2} \text{ of } 24$$
$$= \frac{1}{2} \times 24$$
$$= 12$$

$\frac{125}{100}$

$$125\% \text{ of } 36 = \frac{5}{4} \text{ of } 36$$
$$= \frac{5}{4} \times 36$$
$$= 45$$

EXERCISES

Solve.

1. 75% of 40 = n
2. 25% of 56 = n
3. 50% of 38 = n
4. 40% of 45 = n
5. 80% of 35 = n
6. 10% of 40 = n
7. 60% of 35 = n
8. 150% of 48 = n
9. 100% of 36 = n
10. 175% of 80 = n
11. 5% of 120 = n
12. 30% of 100 = n
13. 90% of 150 = n
14. 125% of 64 = n
15. 20% of 60 = n

125% of 20 = n

Use the shortcut to solve these equations.

16. 125% of 36 = n
17. 125% of 48 = n
18. 150% of 18 = n
19. 150% of 30 = n
20. 175% of 24 = n
21. 175% of 36 = n
22. 120% of 50 = n
23. 120% of 80 = n
24. 160% of 45 = n
25. 160% of 75 = n
26. 105% of 80 = n
27. 105% of 100 = n

Solve.

28. A 3-speed bicycle is on sale for 80% of the regular price. The regular price is $96. What is the sale price?

29. In a survey of 120 students it was found that 85% owned a bicycle. How many students was that?

30. The price of a certain lightweight touring bicycle is $189 plus 5% sales tax. What is the total cost?

★ 31. Terry wants to trade in her old bicycle for a new 10-speed bicycle that sells for $169. The dealer will allow her $45 for her old bicycle. How much money will she need if she has to pay a 5% sales tax on the difference?

More about finding a percent of a number

In a survey of 120 students it was found that 20% have stereo sets. How many students have stereos?

Here are 3 ways to solve the problem.

(1) Multiply by an equivalent fraction.

$$\frac{20}{100} = \frac{1}{5} \qquad \frac{1}{\cancel{5}_{1}} \times \cancel{120}^{24} = 24$$

20% of 120 = 24

(2) Multiply by an equivalent decimal.

$$20\% = 0.20 \qquad \begin{array}{r} 120 \\ \times 0.20 \\ \hline 24.00 \end{array}$$

20% of 120 = 24

(3) Solve a proportion.

$$\frac{20}{100} = \frac{n}{120}$$

students with stereos (n)
students in survey (120)

$$100n = 2400$$
$$n = 24$$

20% of 120 = 24

So, 24 students have stereos.

EXERCISES

Solve by multiplying by an equivalent fraction.

1. 25% of 32 = n **2.** 75% of 36 = n **3.** 50% of 60 = n

4. 40% of 40 = n **5.** 80% of 35 = n **6.** 60% of 150 = n

7. 150% of 60 = n **8.** 10% of 80 = n **9.** 120% of 75 = n

Solve by multiplying by an equivalent decimal.

10. 18% of 25 = n

11. 16% of 38 = n

12. 22% of 19 = n

13. 25% of 56 = n

14. 36% of 74 = n

15. 17% of 58 = n

16. 8% of 83 = n

17. 3% of 65 = n

18. 2% of 75 = n

19. 6% of 142 = n

20. 9% of 108 = n

21. 7% of 256 = n

22. 118% of 325 = n

23. 126% of 378 = n

Solve by solving a proportion.

24. 8% of 51 = n

25. 6% of 63 = n

26. 3% of 54 = n

27. 15% of 29 = n

28. 29% of 78 = n

29. 35% of 125 = n

30. $37\frac{1}{2}$% of 77 = n

31. $42\frac{1}{2}$% of 61 = n

32. $18\frac{1}{4}$% of 59 = n

33. $8\frac{1}{3}$% of 63 = n

34. $6\frac{3}{4}$% of 75 = n

35. $9\frac{1}{2}$% of 95 = n

In another survey, 160 students were asked to estimate how many minutes per day they spent listening to their favorite radio station. The table shows the results of the survey.

TIME SPENT LISTENING TO FAVORITE STATION	
Time in Minutes	**Percent of Students Surveyed**
0	$12\frac{1}{2}$
1–30	$6\frac{1}{4}$
31–60	$37\frac{1}{2}$
61–90	10
91–120	15
121–150	$7\frac{1}{2}$
151–180	$6\frac{1}{4}$
181–210	5

How many students listened

36. between 61 and 90 minutes?

37. more than 3 hours?

38. between half an hour and an hour?

★ **39.** more than 30 minutes?

Finding the number when a percent is known

We can solve a proportion to find the number when a percent of the number is known.

Seventy-five percent of Felicia's records are jazz. If she has 15 jazz records, how many records does she have in all?

Equation.

Percent		Number		Percent of Number
↓		↓		↓
75%	of	n	=	15

Solution.

$$\begin{matrix}\text{Part} \rightarrow \\ \text{Whole} \rightarrow\end{matrix} \frac{75}{100} = \frac{15}{n} \begin{matrix}\leftarrow \text{Part} \\ \leftarrow \text{Whole}\end{matrix}$$

$$75 \times n = 1500$$

$$n = 20$$

Felicia has 20 records in all.

EXERCISES

Solve by using a proportion.

$\frac{25}{100} = \frac{8}{n}$

1. 25% of n = 8
2. 20% of n = 16
3. 30% of n = 24
4. 75% of n = 36
5. 60% of n = 9
6. 25% of n = 63
7. 30% of n = 42
8. 75% of n = 96
9. 15% of n = 200
10. $16\frac{2}{3}$% of n = 45
11. $37\frac{1}{2}$% of n = 12
12. $62\frac{1}{2}$% of n = 45
13. $87\frac{1}{2}$% of n = 63
14. $12\frac{1}{2}$% of n = 15
15. $6\frac{1}{4}$% of n = 28
16. $37\frac{1}{2}$% of n = 33
17. 175% of n = 49
18. 300% of n = 81

Solve.

19. Beth bought a record on sale for 75% of the regular price. She paid \$3.60 for the record. What was the regular price?

20. One week Lynn spent 60% of her allowance for a new record. The record cost \$3.60. How much was her allowance?

21. John paid \$24 for new earphones. They were on sale for 20% off the regular price.

a. What percent of the regular price did he pay?

b. What was the regular price?

22. George can buy a turntable by making a 12% down payment. If the down payment is \$9.36, what is the full price?

23. Eighteen of Jill's 45 records are rock and roll. What percent are rock and roll and what percent are not?

24. Sally has 72 records. If 25% of her records are Broadway musicals, how many records is that?

★ **25.** Two speakers were on sale for \$75. The regular price was \$90. By what percent were the speakers reduced?

KEEPING SKILLS SHARP

Estimate to find the incorrect answer.

1. a. $\begin{array}{r} 2.65 \\ 1.2\overline{)3.180} \end{array}$

b. $\begin{array}{r} 6.28 \\ 2.4\overline{)1.5072} \end{array}$

Divide. Round each quotient to the nearest hundredth.

2. $0.31\overline{)0.9382}$

3. $0.43\overline{)0.3916}$

4. $0.83\overline{)0.1374}$

5. $0.57\overline{)3.056}$

6. $6.4\overline{)0.2973}$

7. $3.9\overline{)5.901}$

8. $7.8\overline{)2.643}$

9. $4.2\overline{)0.3967}$

10. $7.5\overline{)8.215}$

11. $8.2\overline{)37.19}$

Practice exercises

Complete this table.

	1.	2.	3.	4.	5.	6.	7.	8.
Decimal	0.31	0.17	0.89					0.09
Fraction				$\frac{1}{2}$	$\frac{3}{4}$		$\frac{1}{100}$	
Percent						3%		

Solve.

9. 25% of 24 = n

10. 50% of 38 = n

11. 75% of 60 = n

12. 40% of 55 = n

13. 60% of 65 = n

14. 10% of 150 = n

15. 5% of 62 = n

16. 125% of 38 = n

17. 150% of 82 = n

18. $6\frac{1}{4}$% of 96 = n

19. $16\frac{2}{3}$% of 72 = n

20. $37\frac{1}{2}$% of 45 = n

21. $8\frac{1}{3}$% of 72 = n

22. $12\frac{1}{2}$% of 112 = n

23. $62\frac{1}{2}$% of 270 = n

24. 40% of 110 = n

25. $66\frac{2}{3}$% of 45 = n

26. 75% of 135 = n

27. $87\frac{1}{2}$% of 156 = n

28. 45% of 90 = n

29. $12\frac{1}{2}$% of 124 = n

Solve.

30. 30% of n = 27

31. 25% of n = 32

32. 80% of n = 41

33. 75% of n = 48

34. 70% of n = 42

35. 120% of n = 96

36. 8% of n = 15

37. 100% of n = 135

38. 1% of n = 6

39. $66\frac{2}{3}$% of n = 34

40. $83\frac{1}{3}$% of n = 58

41. $87\frac{1}{2}$% of n = 135

42. $37\frac{1}{2}$% of n = 150

43. $33\frac{1}{3}$% of n = 216

44. $62\frac{1}{2}$% of n = 352

45. $33\frac{1}{3}$% of n = 124

46. 120% of n = 36

47. 43% of n = 19

48. $112\frac{1}{2}$% of n = 54

49. 110% of n = 22

50. 60% of n = 12

Problem solving

1. Sharon holds the school counting record. She counted by ones to 160 in one minute. Calvin's best effort was only 65% of the record. How far did Calvin count in one minute?

2. The old bubble gum blowing record was a bubble measuring 8 inches in diameter. The diameter of the bubble Todd blew was 120% of the old record. How large a bubble did Todd blow?

3. Brenda set a new record. She flipped a penny 12 times and it came up heads each time. The old record was $66\frac{2}{3}\%$ of the new record. How many consecutive heads came up on the old record?

4. Janice spun a nickel for 24 seconds. Her time was 75% of the record. What was the record time for spinning a nickel?

5. Cindy named 68% of the 50 state capitals. This was 16 less than the school record. What was the school record for naming state capitals?

★ 6. Sidney can read 450 words per minute. If he could read 12% faster, he would tie the school record. How many words per minute is the school record?

Problem solving

1. To raise money for a canoe camping trip, a club decided to have a car wash. During the first 2 hours they washed 5 cars. At that rate, how many cars could they wash in 7 hours?

2. During the day they washed 19 cars. They charged $1.85 a car. If they spent $4.39 on supplies, how much profit did they make?

3. The fourteen boys decided that they would need $5.25 each to go on the canoe trip. If they earned $30.76 washing cars, how much more did they need?

4. Fourteen boys went on the trip. They were charged $7.35 rent for each canoe. If not more than 3 boys could be in a canoe, what was each boy's share of the canoe rental? Round your answer to the nearest cent.

5. When planning the trip, the club decided that each boy could take 40% of his weight in supplies. Sam weighed 52 kilograms. How many kilograms of supplies could he take?

6. It was 17 kilometers to where they decided to camp. They canoed 5 kilometers before taking a break. What percent of the trip had they completed?

7. At the end of 3 hours they had canoed 7 kilometers. At that rate, how long would it take them to canoe 11 kilometers?

8. For their last meal they made chili. The recipe called for 2 cups of tomatoes and 3 cups of beans. They used $7\frac{1}{2}$ cups of tomatoes. How many cups of beans should they have used?

9. The boys took a different return route. After canoeing 6 kilometers, they had traveled $37\frac{1}{2}\%$ of the way. How many kilometers long was the return trip?

10. The first 9 kilometers of the return trip took 4 hours. At that rate, how long would it take them to return? [*Hint:* For length of return trip, see exercise 9.]

11. HOOK, LINE, AND SINKER

Bob paid Sue 25¢ for every fish she caught. Sue paid Bob 10¢ for every fish he caught. Bob caught twice as many fish as Sue, but Sue had 15¢ more than Bob. How many fish did Bob catch?

CHAPTER CHECKUP

Give the ratio. [pages 266–267]

1. triangles to circles
2. circles to squares
3. squares to circles
4. red figures to green figures
5. blue squares to red circles
6. red triangles to green figures

Solve each proportion. [pages 268–269, 276]

7. $\frac{3}{4} = \frac{n}{32}$ 8. $\frac{5}{9} = \frac{4}{n}$ 9. $\frac{n}{6} = \frac{3}{10}$ 10. $\frac{6}{6} = \frac{n}{7}$

11. $\frac{n}{3} = \frac{5}{8}$ 12. $\frac{5}{n} = \frac{3}{7}$ 13. $\frac{n}{4} = \frac{15}{23}$ 14. $\frac{7}{n} = \frac{29}{18}$

Solve by using a proportion. [pages 270–275]

15. Oranges cost $1.79 per dozen. How much do 20 oranges cost?

16. A passenger train traveled 215 km in $2\frac{1}{4}$ hours. At that rate, how far would it travel in 4 hours?

Change each fraction to a percent. [pages 278–283]

17. $\frac{1}{4}$ 18. $\frac{3}{4}$ 19. $\frac{5}{2}$ 20. $\frac{1}{3}$ 21. $\frac{3}{8}$ 22. $\frac{5}{16}$

Solve. [pages 284–290]

23. 25% of $72 = n$ 24. 40% of $60 = n$ 25. 8% of $75 = n$

26. $33\frac{1}{3}\%$ of $50 = n$ 27. $87\frac{1}{2}\%$ of $94 = n$ 28. $112\frac{1}{2}\%$ of $150 = n$

29. 50% of $n = 19$ 30. 18% of $n = 9$ 31. 72% of $n = 45$

32. 175% of $n = 25$ 33. $83\frac{1}{3}\%$ of $n = 28$ 34. $16\frac{2}{3}\%$ of $n = 20$

Solve. [pages 291–293]

35. An ecology club collected a total of 8000 kg of newspaper. One team collected $33\frac{1}{3}\%$ of the total. How many kg did that team collect?

36. Some hiking boots are on sale for 75% of the regular price. What is the regular price if the sale price is $28.35?

CHAPTER PROJECT

1. List the titles of the six songs that you think are the most popular. Order the titles from most popular to least popular.
2. Survey 20 or more students and have them pick their favorite song from your list. Be sure to keep a tally.
3. Compare the way you ordered the titles with your survey. For example, did your number 1 song get the greatest number of votes?
4. Use your survey and make a bar graph.

CHAPTER REVIEW

$$\frac{5}{3} = \frac{n}{8}$$
$$5 \times 8 = 3n$$
$$40 = 3n$$
$$13\frac{1}{3} = n$$

Solve each proportion.

1. $\frac{5}{9} = \frac{n}{27}$
2. $\frac{7}{2} = \frac{n}{5}$
3. $\frac{3}{11} = \frac{n}{8}$
4. $\frac{6}{5} = \frac{9}{n}$
5. $\frac{8}{4} = \frac{5}{n}$
6. $\frac{9}{6} = \frac{12}{n}$
7. $\frac{n}{4} = \frac{12}{5}$
8. $\frac{n}{9} = \frac{6}{5}$
9. $\frac{n}{7} = \frac{8}{3}$
10. $\frac{9}{n} = \frac{7}{14}$
11. $\frac{6}{n} = \frac{15}{3}$
12. $\frac{6}{n} = \frac{12}{5}$

$$\frac{2}{3} = n\%$$

$$\begin{array}{r} 0.66\frac{2}{3} \\ 3\overline{)2.00} \\ -1\,8 \\ \hline 20 \\ -18 \\ \hline 2 \end{array}$$

So, $\frac{2}{3} = 66\frac{2}{3}\%$

Change each fraction to a percent.

13. $\frac{1}{3}$
14. $\frac{3}{8}$
15. $\frac{7}{4}$
16. $\frac{2}{9}$
17. $\frac{5}{6}$
18. $\frac{9}{8}$

$$37\frac{1}{2}\% \text{ of } 60 = n$$
$$\frac{37\frac{1}{2}}{100} = \frac{n}{60}$$
$$37\frac{1}{2} \times 60 = 100n$$
$$2250 = 100n$$
$$22\frac{1}{2} = n$$

Solve.

19. 20% of $60 = n$
20. 25% of $48 = n$
21. 50% of $38 = n$
22. 75% of $36 = n$
23. 60% of $45 = n$
24. 150% of $72 = n$
25. $33\frac{1}{3}\%$ of $54 = n$
26. $37\frac{1}{2}\%$ of $72 = n$
27. $66\frac{2}{3}\%$ of $96 = n$
28. $12\frac{1}{2}\%$ of $18 = n$

$$33\frac{1}{3}\% \text{ of } n = 8$$
$$\frac{33\frac{1}{3}}{100} = \frac{8}{n}$$
$$33\frac{1}{3}n = 100 \times 8$$
$$33\frac{1}{3}n = 800$$
$$n = 24$$

Solve.

29. 25% of $n = 7$
30. 20% of $n = 12$
31. 40% of $n = 20$
32. 75% of $n = 36$
33. 125% of $n = 75$
34. 120% of $n = 72$
35. $37\frac{1}{2}\%$ of $n = 33$
36. $33\frac{1}{3}\%$ of $n = 23$
37. $62\frac{1}{2}\%$ of $n = 30$
38. $6\frac{1}{4}\%$ of $n = 15$

CHAPTER CHALLENGE

The painter wants to mix a special shade of green using this ratio:

3	:	5	:	2
yellow		blue		white

She has *2* liters of yellow paint. How much of the other colors should she use?

Solution.

yellow: $\frac{3}{5} = \frac{2}{b}$:blue

$3b = 10$

$b = 3\frac{1}{3}$

yellow: $\frac{3}{2} = \frac{2}{w}$:white

$3w = 4$

$w = 1\frac{1}{3}$

She should mix 2 liters of yellow, $3\frac{1}{3}$ liters of blue, and $1\frac{1}{3}$ liters of white.

Tell how much of each color should be used.

1. Ratio: 2 red : 3 white : 4 green

   Use 1 liter of red.

2. Ratio: 3 black : 4 red : 2 yellow

   Use 2 liters of black.

3. Ratio: 3 blue : 4 white : 5 green

   Use 4 liters of green.

4. Ratio: 2 brown : 3 orange : 4 yellow

   Use 4 liters of orange.

MAJOR CHECKUP
STANDARDIZED FORMAT

Choose the correct letter.

1. Six billion sixty million is
   - a. 6,060,000
   - b. 60,006,000
   - c. 660,000,000
   - d. none of these

2. Find the difference.

   62.8 − 3.54
   - a. 59.26
   - b. 2.74
   - c. 27.4
   - d. none of these

3. Multiply.

   $$\begin{array}{r} 3.26 \\ \times 1.5 \\ \hline \end{array}$$
   - a. 0.489
   - b. 4.89
   - c. 48.9
   - d. none of these

4. Find the average of 6.8, 3.5, 9.2, 7.45, and 6.05.
   - a. 6.0
   - b. 66
   - c. 6.6
   - d. none of these

5. The prime factorization of 24 is
   - a. 6×4
   - b. $2 \times 3 \times 4$
   - c. 1×24
   - d. none of these

6. Give the sum.

   $\frac{5}{9} + \frac{5}{6}$
   - a. $\frac{2}{3}$
   - b. $\frac{5}{27}$
   - c. $1\frac{7}{18}$
   - d. none of these

7. Give the difference.

   $\frac{9}{5} - \frac{2}{3}$
   - a. $3\frac{1}{2}$
   - b. $1\frac{2}{15}$
   - c. $\frac{7}{15}$
   - d. none of these

8. Give the product.

   $2\frac{1}{2} \times 4\frac{1}{4}$
   - a. $\frac{10}{17}$
   - b. $10\frac{5}{8}$
   - c. $1\frac{7}{10}$
   - d. none of these

9. $\frac{4}{5} = $ ___?___
   - a. 80%
   - b. 8%
   - c. 125%
   - d. none of these

10. 75% of 48 = ___?___
    - a. 60
    - b. 36
    - c. 3600
    - d. none of these

11. A wall is 4.2 m long and 1.8 m high. Which estimate is closest to the area of the wall?
    - a. 6 m^2
    - b. 7 m^2
    - c. 8 m^2
    - d. 12 m^2

12. 1 cm on a map stands for 30 m. What length does 6 cm on the map stand for?
    - a. 5 m
    - b. 5 cm
    - c. 180 m
    - d. none of these

Consumer Mathematics

10

READY OR NOT?

Solve.

1. $\frac{3}{4}$ of \$80 = n
2. $\frac{2}{3}$ of \$48 = n
3. $\frac{5}{6}$ of \$84 = n
4. 25% of \$16 = n
5. 60% of \$25 = n
6. $12\frac{1}{2}$% of \$36 = n

Solve.

7. $\frac{1}{3}$ of n = \$13
8. $\frac{3}{8}$ of n = \$24
9. $\frac{4}{5}$ of n = \$56
10. 20% of n = \$64
11. $33\frac{1}{3}$% of n = \$35
12. $137\frac{1}{2}$% of n = \$121

Earning money

The examples show some different ways in which people are paid.

Joan baby-sits to earn money. She is paid \$1.50 an hour.

David sells peanuts at a ball park. He is paid a commission of 10% of his sales.

Debbie is paid 35¢ a week for each subscriber on her paper route.

EXERCISES

Joan is paid $1.50 an hour. This chart shows Joan's record of her hours worked and the pay received from 6 baby-sitting jobs.

Study the example. Then copy and complete the chart.

	From	*To*	*Number of Hours*	*Charge*
1.	8:00 P.M.	10:30 P.M.	$2\frac{1}{2}$	$3.750
2.	6:30 P.M.	9:30 P.M.		
3.	7:00 P.M.	9:30 P.M.		
4.	8:15 P.M.	9:30 P.M.		
5.	5:15 P.M.	8:45 P.M.		
6.	10:45 A.M.	2:15 P.M.		

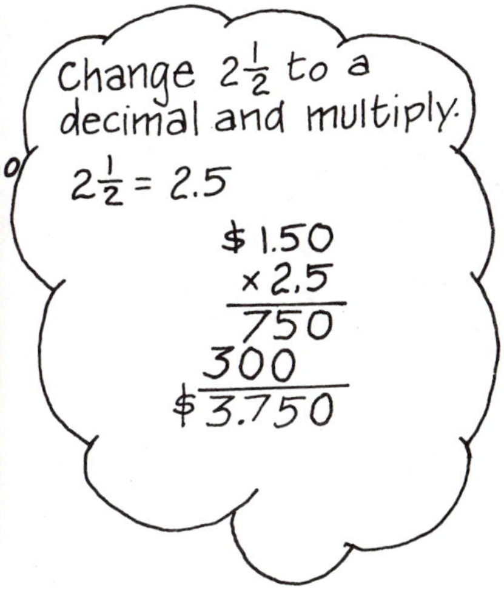

David is paid a commission of 10% of his sales. Find David's commission on the following sales.

7. David sold $120 worth of peanuts during one game last week?

8. David sold 264 bags of peanuts for 60¢ a bag during a Sunday double-header.

Debbie is paid 35¢ a week for each subscriber on her paper route. Find Debbie's pay.

9. How much would she earn if she had 88 subscribers?

10. If she increased her route to 104 subscribers, how much would she earn in 12 weeks?

11. **UPS AND DOWNS**

The price of a $200 bicycle is reduced 10% and later raised 10%. What is the final price? (The answer is not $200.) Answer the same question if the price is raised 10% and later reduced 10%.

Budgets

A **budget** is a plan for using your money.
Here is Jeremy's budget:

Here are some reasons why he has a budget:

(1) To spread out spending so that he won't run out of money between paychecks.

(2) To be sure that he puts some money into savings each payday.

(3) To control "impulsive" spending so that all necessities can be taken care of.

EXERCISES

In exercises 1–6, use the budget shown at the top of the page.

Hint: $\frac{5}{15} = 15\overline{)5.00} = ?\%$ Write the quotient as a percent.

1. What percent of his earnings did Jeremy save?
2. What percent of his earnings did he spend for lunches?
3. What percent did he spend for movies?
4. What percent did he spend for food?
5. One week Jeremy saw two movies that cost $3.50 each. The extra money for movies came from his savings. What percent did he save that week?
6. One week he earned an extra $5. He spent $3 of that and saved the rest. What percent of his total earnings did he save?

Remember that you can find a fraction or percent of a quantity by multiplying. For example, to find the amount budgeted for utilities, you might use this method:

8% of \$24,000 = ?

↑ equivalent ⟶ 0.08

$$\begin{array}{r} \$24{,}000 \\ \times\, 0.08 \\ \hline \$1920.00 \end{array}$$

Refer to the graph of the family budget to answer the questions below.

7. How many dollars did the family budget for housing?

8. How many dollars did the family budget for savings?

9. What fraction of the total budget is for food and housing? Give answer in lowest terms.

10. How much more money was budgeted for clothing than for life insurance?

11. How much money is budgeted for both recreation and automobile?

12. What is the total percent budgeted for clothing, housing, and food?

13. If life insurance is considered as savings, how much money is budgeted to be saved?

14. The monthly car payment is \$84.96. How much budgeted money a month is left for other car expenses?

15. Keep track of all your expenses for one week. Make a record of all these expenses.

16. Make a budget in the form of a circle graph to show how you would like to use your money.

Buying wisely

Suppose that it is your job to buy the food supplies for a camping trip. Before buying an item, you should ask yourself such questions as:

(1) Do we really need the item?
(2) Is the price within our budget?
(3) If two similar items are available, are they of the same quality?
(4) If two similar items are available, is one a better buy than the other?

If you answer yes to questions (1) through (3), then you could answer question (4) by computing the **unit price** of each item and comparing the unit prices.

To compute the unit price of the peanut butter, divide the price in cents by the weight in grams.

Large Jar

```
             0.679
grams → 234)159.000 ← cents
           -1404
             1860
            -1638
              2220
             -2106
               114
```

Small Jar

```
             0.741
grams → 120)89.000 ← cents
           -840
             500
            -480
              200
             -120
               80
```

Round each quotient to hundredths.

Large Jar	Small Jar
0.68 cent per gram	0.74 cent per gram

The unit price for the larger jar is less than the unit price for the smaller jar. So the larger jar is the better buy.

EXERCISES

In many supermarkets the unit prices are on display. Below are some examples. For each item give the amount, the price, and the unit price.

1.

UNIT PRICE	50 g	RETAIL PRICE
1.48		$.74
CNTS/GRAM	GROUND ALMONDS	

2.

UNIT PRICE	450 g	RETAIL PRICE
1.47		$.66
DOLS/KG	SPAGHETTI SAUCE	

3.

UNIT PRICE	0.475 L	RETAIL PRICE
1.03		$.49
DOLS/LITER	CURRANT JELLY	

4.

UNIT PRICE	260 g	RETAIL PRICE
4.50		$1.17
DOLS/KG	TUNA	

Suppose the following items are on your shopping list. Compute the unit price of each and tell which is the better buy.

5. Ground beef
1 kg for $4.50
0.850 kg for $3.65

6. Pear halves
822 g for $1.09
454 g for 49¢

7. Pickles
822 g for $1.59
454 g for $1.19

★ **8.** Tomato soup
305 g for 28¢
454 g for 41¢

KEEPING SKILLS SHARP

Change to a percent.

1. $\frac{3}{4}$ **2.** $\frac{5}{2}$

3. $\frac{2}{5}$ **4.** $\frac{1}{3}$

5. $\frac{5}{8}$ **6.** $\frac{9}{16}$

7. 1 **8.** $\frac{5}{6}$

9. $\frac{3}{10}$ **10.** $\frac{3}{8}$

11. $\frac{6}{5}$ **12.** $\frac{9}{8}$

Solve. (*Hint:* Change the percent to a decimal and multiply.)

13. 20% of 60 = n

14. 35% of 28 = n

15. 42% of 74 = n

16. 85% of 95 = n

17. 62% of 153 = n

18. 34% of 276 = n

19. 72% of 72 = n

20. 6% of 100 = n

21. 100% of 83 = n

22. 130% of 28 = n

23. 1% of 57 = n

24. 8% of 20 = n

Computing prices

The Bates Junior High School Student Council raises money by selling school supplies.

The examples below show how Ramón computed prices for some sales.

Problem: What is the price of one pencil?
Pencils: 2 for 25¢
One pencil sold

Solution: Compute the exact price of 1 pencil and *round up* to the next cent.

$$2\overline{)25}\ \ 12\frac{1}{2}$$

The price is 13¢.

Problem: What is the price of 2 erasers?
Erasers: 3 for 59¢
2 erasers sold

Solution: Compute the exact price of 2 erasers and *round up* to the next cent.

$$\frac{3}{59} = \frac{2}{n}$$

$$3n = 118$$

$$n = 39\frac{1}{3}$$

The price is 40¢.

Problem: What is the price of 5 erasers?

Solution: Add the price of 3 erasers and the price of 2 erasers.

$59 + 40 = 99$ The price is 99¢.

EXERCISES

Refer to the prices on page 306 to compute these total prices.

1. 2 pencils and 1 large eraser
2. 1 pencil and 2 large erasers
3. 3 pencils
4. 1 protractor, 1 ruler, and 2 pencil erasers
5. 1 pencil, 1 compass, and 1 pencil eraser
6. 4 pencils and 5 pencil erasers

Solve.

7. One student buys 7 pencils and gives you a dollar. How much change should you give?
8. Another student buys 3 packages of notebook paper and 5 pencils. He gives you $5. How much change should you give?
9. Later, another student returned 1 compass and 2 pens. He then bought 4 packages of notebook paper and gave Ramón $2. How much change did Ramón give him?

★ 10. A student returned three different items and Ramón refunded her money. Ramón gave her 2 quarters, 2 dimes, and four pennies. What did the student return?

Discounts

OUTDOOR AMERICA, a sporting goods store, is having a sale.

The price of the pack frame was decreased by 20%. The *amount* a price is decreased is called the **discount.**

Here are 3 ways to find the sale price.

$$\frac{20}{100} = \frac{x}{\$28.60}$$
$$100x = \$572$$
$$x = \$5.72 \text{ (discount)}$$

$$\begin{array}{r} \$28.60 \\ -5.72 \\ \hline \$22.88 \end{array} \text{ (sale price)}$$

$$\begin{array}{r} \$28.60 \\ \times 0.20 \\ \hline \$5.7200 \end{array} \text{ (discount)}$$

$$\begin{array}{r} \$28.60 \\ -5.72 \\ \hline \$22.88 \end{array} \text{ (sale price)}$$

$$\begin{array}{r} \$28.60 \\ \times 0.80 \\ \hline \$22.8800 \end{array} \text{ (sale price)}$$

If the discount is 20%, then the sale price is 80% of the regular price.

The *percent* a price is decreased is called the **percent of discount.**

This pair of hiking boots was discounted for a sale.

To find the percent of discount, find what percent of the original price the discount was.

amount of discount → 6; original price → 48

$$\frac{6}{48} = \frac{n}{100}$$
$$48n = 600$$
$$n = 12\tfrac{1}{2}$$

The percent of discount was $12\frac{1}{2}\%$.

EXERCISES

The regular price and the percent of discount are given. Compute the discount.

1. tent: \$36.50
   discount: 25%
2. mess kit: \$12.50
   discount: 10%
3. pack: \$27.60
   discount: $33\frac{1}{3}\%$
4. sleeping bag: \$56.80
   discount: $12\frac{1}{2}\%$

The regular price and the percent of discount are given. Find the total cost, including a sales tax of 5%. Refer to the table for the sales tax.

5. flashlight: \$8.50
   discount: 10%
6. poncho: \$16.60
   discount: 25%
7. pack frame: \$29.50
   discount: 12%
8. stove: \$24.75
   discount: $33\frac{1}{3}\%$
9. cooler: \$17.94
   discount: $16\frac{2}{3}\%$
10. tools: \$21.95
    discount: 20%

Find the percent of discount.

11.

★ 12.

Sale	Tax	Sale	Tax
6.90– 7.09	.35	18.90–19.09	.95
7.10– 7.29	.36	19.10–19.29	.96
7.30– 7.49	.37	19.30–19.49	.97
7.50– 7.69	.38	19.50–19.69	.98
7.70– 7.89	.39	19.70–19.89	.99
7.90– 8.09	.40	19.90–20.09	1.00
8.10– 8.29	.41	20.10–20.29	1.01
8.30– 8.49	.42	20.30–20.49	1.02
8.50– 8.69	.43	20.50–20.69	1.03
8.70– 8.89	.44	20.70–20.89	1.04
8.90– 9.09	.45	20.90–21.09	1.05
9.10– 9.29	.46	21.10–21.29	1.06
9.30– 9.49	.47	21.30–21.49	1.07
9.50– 9.69	.48	21.50–21.69	1.08
9.70– 9.89	.49	21.70–21.89	1.09
9.90–10.09	.50	21.90–22.09	1.10
10.10–10.29	.51	22.10–22.29	1.11
10.30–10.49	.52	22.30–22.49	1.12
10.50–10.69	.53	22.50–22.69	1.13
10.70–10.89	.54	22.70–22.89	1.14
10.90–11.09	.55	22.90–23.09	1.15
11.10–11.29	.56	23.10–23.29	1.16
11.30–11.49	.57	23.30–23.49	1.17
11.50–11.69	.58	23.50–23.69	1.18
11.70–11.89	.59	23.70–23.89	1.19
11.90–12.09	.60	23.90–24.09	1.20
12.10–12.29	.61	24.10–24.29	1.21
12.30–12.49	.62	24.30–24.49	1.22
12.50–12.69	.63	24.50–24.69	1.23
12.70–12.89	.64	24.70–24.89	1.24
12.90–13.09	.65	24.90–25.09	1.25
13.10–13.29	.66	25.10–25.29	1.26
13.30–13.49	.67	25.30–25.49	1.27
13.50–13.69	.68	25.50–25.69	1.28
13.70–13.89	.69	25.70–25.89	1.29
13.90–14.09	.70	25.90–26.09	1.30
14.10–14.29	.71	26.10–26.29	1.31
14.30–14.49	.72	26.30–26.49	1.32
14.50–14.69	.73	26.50–26.69	1.33
14.70–14.89	.74	26.70–26.89	1.34
14.90–15.09	.75	26.90–27.09	1.35
15.10–15.29	.76	27.10–27.29	1.36
15.30–15.49	.77	27.30–27.49	1.37
15.50–15.69	.78	27.50–27.69	1.38
15.70–15.89	.79	27.70–27.89	1.39
15.90–16.09	.80	27.90–28.09	1.40
16.10–16.29	.81	28.10–28.29	1.41
16.30–16.49	.82	28.30–28.49	1.42
16.50–16.69	.83	28.50–28.69	1.43
16.70–16.89	.84	28.70–28.89	1.44
16.90–17.09	.85	28.90–29.09	1.45
17.10–17.29	.86	29.10–29.29	1.46
17.30–17.49	.87	29.30–29.49	1.47
17.50–17.69	.88	29.50–29.69	1.48
17.70–17.89	.89	29.70–29.89	1.49
17.90–18.09	.90	29.90–30.09	1.50
18.10–18.29	.91	30.10–30.29	1.51
18.30–18.49	.92	30.30–30.49	1.52
18.50–18.69	.93	30.50–30.69	1.53
18.70–18.89	.94	30.70–30.89	1.54

A checking account

People put their money into checking accounts in banks to keep it safe. Whenever they wish to use some of the money, they write a check. Checks are just notes that tell the bank to give some of the money in an account to a company or person.

This check tells the First National Bank of Hometown, USA, to pay Mary R. Cooper $18.25 from the checking account of Sarah Smith.

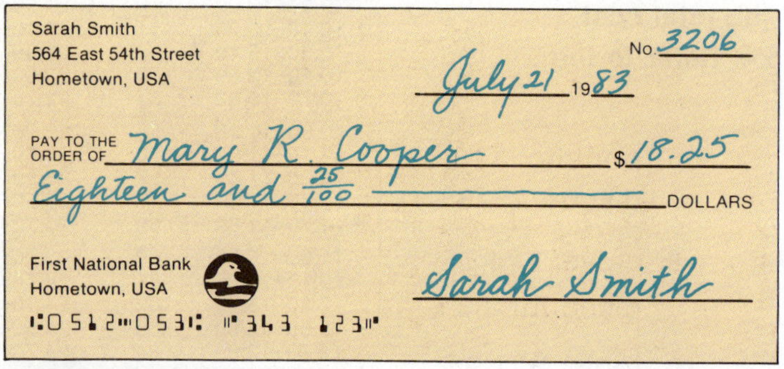

Mary R. Cooper will have to sign her name on the back of the check before she can cash the check. This is called **endorsing** the check. The endorsed check is proof that Sarah Smith paid the money to Mary R. Cooper.

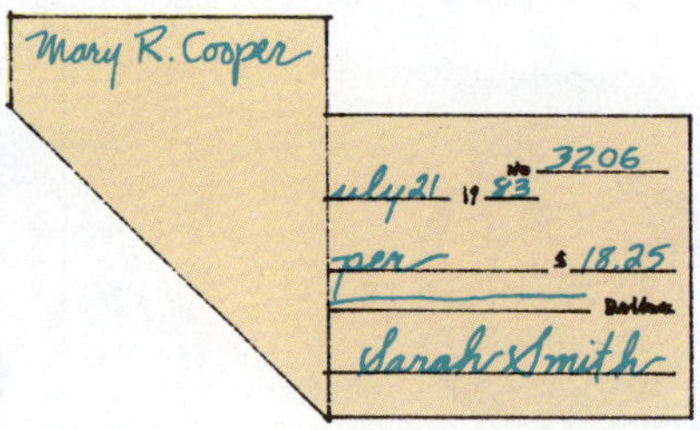

EXERCISES

1. Why do you think that a check should be written in ink?
2. What is the number of the check that was written to Mary Cooper?
3. Why do you think the amount is written in both numerals and words?

A checkbook has a section in which a record of the account is kept. Study this part of Sarah Smith's record:

CHECK NO.	DATE	PAY TO	AMOUNT	DEPOSIT	BALANCE	
					26	42
3204	July 11	LP Records	14.20		12	22
	July 15			50.00	62	22
3205	July 18	Allan's Bookstore	8.57		53	65
3206	July 21	Mary R. Cooper	18.25		35	40

Use the record to complete.

4. On what date was a $50 deposit made?
5. Check **3204** was written for what amount?
6. How much was the check written to Allan's Bookstore?
7. What was the balance in the account after check **3206** was written?
8. To whom was a check of $14.20 written?
9. ★ On July 23, Sarah deposited $65.00 and wrote a check for $32.46. What was her balance after this?

Write in words, as you would on a check.

10. $37.50 **11.** $42.75 **12.** $60.00 **13.** $135.45 **14.** $203.04

PROJECT

Most probably, your parents have a checking account. Ask them to help you write a short answer for each question.

1. What does "overdrawn" mean?
2. What is a "service charge"?
3. What is an "automatic payment"?
4. What does it mean to make a check out to "cash"?
5. What is a "cashier's check"?

Borrowing money

An expensive item such as a stereo, motorcycle, or automobile is usually bought with money borrowed from a bank, credit union, or loan company. The borrower pays **interest** (rent) for using the money. The amount of interest depends on the **principal** (the amount borrowed), the **rate** (percent) of interest charged, and the **time** for which the money is borrowed.

Mr. Kratz will have to sign a note saying he will pay back the $600 plus the interest in 1 year.

The amount of interest can be computed by using this formula:

$$\text{Interest} = \text{principal} \times \text{rate} \times \text{time}$$

$$I = prt$$

principal: $600	$I = prt$
rate: 18% per *year*	$I = \$600 \times \frac{18}{100} \times 1$
time: 1 *year*	$I = \$108$

The time units must be the same when using the formula.

The interest is $108.

He will repay $708 at the end of 1 year.

Important: Before you borrow, you should compare the rates of several banks.

EXERCISES

Compute the interest to the nearest cent.

	Interest	Principal	Interest rate per year	Time
1.		\$800	16%	1 year
2.		\$1000	14%	1 year
3.		\$2000	12%	$\frac{1}{2}$ year 6 months
4.		\$750	15%	9 months
5.		\$500	18%	8 months
6.		\$1250	$12\frac{1}{2}$%	1 year
7.		\$500	$10\frac{1}{2}$%	6 months
8.		\$850	$14\frac{1}{2}$%	8 months
★ 9.		\$1250	$12\frac{3}{4}$%	5 months
★ 10.		\$1875	$15\frac{1}{4}$%	2 months

Solve.

11. Jane borrowed \$1800 for 1 year to buy a car. If the yearly rate was $10\frac{1}{2}$%, how much interest would she owe at the end of the year?

12. **a.** A motorcycle is on sale for \$1350. A bank will loan 80% of the money. How much money is that?
    b. If the bank will loan the money at 11% for 1 year, how much interest will be charged?

★ 13. John saved \$120 to buy a stereo set. A stereo that generally sells for \$379.00 is on sale for 25% off.
    a. How much money will he have to borrow?
    b. If he borrows the money for 6 months at 12%, how much interest will he pay?

KEEPING SKILLS SHARP

Solve.

1. 25% of $n = 32$
2. 50% of $n = 20$
3. $33\frac{1}{3}$% of $n = 18$
4. $66\frac{2}{3}$% of $n = 14$
5. 15% of $n = 7.5$
6. 28% of $n = 42$
7. 16% of $n = 4$
8. $37\frac{1}{2}$% of $n = 27$
9. $62\frac{1}{2}$% of $n = 31$
10. 115% of $n = 69$
11. $112\frac{1}{2}$% of $n = 68.4$
12. 120% of $n = 105$
13. 40% of $n = 100$
14. 80% of $n = 96$
15. 75% of $n = 61.5$
16. 18% of $n = 36$
17. 4% of $n = 0.992$
18. 1% of $n = 6.3$
19. 100% of $n = 191$
20. $87\frac{1}{2}$% of $n = 54.6$

Installment buying

Suppose you wanted to buy this television set but did not have $180. You could buy it on the **installment plan.**

This means that you could put $30 down and sign an agreement to pay $14 a month for 12 payments. Then you could take the television home.

(Until you are older and have "established credit," you will probably have to have someone sign the agreement with you. Such a person is called a **cosigner.**

When you buy an item on an installment plan, you usually have to pay more than when you pay cash.

Cash	*Installment Plan*
$180	$\begin{array}{r} \$14 \\ \times 12 \\ \hline 28 \\ 14 \\ \hline \$168 \\ +30 \\ \hline \$198 \end{array}$

The extra $18 that you are charged on the installment plan is interest that you must pay because you borrowed from the television store.

Before buying on an installment plan, check to see if you could save money by borrowing the money from a bank or credit union.

EXERCISES

How much more does the installment plan cost than paying cash?

1.

Cash: $68
Installment plan:
$20 down
$4.48 per month for 12 months

2.

Cash: $156
Installment plan:
$30 down
$12.06 per month for 12 months

3.

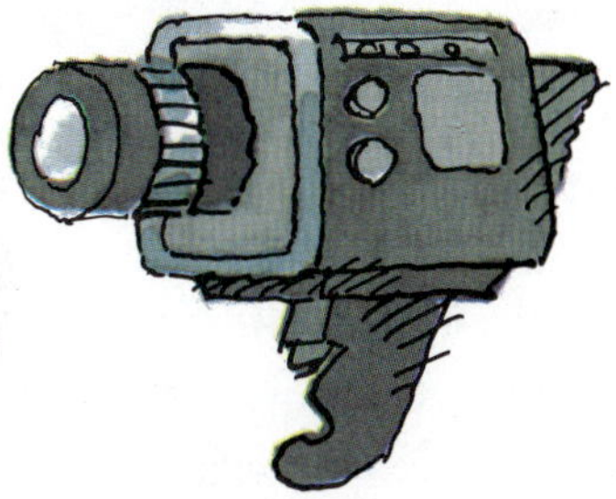

Cash: $150
Installment plan:
$35 down
$20.60 per month for 6 months

4.

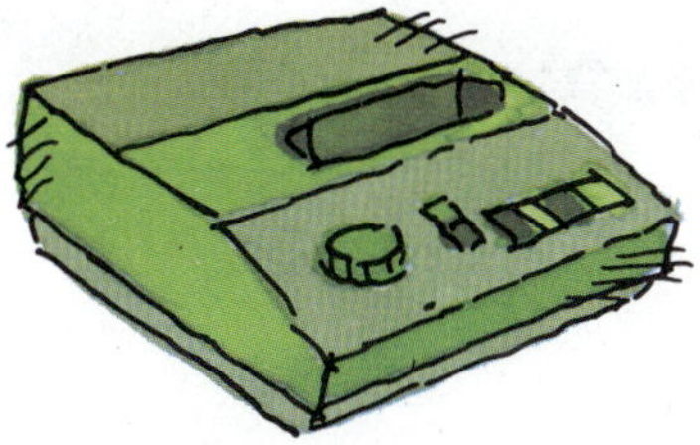

Cash: $72
Installment plan:
$15 down
$4.37 per month for 15 months

5.

Cash: $248
Installment plan:
$50 down
$14.36 per month for 16 months

6.

Cash: $159
Installment plan:
no down payment
$19.25 per month for 9 months

Solve.

7. Suppose that it costs $14.88 interest to finance a $120 purchase for 1 year on an installment plan. How much could be saved by borrowing the money at a bank for $10\frac{1}{2}\%$?

8. Suppose that a merchant offers to finance $460 for 18 months at $30.35 per month. You can also borrow the money for $10\frac{1}{2}\%$ at a bank. Which would cost less? How much less?

Mathematics in careers

Besides selling gasoline, most gasoline station operators also sell such items as oil, tires, and batteries. They also provide such services as oil changes, wheel balancing, minor repairs, and tune-ups.

In a certain state, gasoline stations charge a 5% sales tax on the sale of all items except gasoline. The sales tax is not charged on labor.

EXERCISES

Compute the total. For the sales tax, refer to the table on page 309.

1.

CRAIG'S SERVICE STATION			
Labor			
Tune-up	$24.50	Gasoline	$15.20
		5L oil	
		($2.19/L)	
		oil filter	$8.15
		Sales Tax	
Labor Total		→	
		Total	

2.

CRAIG'S SERVICE STATION			
Labor			
Balance	$12.00	Gasoline	$13.45
Wheels		4 L Oil	
		($2.19/L)	
		Sales Tax	
Labor Total		→	
		Total	

3.

CRAIG'S SERVICE STATION			
Labor			
Lubrication	$5.95	Gasoline	$14.95
		Oil Filter	$7.85
		Air Filter	$6.95
		4L oil	
		($1.79/L)	
		Sales Tax	
Labor Total		→	
		Total	

Computers in stores

Some stores give sales receipts like the one shown. Stores can use the Universal Product Code to *automatically* list the name of each item and its price on the sales receipt.

The Universal Product Code (UPC) is explained on page 71.

BROCCOLI SPRS	.61
COTTAGE CHZ	1.15
MARGARINE	1.07
TOTAL	2.83
CASH TEND	5.00
2.17 CHANGE	

EXERCISES

Use the simplified bar code on page 71 to determine the number for each of the six items. Then copy and complete the table below.

BROCCOLI .61

CEREAL 1.45

MILK .93

MARGARINE 1.07

COTTAGE CHEESE 1.15

COOKIES 1.39

	Number	*Name of Item*	*Quantity*	*Price of Item*	*Cost*
1.	416		2		
2.	502		4		
3.	487		1		
4.	509		3		
5.	538		4		
6.	413		2		

CHAPTER CHECKUP

Solve. [pages 300–301, 310–311]

1. John Roberto worked 3 hours for the amount shown on the check. How much did he earn per hour?

2. The balance of the account before the check was written was $142.13. What was the balance after the check was written?

3. Who wrote the check?

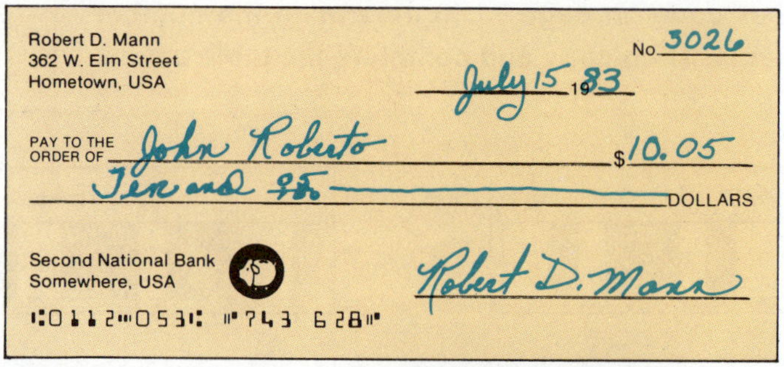

[pages 306–307]

Find the price of

4. 1 book.

5. 2 books.

6. 5 books.

Solve. [pages 308–309]

7. A $112 bicycle was put on sale at a discount of 20%. What was the sale price?

8. A $48 camera was put on sale for $36. What was the percent of discount?

Solve. [pages 312–315]

9. If you borrowed $85 for 1 year and paid 12% interest, how much interest would you pay?

10. A $156 canoe can be bought for $30 down and $12.39 a month for 12 months. Find the amount of interest.

CHAPTER PROJECT

1. From a classified ad in a newspaper, compute the average prices of 1-year-old Chevrolets, 2-year-old Chevrolets, 3-year-old Chevrolets, and 4-year-old Chevrolets.

2. Show your findings on a bar graph like this:

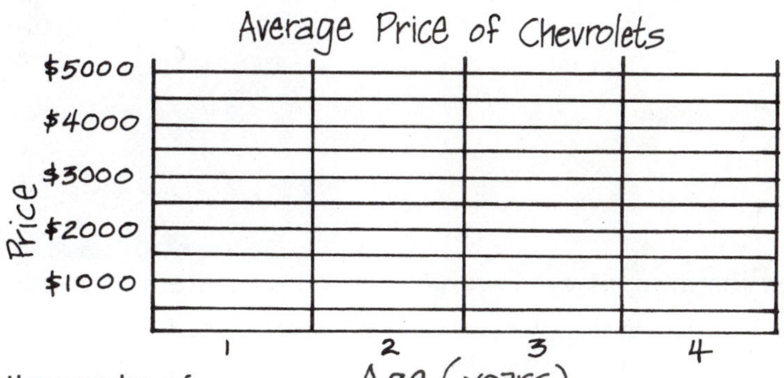

3. Repeat the project for another make of automobile. Compare your results.

CHAPTER REVIEW

To find the price of 1 or 2 balls, compute the exact price and round up to the next cent.

Find the price of

1. 1 tennis ball.
2. 2 tennis balls.
3. 5 tennis balls.

The discount is 20% of \$18.50. The sale price is \$18.50 minus the discount.

4. How much is the discount?
5. What is the sale price?

$I = prt$

Interest = principal × rate × time

Find the simple interest for a loan of

6. \$180 at 12% for 1 year.
7. \$640 at 16% for 6 months.

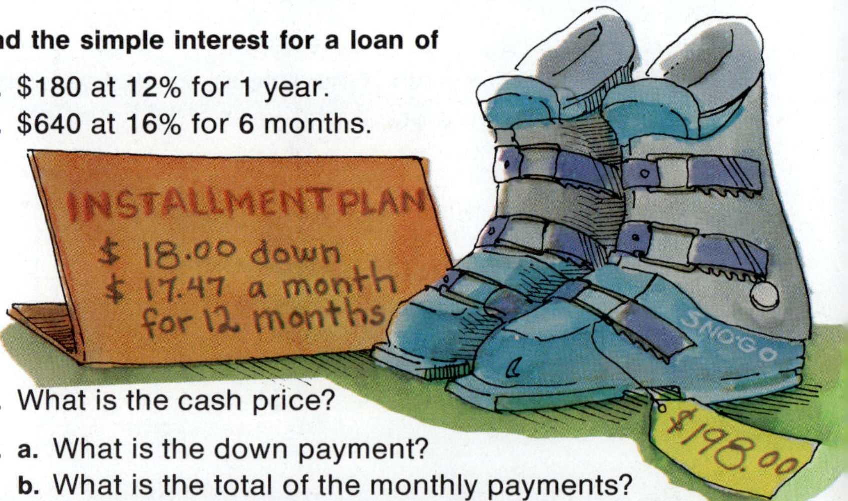

The installment cost is the down payment plus the total of the monthly payments.

8. What is the cash price?
9. a. What is the down payment?
   b. What is the total of the monthly payments?
10. How much more does the installment plan cost?

CHAPTER CHALLENGE

Perhaps you have a savings account. If you do, you probably know that the interest your account earns is added to your account several times during the year.

Study this example.

On January 1, \$100 is deposited in a savings account that pays 6% interest per year.

At the end of 3 months $\left(\frac{1}{4}\text{ year}\right)$ the interest is added to the account.

$I = p \times r \times t$

$I = \$100 \times \frac{6}{100} \times \frac{1}{4}$

$I = \$1.50$

The new balance is \$101.50.

At the end of the next 3 months, the interest is figured on \$101.50 and added to the account.

$I = p \times r \times t$

$I = \$101.50 \times \frac{6}{100} \times \frac{1}{4}$

$I = \$1.52$ (rounded to nearest cent)

The new balance is \$103.02.

Notice that "interest on interest" is being paid. This is called **compounding** the interest.

Assume that you deposit \$100 in a savings account that pays 6% interest compounded every 3 months. Find how much money you would have in your account after 2 years.

MAJOR CHECKUP
STANDARDIZED FORMAT

Choose the correct letter.

1. Round 52.6396 to the nearest thousandth.

a. 52.640
b. 52.639
c. 52.649
d. none of these

2. Divide.

$0.79\overline{)21.646}$

a. 2.74
b. 274.0
c. 27.4
d. none of these

3. Complete.

5.38 kg = ___?___ g

a. 0.538
b. 53.8
c. 538
d. none of these

4. Substitute and simplify.

g	T
7	3

$6g - T$

a. 24
b. 39
c. 64
d. none of these

5. Add.

$$9\frac{2}{3} + 8\frac{3}{8}$$

a. $17\frac{1}{24}$
b. $17\frac{7}{24}$
c. $18\frac{1}{24}$
d. none of these

6. Subtract.

$$5\frac{1}{3} - 3\frac{3}{4}$$

a. $7\frac{5}{12}$
b. $2\frac{7}{12}$
c. $1\frac{7}{12}$
d. none of these

7. Give the product.

$\frac{7}{3} \times \frac{3}{5}$

a. $1\frac{2}{5}$
b. $3\frac{8}{9}$
c. $\frac{9}{35}$
d. none of these

8. Find the area.

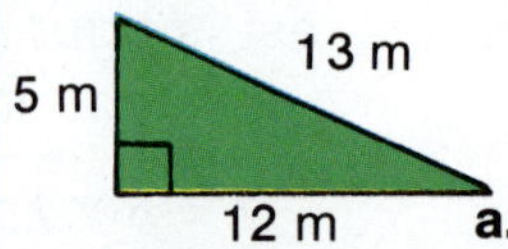

a. 30 sq m
b. 17 sq m
c. 60 sq m
d. none of these

9. Solve.

$\frac{n}{15} = \frac{13}{8}$

a. $24\frac{3}{8}$
b. $6\frac{14}{15}$
c. $9\frac{3}{13}$
d. none of these

10. 75% of $n = 24$

$n = \underline{?}$

a. 36
b. 18
c. 6
d. none of these

11. Pencils are 3 for 40¢. How much will 2 pencils cost?

a. 27¢ **b.** 26¢
c. 80¢ **d.** none of these

12. The TV program was 1 hour long. 9 minutes were used for ads. How many minutes were not used for ads?

a. 91 **b.** 51
c. 21 **d.** none of these

Geometry

11

READY OR NOT?

Match.

1. point — a.
2. segment — b.
3. ray — c.
4. line — d.
5. angle — e.

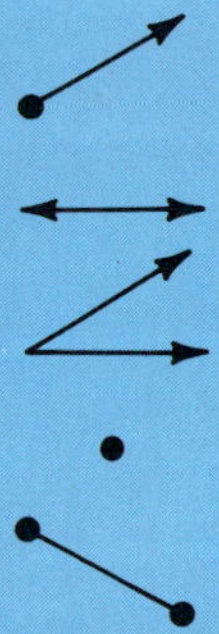

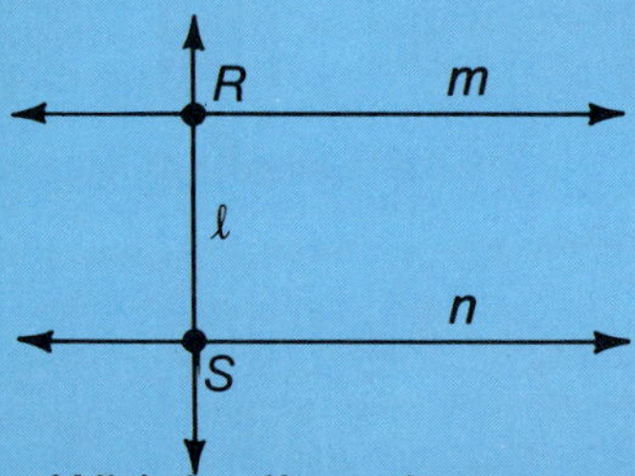

6. Which line intersects line ℓ at point *R*?
7. Which line ∥ (is parallel to) line *m*?
8. Which line ⊥ (is perpendicular to) line *m*?

Constructions

A **compass** and a **straightedge** are used to construct segments.

Follow these steps to copy a segment, $\overline{AB}$.

Step 1. Set the compass legs on *A* and *B*.

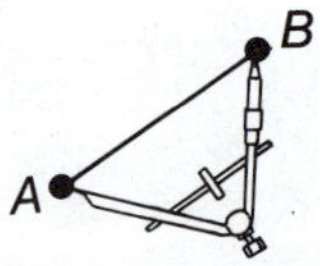

Step 2. Mark the paper with both the pencil and the point.

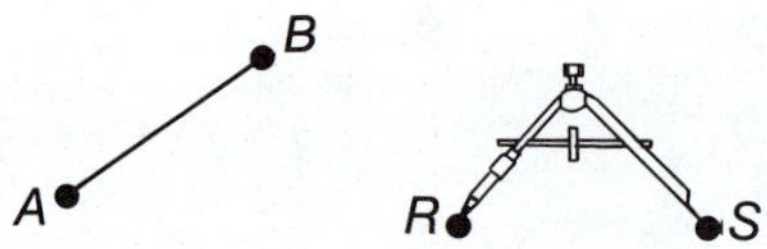

Step 3. Draw the segment.

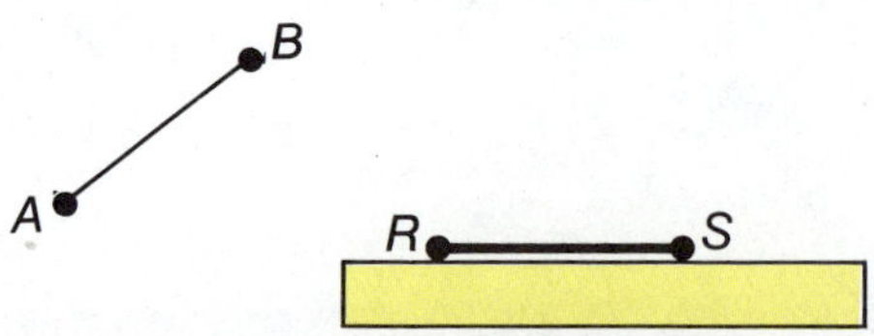

$\overline{AB}$ and $\overline{RS}$ are the same length.

Follow these steps to copy an angle, $\angle B$.

Step 1. Use your straightedge to draw $\overrightarrow{TR}$.

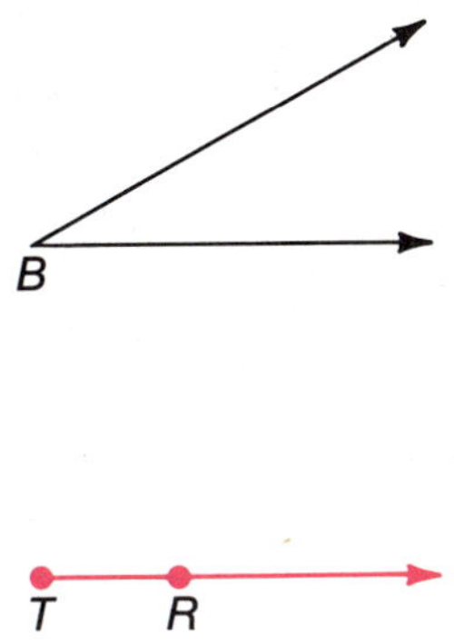

Step 2. Draw an arc that intersects the sides of $\angle B$.

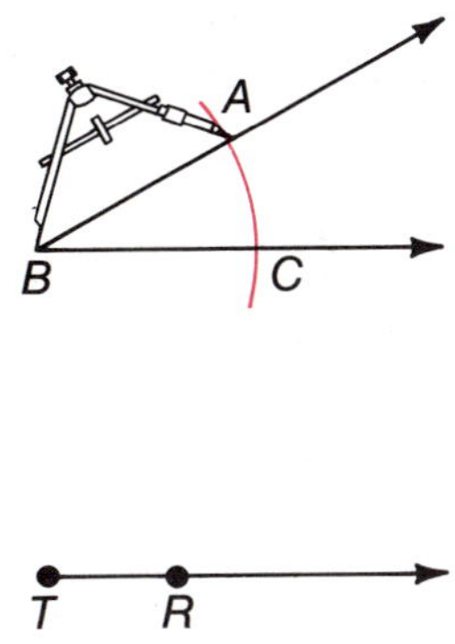

Step 3. Use the same compass setting to draw an arc as shown.

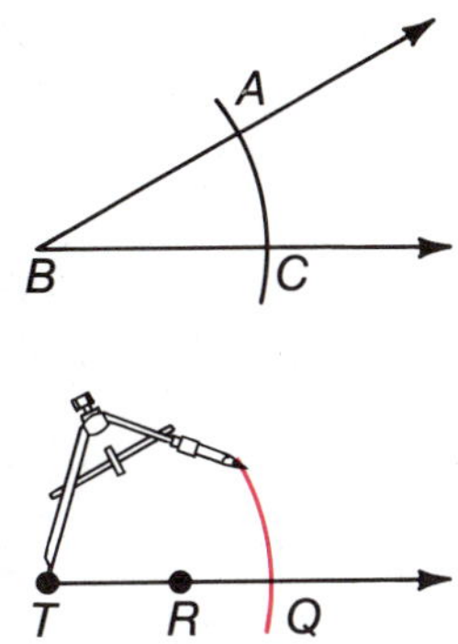

Step 4. Set the compass as shown.

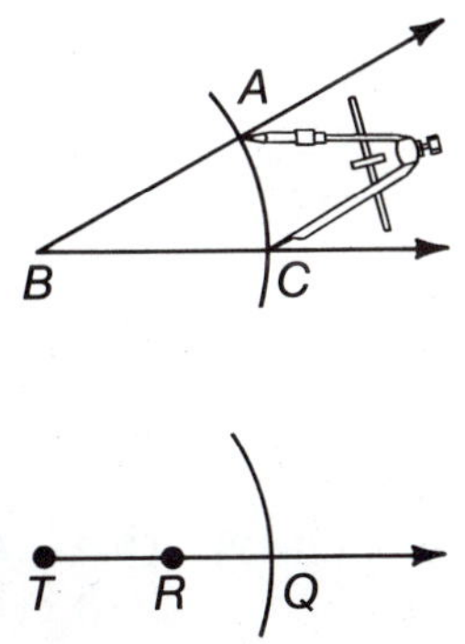

Step 5. Use the same setting to draw an arc as shown.

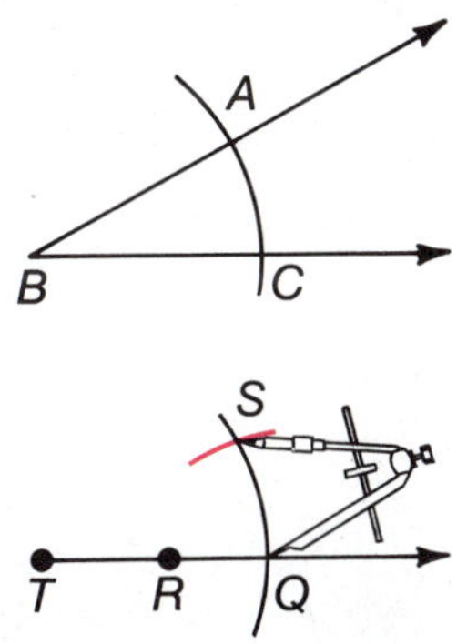

Step 6. Draw $\overrightarrow{TS}$, as shown.

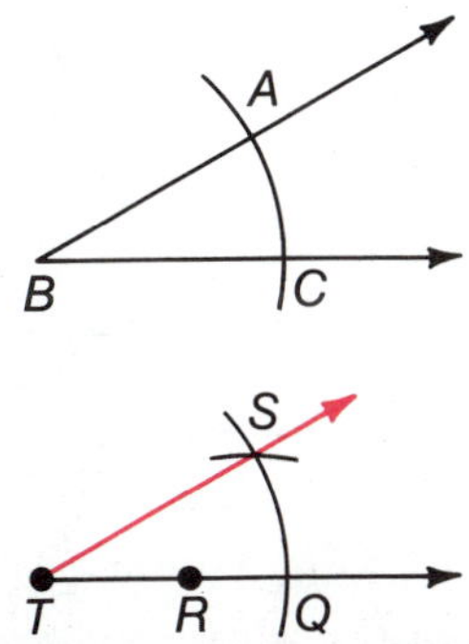

$\angle B$ and $\angle T$ are the same size.

EXERCISES

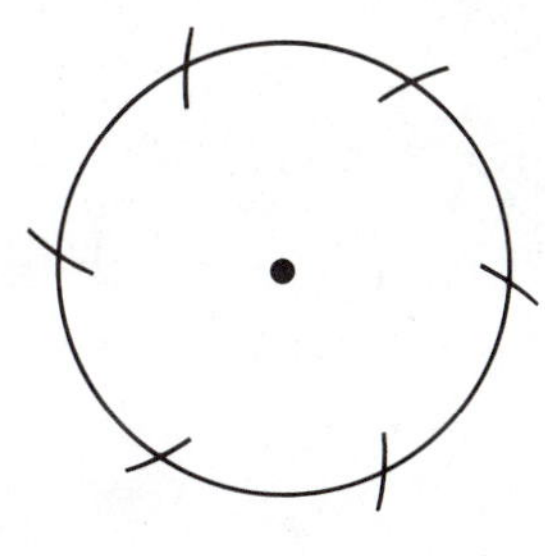

1. First draw some segments. Then use a compass and straightedge to copy each segment.
2. First draw some angles. Then use a compass and straightedge to copy each angle.
3. Draw a circle. Do not change the compass setting. Mark off 6 equal arcs on the circle.

Bisecting segments and angles

Here is how to use your compass and straightedge to **bisect** (find the midpoint of) a segment.

Step 1. Using point *A* as the center, draw an arc as shown.

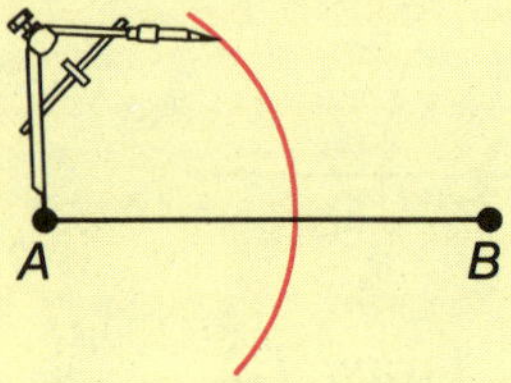

Step 2. Using point *B* as the center and the same compass setting, draw another arc.

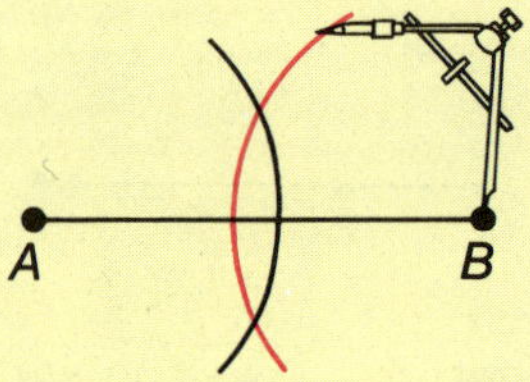

Step 3. Put your straightedge on the points where the arcs cross. Mark where your straightedge crosses the segment.

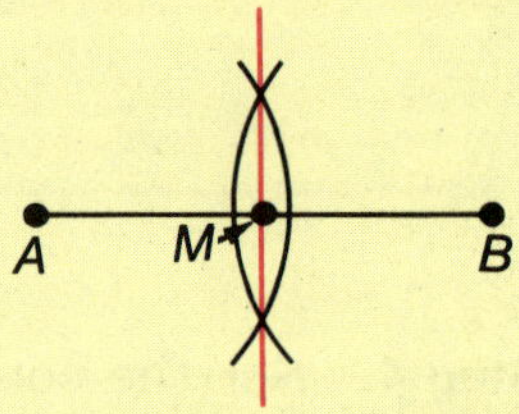

M is the midpoint of $\overline{AB}$.

You can also bisect an angle with a compass and straightedge.

Step 1. Using the vertex as the center, draw an arc as shown.

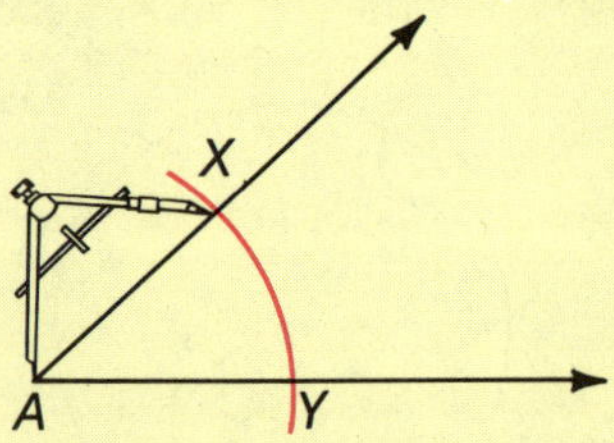

Step 2. Use points *X* and *Y* as centers. Draw two crossing arcs, using the same compass setting for both.

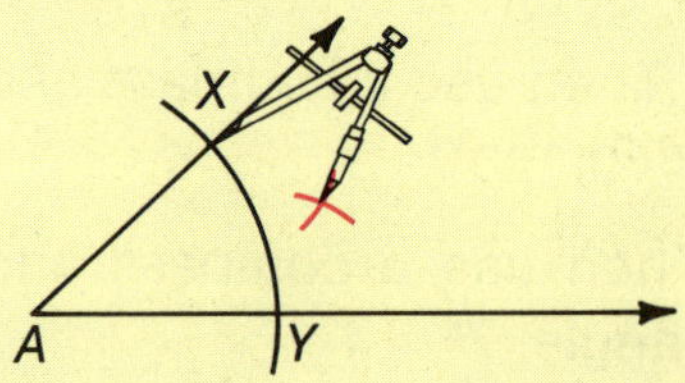

Step 3. Draw the ray from the vertex to the point where the arcs cross.

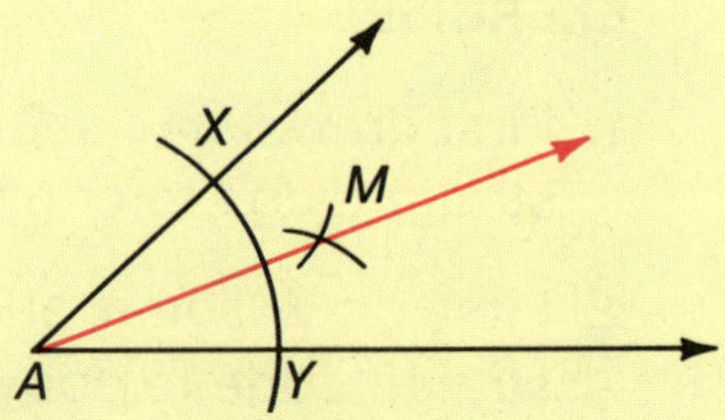

$\overrightarrow{AM}$ is the bisector of $\angle A$.

EXERCISES

1. Draw a segment and bisect it.
2. Draw an angle and bisect it.
3. Draw a large triangle. Draw the bisector of each angle. If you are very careful, the three bisectors will all cross at the same point.

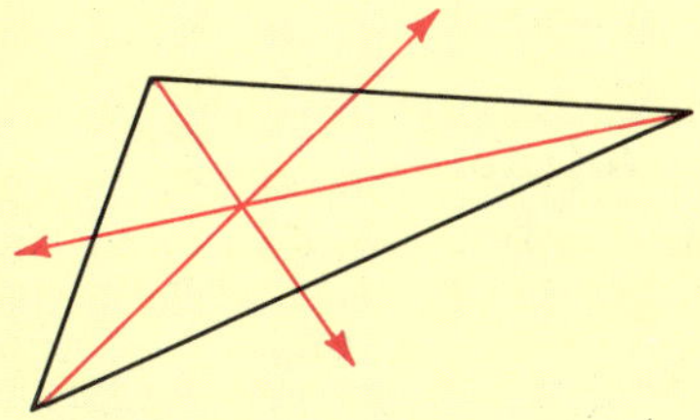

4. Draw a large triangle. Bisect each side. Draw a segment from each vertex to the midpoint of the opposite side. These segments are called **medians** of the triangle. If you are very careful, the three medians will all cross at the same point.

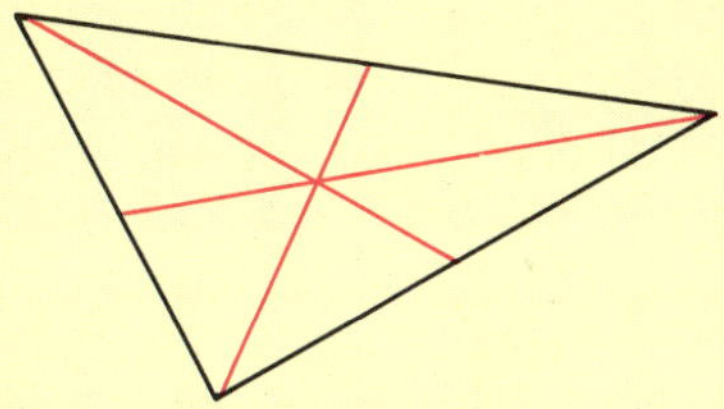

5. Draw a large triangle. Bisect each side. Connect the midpoints of the sides by drawing three segments. Look at the four small triangles. Do you think they are the same size and shape?

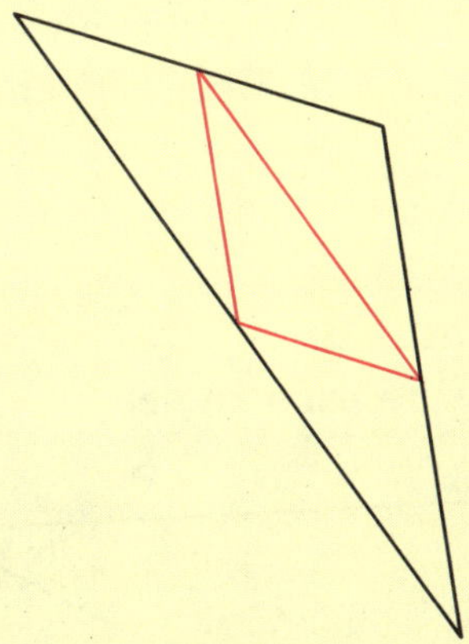

6. Measure the length of each side of the large triangle in exercise 5. Measure the length of each segment that connects two midpoints. Do you see a relationship among the segments?

7. COMPASS CAPERS

Draw a large triangle. See if you can find a way to construct a circle that is inside the triangle, with each side of the triangle touching the circle in just one point. The circle is called an **inscribed circle.** (*Hint:* First construct the angle bisectors.)

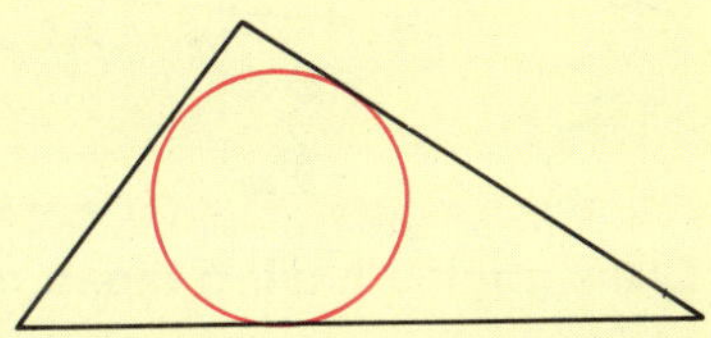

Measuring angles

A **protractor** is used to measure angles. Here is how to use a protractor:

(1) Place the center of the protractor at the vertex of the angle.

(2) Place the 0 mark on one side of the angle.

(3) Read the measure of the angle where the other side crosses the protractor.

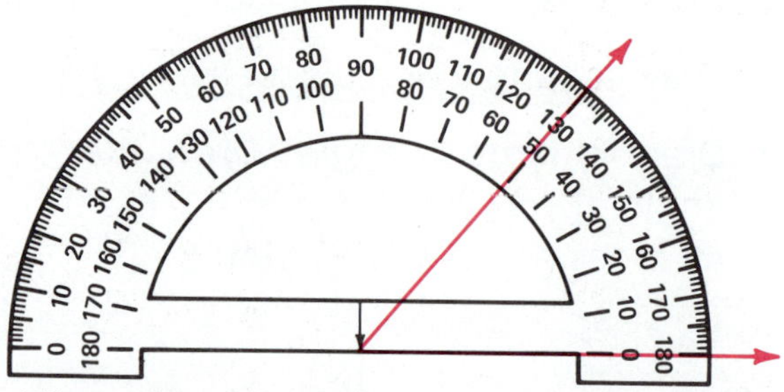

The measure of this angle is 50°. Read "50°" as "fifty degrees."

An **acute angle** is an angle that measures between 0° and 90°.

A **right angle** is an angle that measures 90°.

An **obtuse angle** is an angle that measures between 90° and 180°.

EXERCISES

Measure each angle.

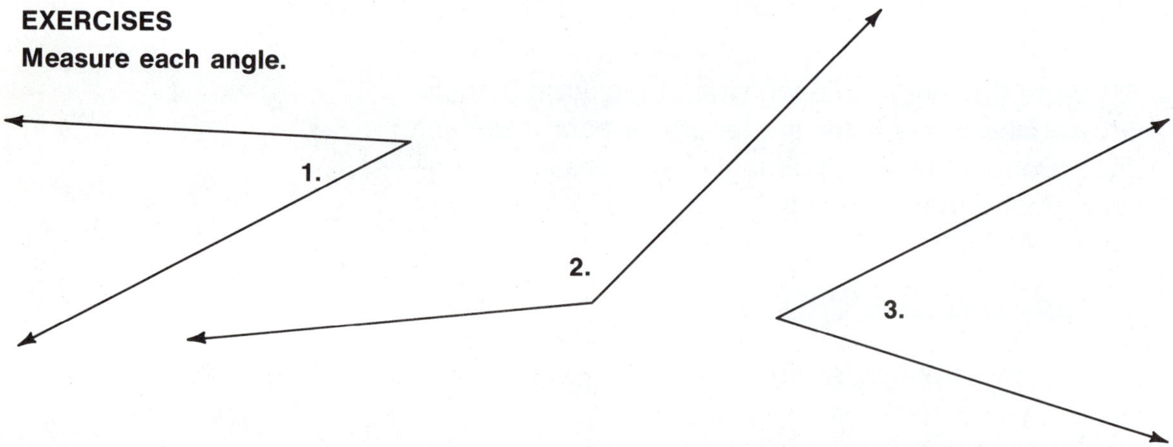

Draw angles having these measures.

4. 35° **5.** 135° **6.** 90° **7.** 75° **8.** 175°

Point B is called the **vertex** of the angle. Here are three ways to name the angle:

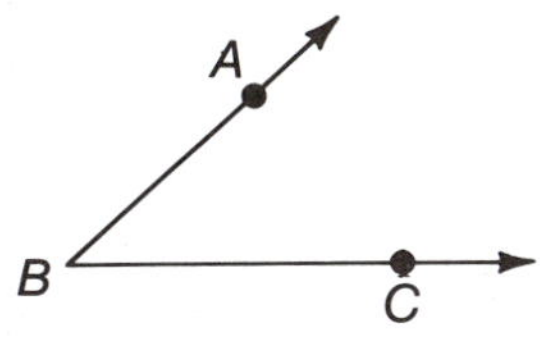

$\angle B$ "angle B"

$\angle ABC$ "angle ABC"

$\angle CBA$ "angle CBA"

Acute, right, or obtuse?

9. $\angle QPT$
10. $\angle SPT$
11. $\angle SPQ$
12. $\angle RPT$

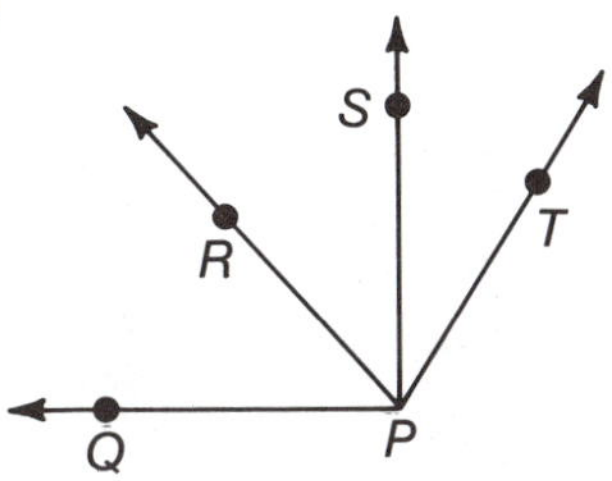

★ 13. Draw a large triangle. Measure each angle. What is the sum of the measures?

★ 14. Repeat exercise 13 for two more triangles. What can you conclude about the sum of the measures for any triangle?

★ 15. A quadrilateral has four sides. Draw three quadrilaterals. Measure each angle. What can you conclude about the sum of the measures for any quadrilateral?

★ 16. A pentagon has five sides. Without drawing or measuring, can you tell the sum of the measures of the angles in a pentagon?

KEEPING SKILLS SHARP

Give each product in simplest form.

1. $4 \times \frac{1}{3}$
2. $\frac{3}{4} \times 3$
3. $\frac{1}{2} \times \frac{4}{5}$
4. $\frac{2}{3} \times \frac{5}{2}$
5. $\frac{3}{8} \times \frac{4}{5}$
6. $\frac{5}{8} \times 8$
7. $6 \times \frac{2}{3}$
8. $\frac{5}{9} \times \frac{9}{5}$
9. $\frac{3}{4} \times \frac{4}{3}$
10. $\frac{5}{6} \times \frac{3}{4}$
11. $\frac{1}{5} \times \frac{1}{3}$
12. $\frac{4}{9} \times \frac{3}{4}$
13. $\frac{1}{8} \times 2$
14. $\frac{5}{6} \times \frac{9}{10}$
15. $\frac{2}{3} \times \frac{3}{4}$

Perpendicular lines and parallel lines

Two lines that intersect to form right angles are called **perpendicular lines.**

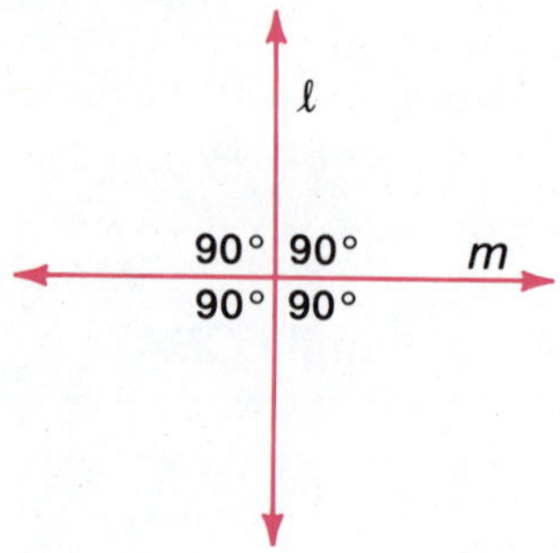

We say:

line ℓ is perpendicular to line m

We write:

$\ell \perp m$

Here is how to use your straightedge and compass to construct two perpendicular lines.

Step 1.
Draw a line.

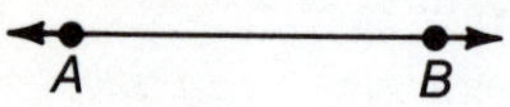

Step 2.
Using point A as center, draw an arc as shown.

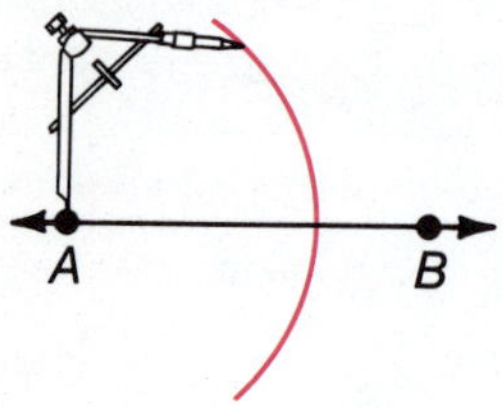

Step 3.
Using the same setting and point B as center, draw an arc as shown.

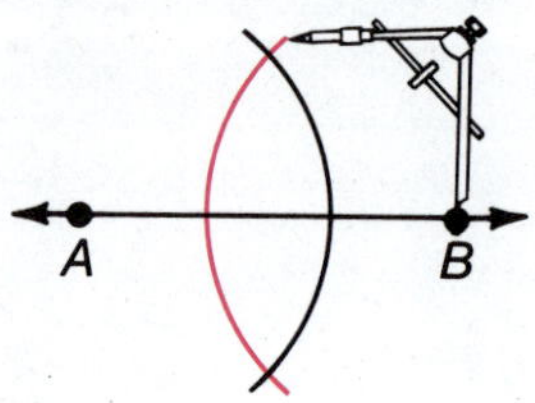

Step 4.
Connect the points where the arcs cross.

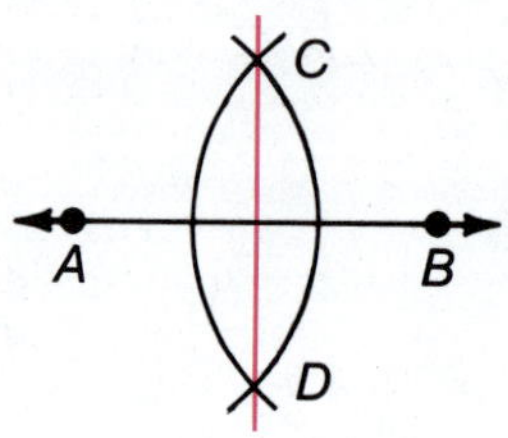

$\overline{CD} \perp \overline{AB}$

Notice that $\overleftrightarrow{CD}$ also bisects $\overline{AB}$. It is called the **perpendicular bisector** of $\overline{AB}$.

Two lines in a plane that do not intersect are called **parallel lines.**

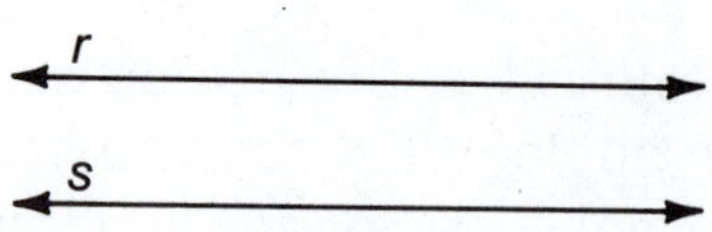

We say:

line r is parallel to line s

We write:

$r \parallel s$

EXERCISES

1. Use a compass and straightedge to construct two perpendicular lines.

2. Follow these steps to construct two parallel lines.

   a. Draw a line ℓ and construct a perpendicular line to line ℓ.

   b. Construct a second perpendicular line to line ℓ.

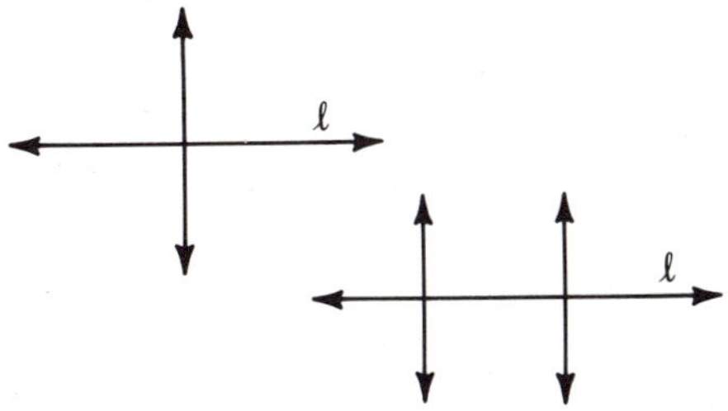

3. Construct a line that is parallel to line ℓ.

4. How many lines do you think there are that are parallel to line ℓ? Perpendicular to line ℓ?

5. Follow these steps to construct a circle that goes through each vertex of a triangle.

   a. Draw a large triangle. Construct the perpendicular bisector of each side.

   b. Put the point of the compass where these bisectors meet. Extend the pencil of the compass to one vertex.

   c. Draw a complete circle with this compass setting. Did the circle pass through each vertex?

   The triangle is called an **inscribed triangle.**

Congruent figures

You have seen pairs of figures that have the same size and shape. Such figures are called **congruent figures.** For example, the two triangles below are congruent because a tracing of $\triangle ABC$ ("triangle ABC") fits $\triangle QPR$.

We write: $\triangle ABC \cong \triangle QPR$

We say: $\triangle ABC$ is congruent to $\triangle QPR$

Notice that for this fitting, we have this matching of the vertices:

$$A \leftrightarrow Q$$
$$B \leftrightarrow P$$
$$C \leftrightarrow R$$

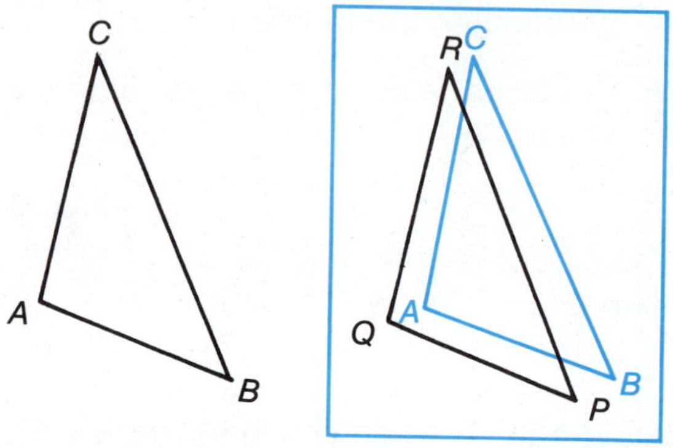

The sides and angles that match are called **corresponding parts.**

Corresponding sides	Corresponding angles
$\overline{AB} \leftrightarrow \overline{QP}$	$\angle A \leftrightarrow \angle Q$
$\overline{AC} \leftrightarrow \overline{QR}$	$\angle B \leftrightarrow \angle P$
$\overline{BC} \leftrightarrow \overline{PR}$	$\angle C \leftrightarrow \angle R$

EXERCISES

In each exercise the two figures are congruent. Name all pairs of corresponding parts for the congruent fitting.

1.

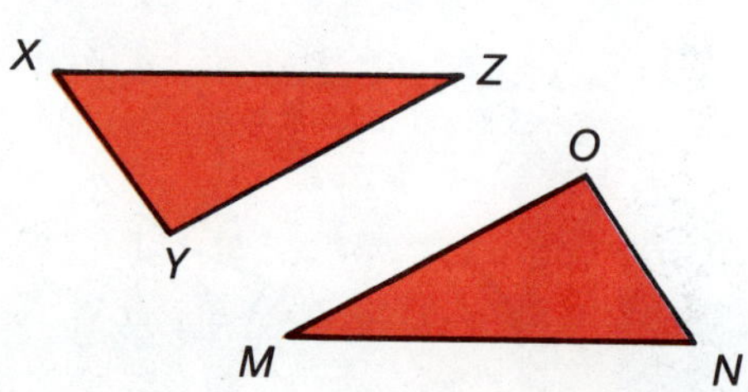

2.

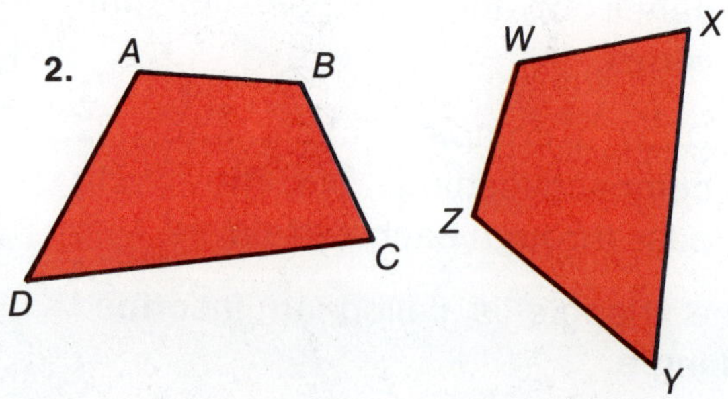

For exercises 3–6, use the figures below. The triangles are congruent.

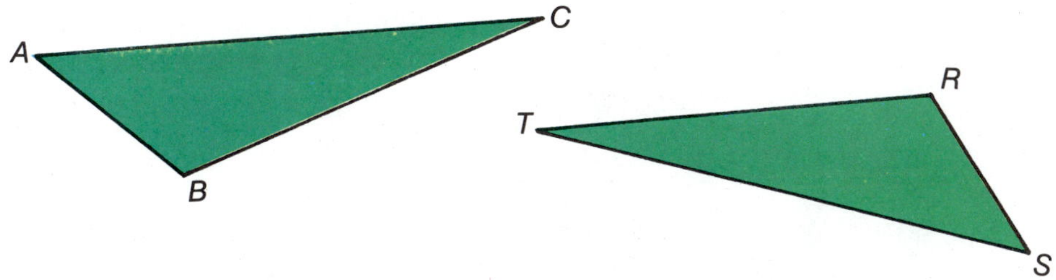

For the congruent fitting,

3. which side of $\triangle RST$ corresponds to $\overline{AB}$?
4. is $\overline{RT}$ congruent to $\overline{AB}$? Why?
5. which angle of $\triangle RST$ is congruent to $\angle C$? How do you know?
6. which side of $\triangle ABC$ is congruent to $\overline{ST}$? How do you know?
7. is it true that the corresponding parts are congruent?

Make tracings to answer exercises 8–11.
True or false?

8. $\triangle AFG \cong \triangle DJI$.
9. $\triangle EFD \cong \triangle BGH$.
10. $\triangle EDC \cong \triangle ABC$.
11. $\triangle AJC \cong \triangle ABC$.

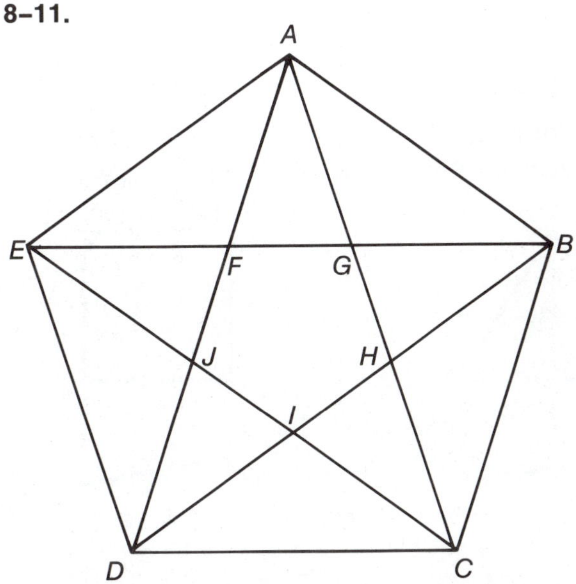

12. **TRIANGLE TANGLE**

How many triangles can you find in this picture?

A line of symmetry

The dotted line seems to divide the leaf "in half." If you fold the leaf along the dotted line, then the two halves match.

The fold line is a **line of symmetry.**

EXERCISES

Is the dotted line a line of symmetry? [*Hint:* If you fold along the dotted line, will the two parts match?]

1.

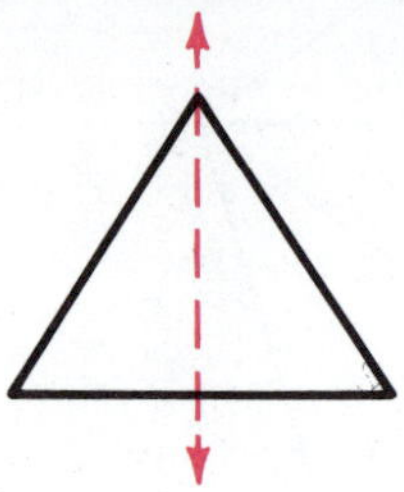

2.

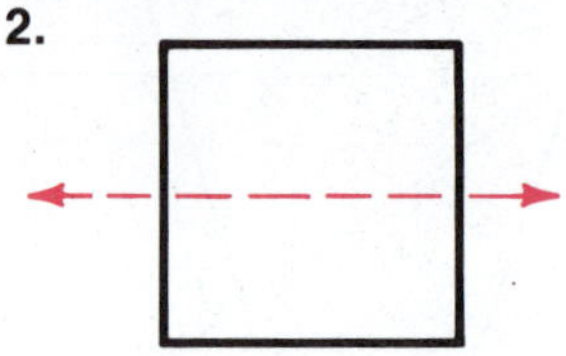

3.

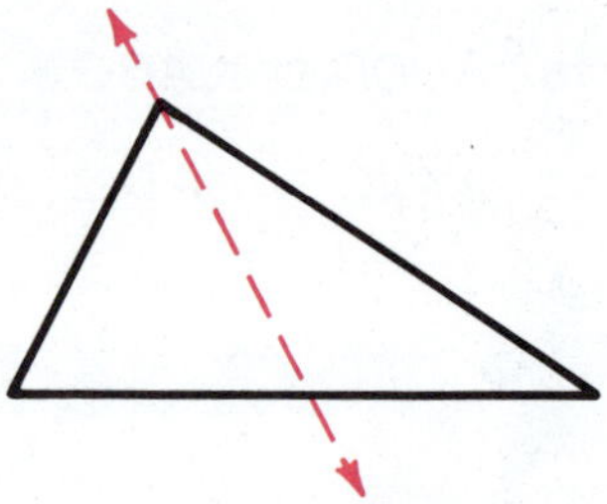

4.

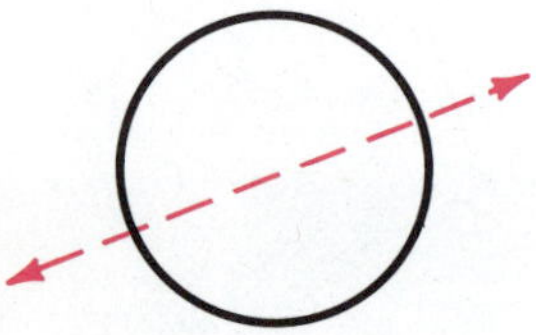

5.

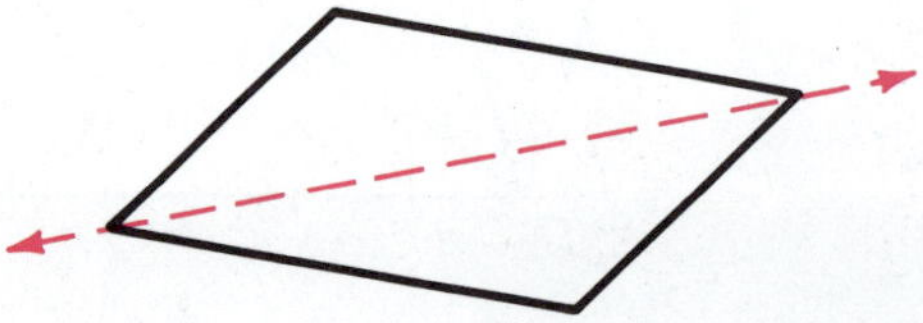

6.

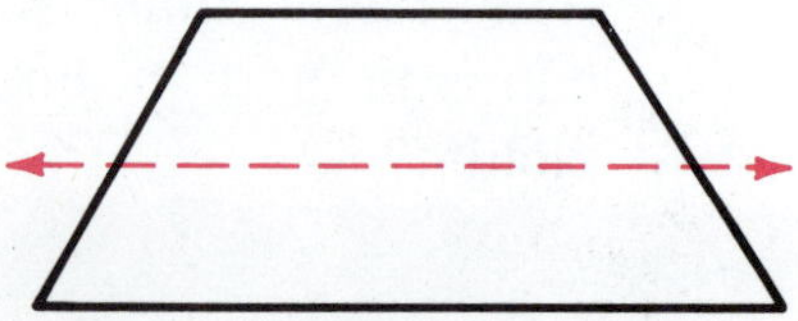

Trace each figure. Draw in all lines of symmetry. Be careful. Some figures may have no lines of symmetry.

7.

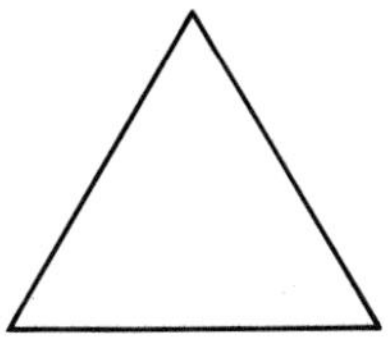

8.

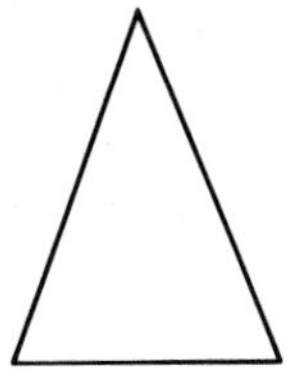

9.

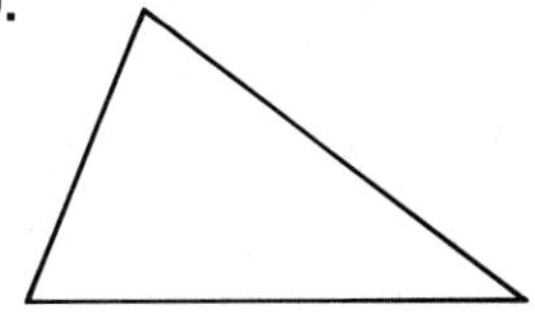

10.

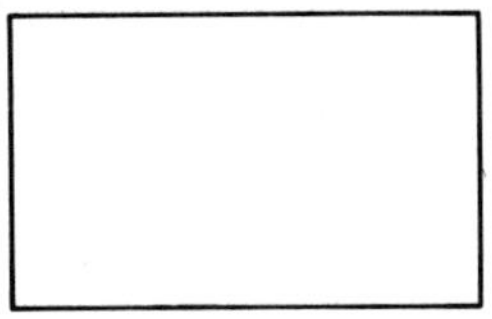

11.

12.

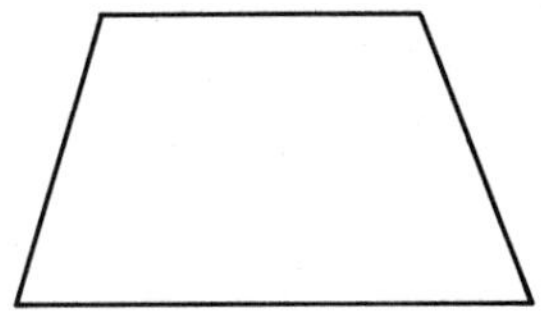

13.

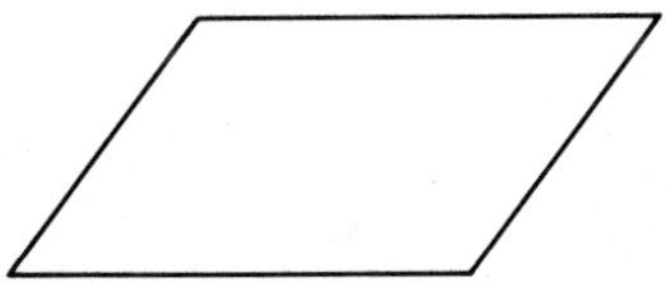

14.

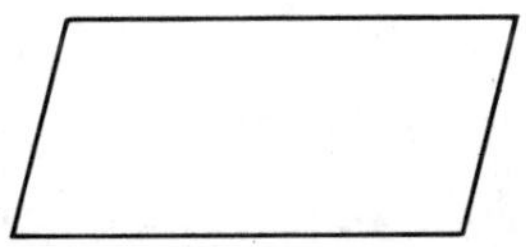

15.

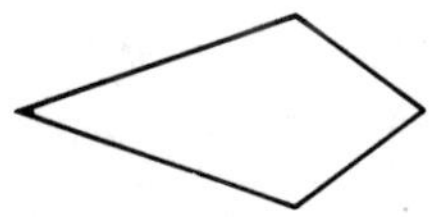

★ **16.** Construct a triangle with no congruent sides. How many lines of symmetry does it have?

★ **17.** Construct a triangle with exactly two congruent sides. How many lines of symmetry does it have?

★ **18.** Construct a triangle with three congruent sides. How many lines of symmetry does it have?

Triangles

Now let's use the idea of a line of symmetry to study the properties of triangles. Any triangle can be classified according to its number of lines of symmetry.

A triangle with 3 lines of symmetry is called an **equilateral triangle.**

A triangle with 1 line of symmetry is called an **isosceles triangle.**

A triangle with 0 lines of symmetry is called a **scalene triangle.**

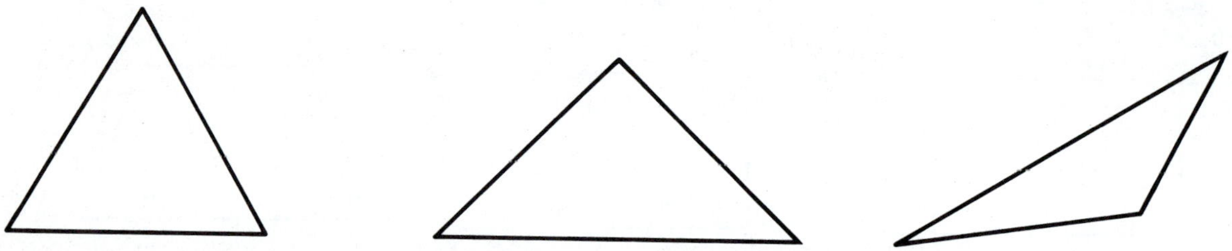

EXERCISES

Trace this scalene triangle.

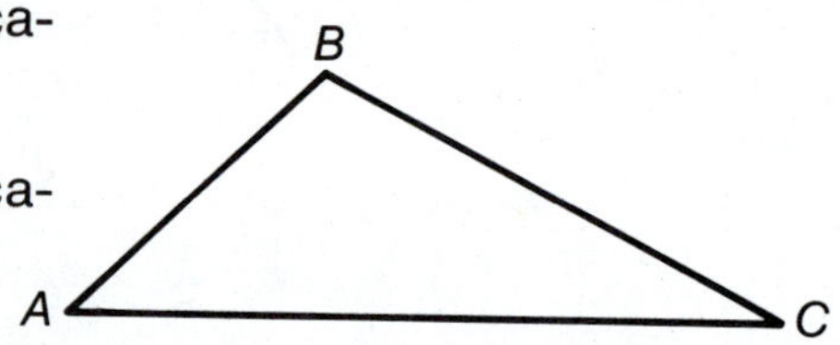

1. Does your tracing of $\overline{AB}$ fit $\overline{BC}$ or $\overline{AC}$? Does a scalene triangle have any congruent sides?
2. Does your tracing of $\angle B$ fit $\angle C$ or $\angle A$? Does a scalene triangle have any congruent angles?

A scalene triangle has no congruent sides.

Trace this isosceles triangle.

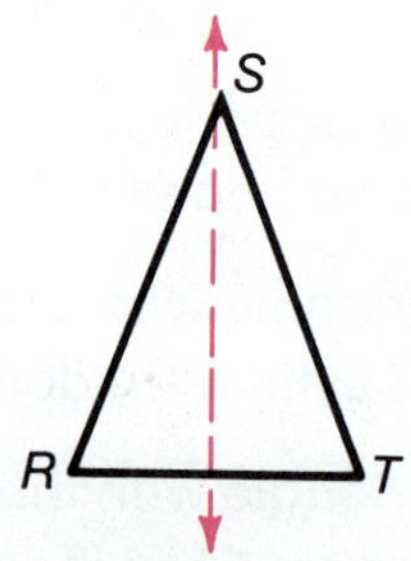

3. Flip your drawing facedown about the line of symmetry. Does $\overline{RS}$ fit $\overline{TS}$? Does $\angle R$ fit $\angle T$?
4. What can you say about 2 sides of an isosceles triangle? About 2 angles of an isosceles triangle?

An isosceles triangle has 2 congruent sides.

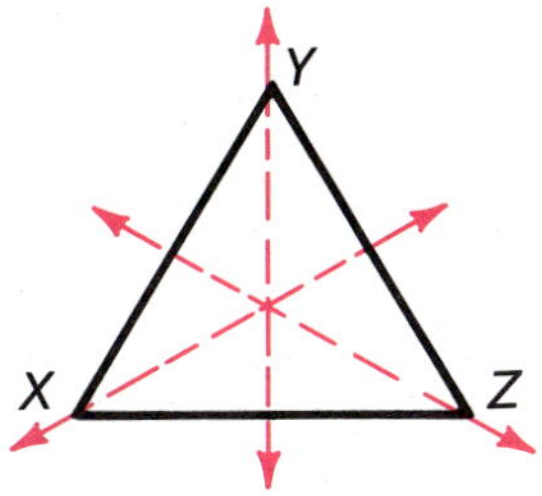

Trace this equilateral triangle.

5. If you flip your tracing facedown, does $\overline{XY}$ fit $\overline{YZ}$ and $\overline{XZ}$? Does $\angle X$ fit $\angle Y$ and $\angle Z$?

6. What can you say about the 3 sides of an equilateral triangle? About the 3 angles?

An equilateral triangle has 3 congruent sides.

7. TRIANGLE TANGLE II

$\overleftrightarrow{AF}$, $\overleftrightarrow{CD}$, and $\overleftrightarrow{BE}$ are lines of symmetry of $\triangle ABC$.

a. Name the 5 equilateral triangles.

b. Name the 12 isosceles triangles.

c. How many scalene triangles can you count? [*Hint:* There are more than 24.]

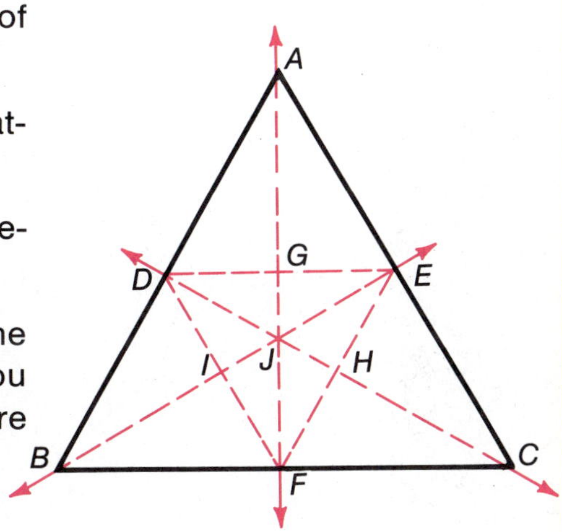

KEEPING SKILLS SHARP

Give each quotient in lowest terms.

1. $3 \div \frac{3}{4}$
2. $\frac{2}{5} \div \frac{2}{3}$
3. $\frac{5}{9} \div \frac{2}{5}$
4. $\frac{8}{5} \div \frac{4}{3}$
5. $\frac{3}{8} \div 5$
6. $\frac{3}{2} \div \frac{9}{4}$
7. $\frac{7}{4} \div \frac{3}{8}$
8. $\frac{9}{4} \div \frac{1}{2}$
9. $\frac{6}{5} \div 3$
10. $\frac{9}{8} \div \frac{3}{4}$
11. $\frac{1}{8} \div \frac{1}{2}$
12. $\frac{2}{5} \div 4$
13. $\frac{1}{3} \div \frac{1}{6}$
14. $12 \div \frac{3}{4}$
15. $\frac{2}{3} \div \frac{3}{2}$

A point of symmetry

Think about taking the lid off each box and turning it halfway around.

Which lid would fit back on the box?

If a figure fits itself after a *half-turn* about a point, the point is a **point of symmetry** of the figure.

EXAMPLES.

Trace.

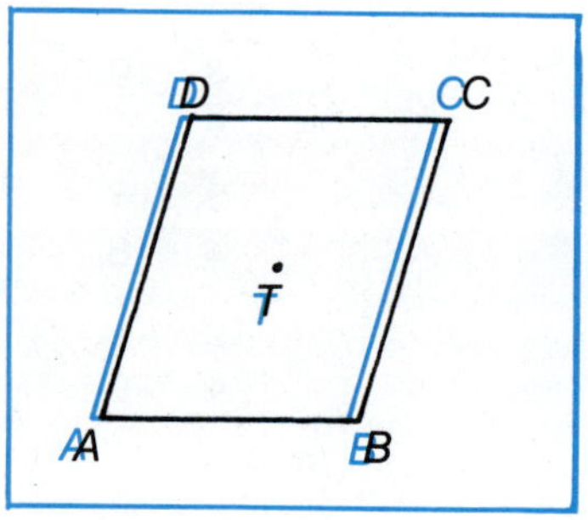

Half-turn. It fits.

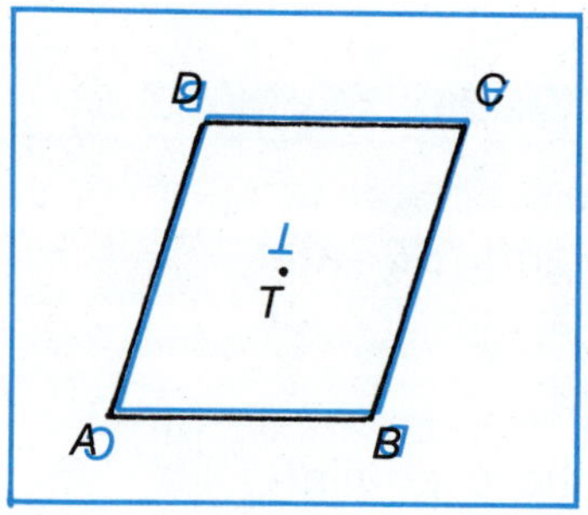

T is a point of symmetry.

Trace.

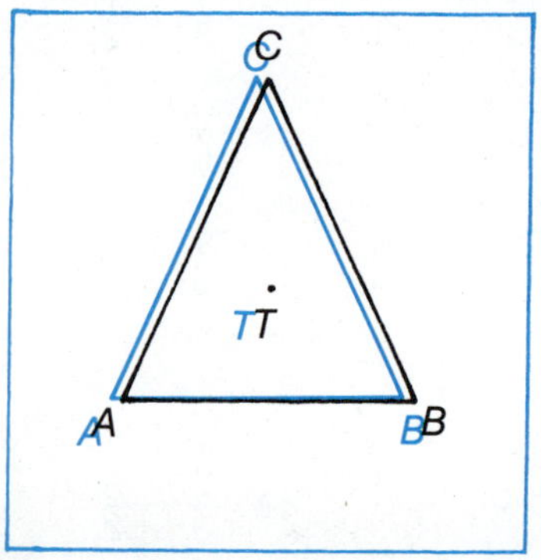

Half-turn. It does not fit.

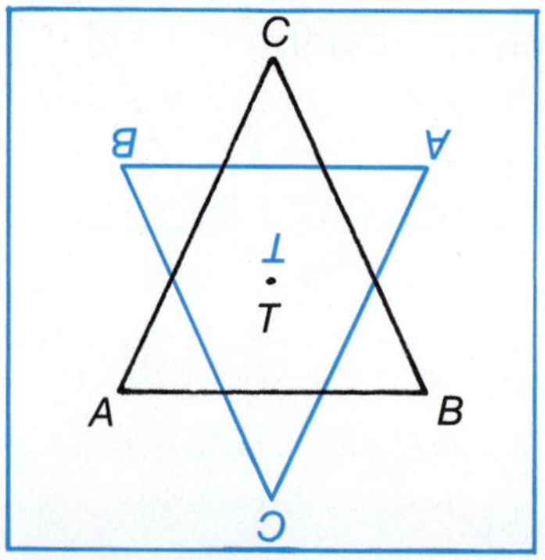

T is *not* a point of symmetry.

EXERCISES

Is the red point a point of symmetry of the figure? [*Hint:* Trace the figure and see if the tracing fits after a *half-turn* about the red point.]

1. 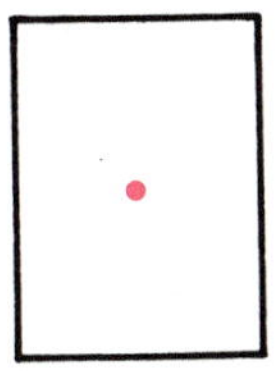

2. 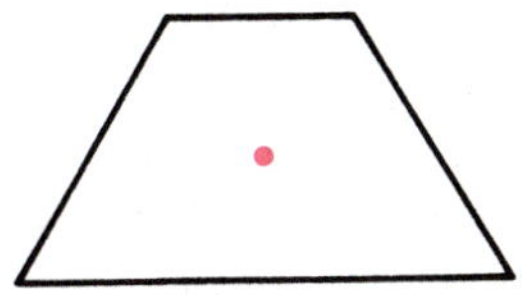

3. 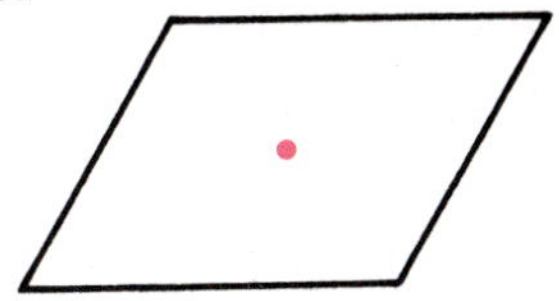

4. 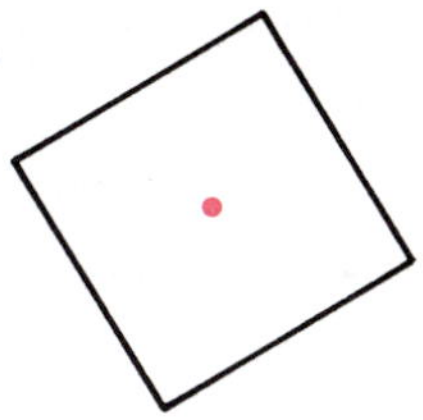

5. 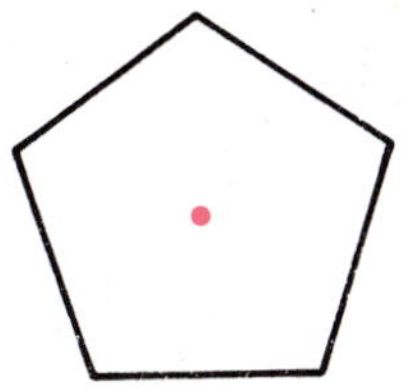

6. 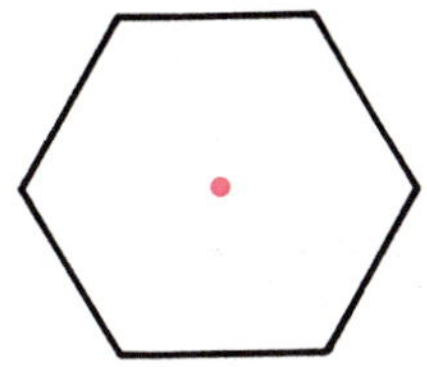

The red point is a point of symmetry.

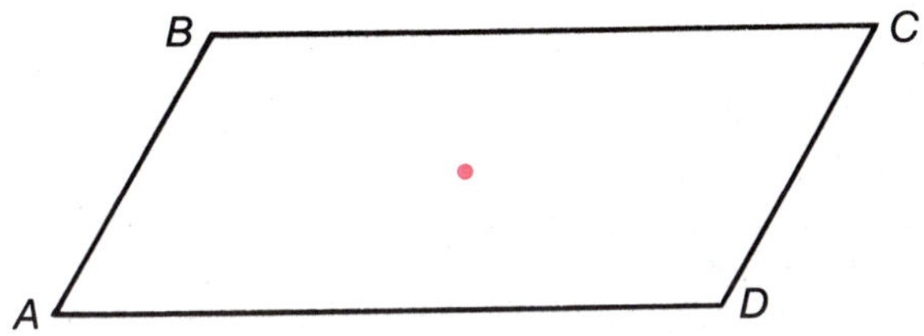

7. a. Under a half-turn about the red point, does $\overline{AB}$ fit $\overline{CD}$?

b. Is $\overline{AB} \cong \overline{CD}$?

c. $\overline{AB}$ and $\overline{CD}$ are called **opposite sides.** Give the other pair of opposite sides.

d. Are the opposite sides congruent?

8. a. Under a half-turn about the red point, does $\angle A$ fit $\angle C$?

b. Is $\angle A \cong \angle C$?

c. $\angle A$ and $\angle C$ are called **opposite angles.** Give the other pair of opposite angles.

d. Are the opposite angles congruent?

9. Since the figure above has a point of symmetry, its opposite sides are parallel. Give the two pairs of parallel sides.

Quadrilaterals

We can use points and lines of symmetry to study properties of some special quadrilaterals (4-sided figures).

A quadrilateral with

a point of symmetry is a **parallelogram.**

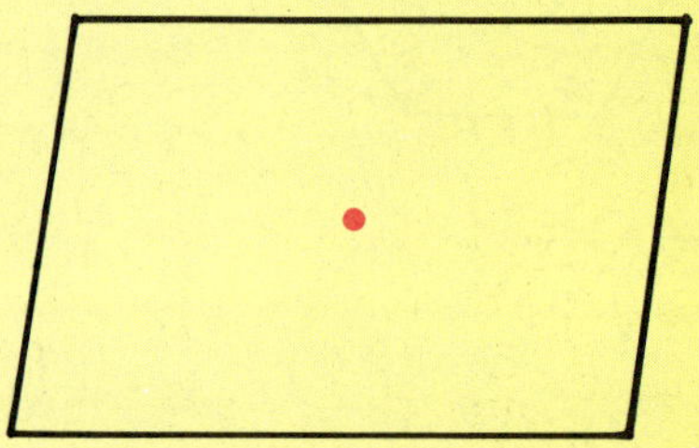

a point of symmetry and two lines of symmetry through opposite vertices is a **rhombus.**

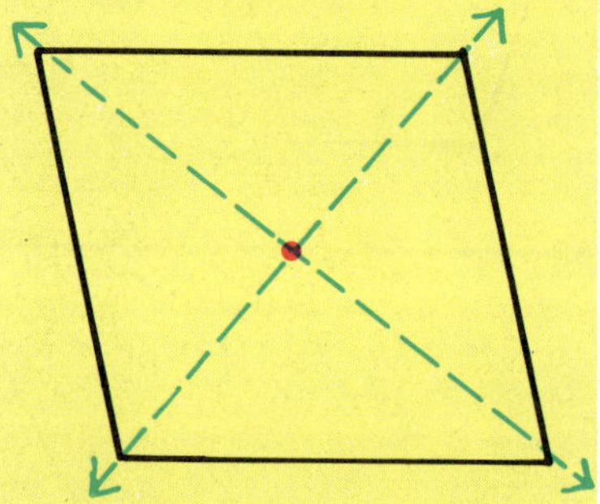

a point of symmetry and two lines of symmetry through opposite sides is a **rectangle.**

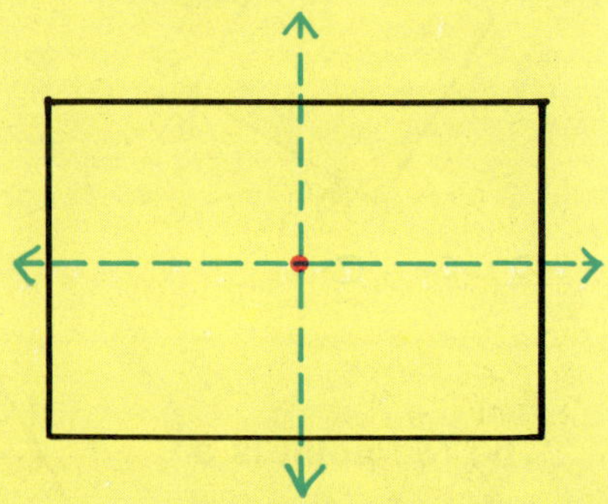

a point of symmetry and four lines of symmetry is a **square.**

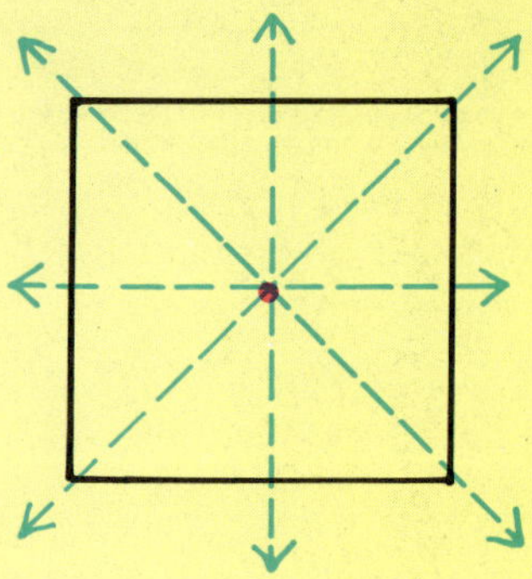

one line of symmetry through opposite sides is an **isosceles trapezoid.**

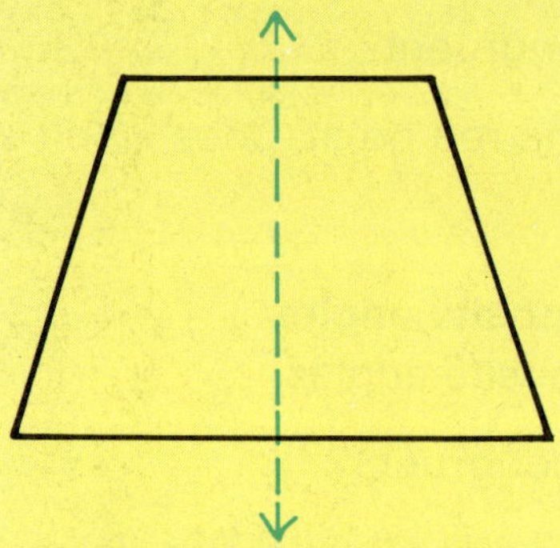

one line of symmetry through opposite vertices is a **kite.**

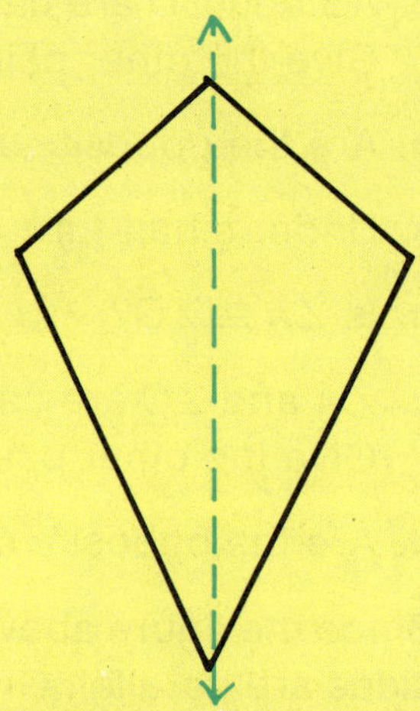

EXERCISES

Which quadrilaterals on page 340 have

1. both pairs of opposite sides parallel? [*Hint:* Which have a point of symmetry?]
2. both pairs of opposite sides congruent?
3. both pairs of opposite angles congruent?
4. all four angles congruent?
5. exactly one pair of parallel sides and exactly one pair of congruent sides?
6. exactly one pair of opposite angles that are congruent?

7. COMPASS CAPERS II

a. Study this method of constructing an equilateral triangle.

b. Construct an equilateral triangle using this method. [*Hint:* Use the radius of the circle to mark off arcs on the circle.]

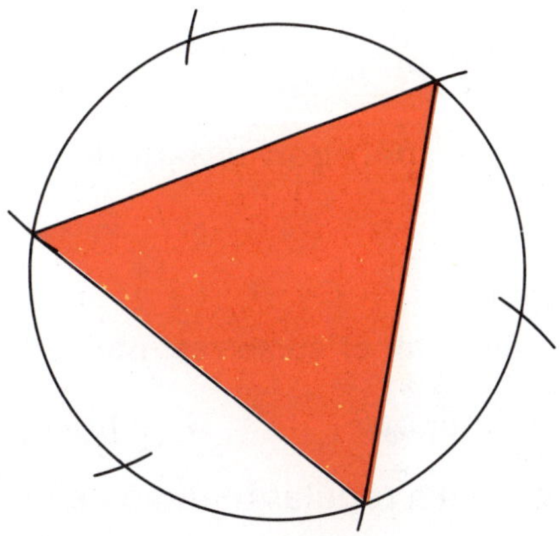

c. Study this method of constructing a square.

d. Construct a square using this method.

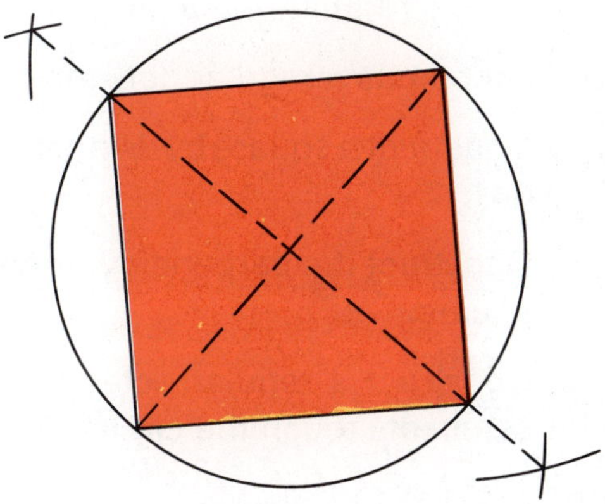

Polygons

These figures are all **polygons.** Their special names are given.

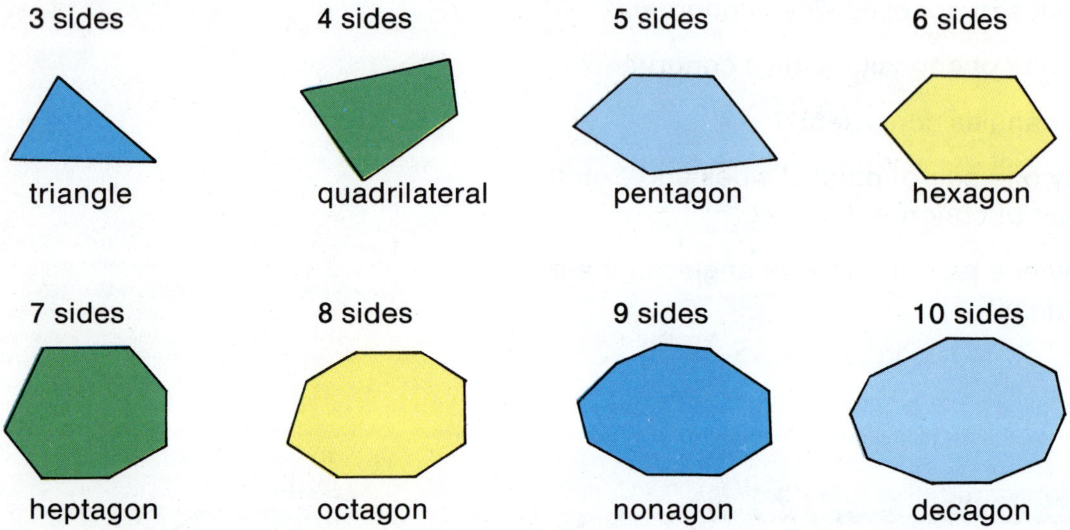

A **regular polygon** has all its sides congruent and all its angles congruent.

EXERCISES

Can you think of some real object that is shaped like

1. a pentagon? **2.** a hexagon? **3.** an octagon?

4. What is a regular triangle called?

5. What is a regular quadrilateral called?

6. Is a rhombus a regular polygon?

7. a. Construct a square by using the "circle" method on page 341.

b. Construct the perpendicular bisector of each side of the square.

c. Construct a regular octagon using the 8 points where segments touch the circle.

Mathematics in careers

A land surveyor determines the shape and size of a part of the earth's surface. He also sets boundaries for property.

The main instrument used by a surveyor is a *transit.* It is used for sighting in a "straight line" and for determining horizontal and vertical angles. Distances are found by measuring with a steel tape. While a survey is being made, the surveyor generally records the data in a field book.

Here are two plots from a field book. Make a scale drawing of each plot. Use a protractor to draw the "corner" angles.

1.

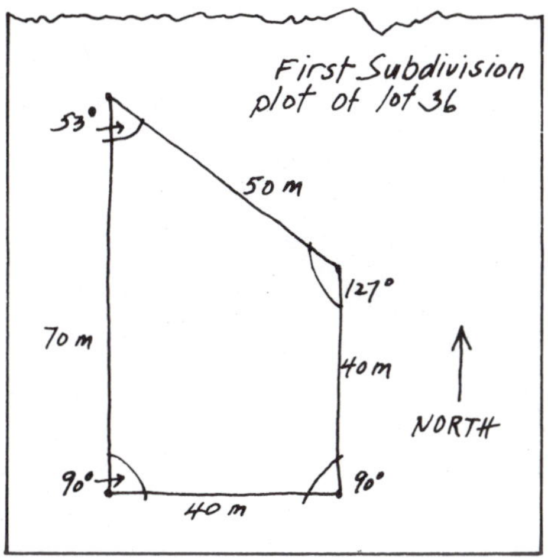

2.

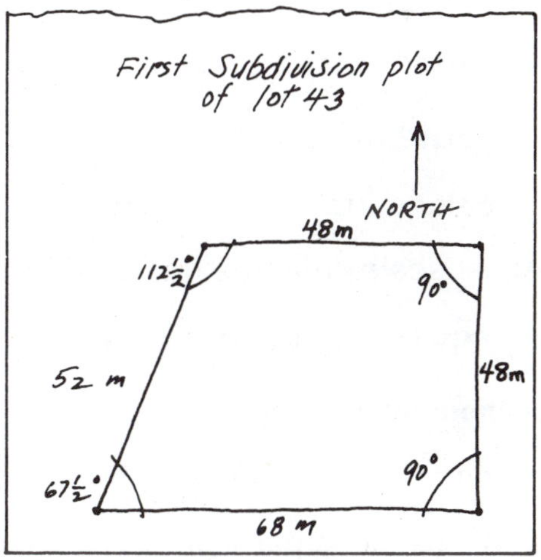

CHAPTER CHECKUP

Acute, right, or obtuse? Complete. [pages 328–329]

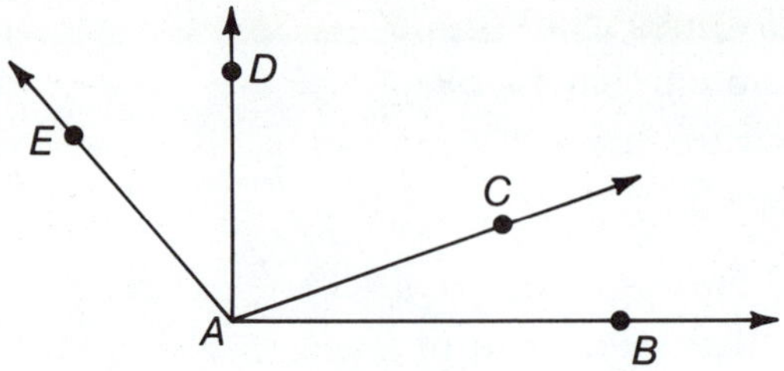

1. ∠*EAB* is ___?___.
2. ∠*BAD* is ___?___.
3. ∠*DAE* is ___?___.
4. An angle that has a measure of 90° is ___?___.

Draw angles with these measures. [pages 328–329]

5. 20° 6. 38° 7. 90° 8. 135°

Which line is [pages 330–331]

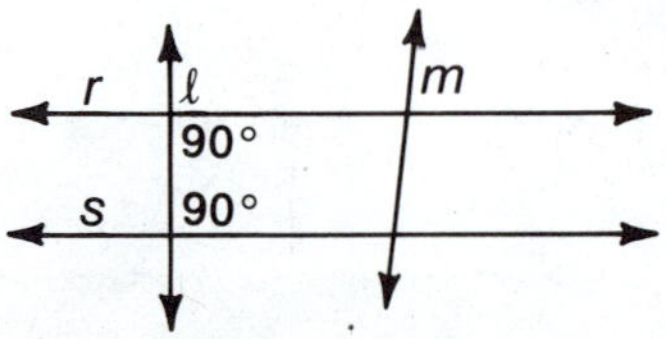

9. perpendicular to line *s*?
10. parallel to line *s*?

These two triangles are congruent. Complete these pairs of corresponding parts for the congruent fitting. [pages 332–333]

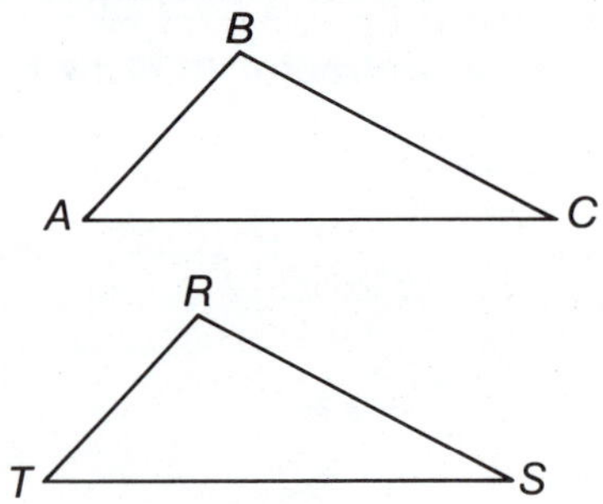

11. $\overline{AB} \leftrightarrow$ ___?___ 12. $\overline{BC} \leftrightarrow$ ___?___
13. $\angle B \leftrightarrow$ ___?___ 14. $\angle C \leftrightarrow$ ___?___
15. Are these corresponding parts congruent?

True or false? [pages 334–337]

16. A scalene triangle has 1 line of symmetry.
17. An isosceles triangle has 2 congruent sides.
18. An equilateral triangle has 3 congruent angles.

Name the quadrilaterals that have [pages 338–341]

19. a point of symmetry and no lines of symmetry.
20. opposite sides parallel. 21. 4 congruent sides.
22. 4 right angles. 23. 4 congruent sides and 4 right angles.

CHAPTER PROJECT

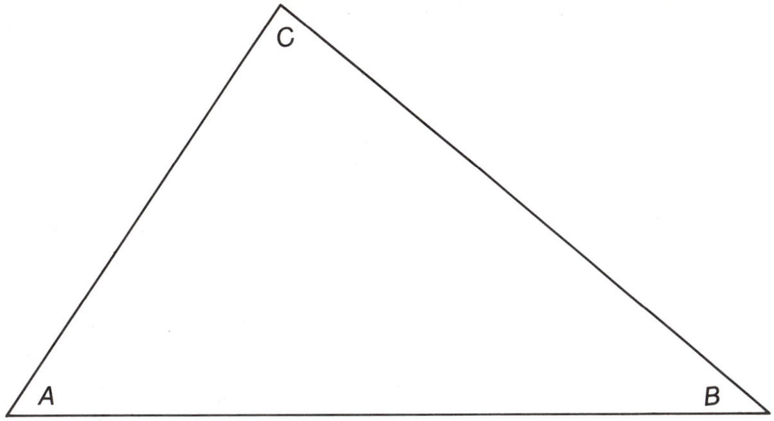

Make 3 copies of this triangle. Cut them out. Use the copies for exercises 1–3.

1. a. Fold the triangle so that points A and B are together. Crease the paper. The line you have folded is the perpendicular bisector of $\overline{AB}$.
   b. Fold the perpendicular bisectors of $\overline{AC}$ and $\overline{BC}$.
   c. What do you notice about the three perpendicular bisectors?

2. a. Fold the triangle so that $\overline{AC}$ falls on $\overline{AB}$. Crease. You have folded the bisector of $\angle A$.
   b. Fold the bisectors of $\angle B$ and $\angle C$.
   c. What do you notice about the three bisectors?

3. a. Fold point A over to point B. Make a small crease where the fold crosses $\overline{AB}$. This is the midpoint of $\overline{AB}$. Do not crease the whole line.
   b. Fold and crease the triangle so that the fold line goes from C to the midpoint of $\overline{AB}$. This is a median of the triangle.
   c. Fold the median that goes from A to the midpoint of $\overline{BC}$ and the median from B to the midpoint of $\overline{AC}$. What do you notice about the three medians?

CHAPTER REVIEW

acute angle 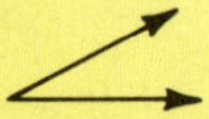

right angle 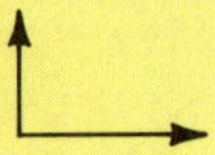

obtuse angle 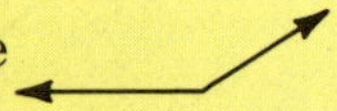

Two lines in a plane that intersect to form right angles are called **perpendicular** (⊥) lines.

Two lines in a plane that do not intersect are called **parallel** (||) lines.

Figures that are the same size and shape are called **congruent figures.**

Triangle	*Lines of Symmetry*
Scalene	0
Isosceles	1
Equilateral	3

Complete.

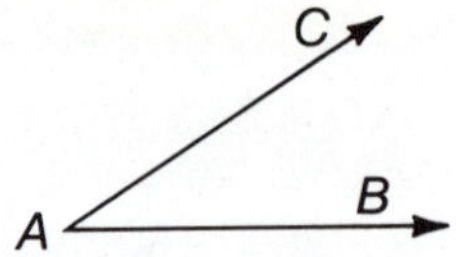

1. The vertex of $\angle BAC$ is point __?__.
2. The measure of $\angle A$ is __?__ 90°. (less than, equal to, greater than)
3. An acute angle measures between 0° and __?__.
4. A right angle measures __?__.
5. An obtuse angle measures between 90° and __?__.

Complete.

6. $\ell \perp$ __?__
7. $\ell \parallel$ __?__

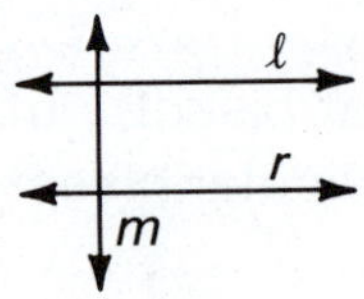

Match congruent figures.

8. 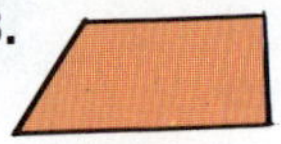
9. 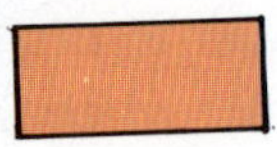
10. 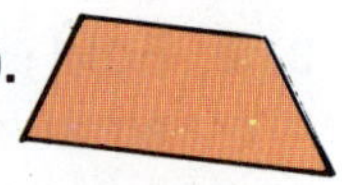

a. 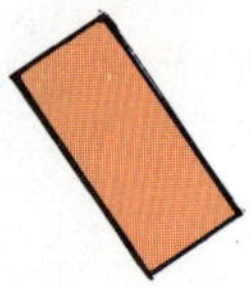
b. 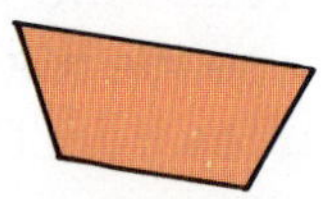
c. 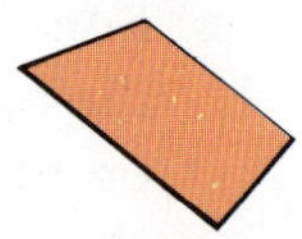

Scalene, isosceles, or equilateral?

11. 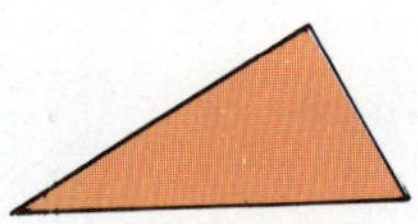
12. 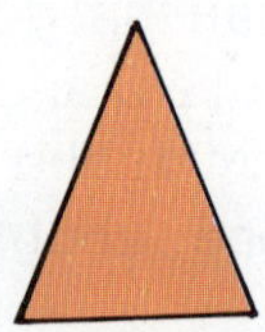
13. 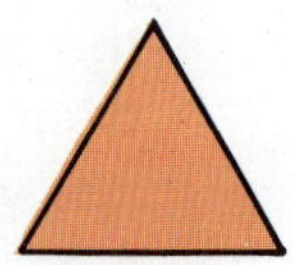

CHAPTER CHALLENGE

A 4-sided figure has 2 diagonals.

1. Draw a 5-sided figure. How many diagonals does it have?

2. Draw a 6-sided figure. How many diagonals does it have?

3. Copy and complete this table. If you need to, draw each figure and its diagonals.

Number of Sides	*Number of Diagonals*
4	
5	
6	
7	
8	

4. Study your completed table. Can you give a rule for finding the number of diagonals if you know the number of sides?

MAJOR CHECKUP
STANDARDIZED FORMAT

Choose the correct letter.

1. The standard numeral for two hundred seven million, eight hundred is

- **a.** 207,800
- **b.** 2,007,800
- **c.** 207,000,800
- **d.** none of these

2. Multiply.

$$\begin{array}{r} 8.76 \\ \times 2.43 \\ \hline \end{array}$$

- **a.** 20.9458
- **b.** 2094.58
- **c.** 21.2868
- **d.** none of these

3. Divide.

$0.235\overline{)14.2645}$

- **a.** 6.7
- **b.** 60.7
- **c.** 607
- **d.** none of these

4. The greatest common factor of 18 and 24 is

- **a.** 3
- **b.** 6
- **c.** 72
- **d.** none of these

5. Subtract.

$$\begin{array}{r} 5 \\ -2\frac{3}{4} \\ \hline \end{array}$$

- **a.** $3\frac{3}{4}$
- **b.** $2\frac{1}{4}$
- **c.** $2\frac{3}{4}$
- **d.** none of these

6. Give the product.

$2\frac{2}{3} \times 1\frac{3}{4}$

- **a.** $4\frac{2}{3}$
- **b.** $\frac{3}{14}$
- **c.** $\frac{21}{32}$
- **d.** none of these

7. Give the area. Use 3.14 for π.

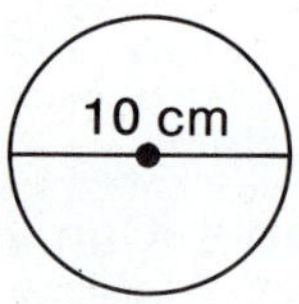

- **a.** 314 cm^2
- **b.** 157 cm^2
- **c.** 78.5 cm^2
- **d.** none of these

8. An item that regularly sells for \$30 was put on sale for \$25. The percent of discount was

- **a.** 25%
- **b.** 20%
- **c.** $16\frac{2}{3}\%$
- **d.** none of these

9. To borrow \$800, you must pay interest at 15% a year. What is the interest for 5 years?

- **a.** \$600
- **b.** \$60
- **c.** \$6
- **d.** none of these

10. How many lines of symmetry does a rectangle have?

- **a.** 1
- **b.** 2
- **c.** 4
- **d.** none of these

11. Doris earned \$23 for working 4 hours. How much did she earn per hour?

- **a.** \$6
- **b.** \$57.50
- **c.** \$92
- **d.** none of these

12. Adele is 6 years younger than Barry. Adele is 11 years old. How many years old is Barry?

- **a.** 66
- **b.** 17
- **c.** 5
- **d.** none of these

Probability and Statistics

12

Equally likely outcomes

If you toss a coin, it can land in one of these two ways:

There are 2 **possible outcomes.**

Since the chance of getting heads or tails is the same, we say that the outcomes are **equally likely.**

If you toss a number cube, it can land in one of six ways:

There are 6 **possible outcomes.** The outcomes are **equally likely.**

If you spin this spinner, there are 4 possible outcomes. The outcomes are *not* equally likely. Why?

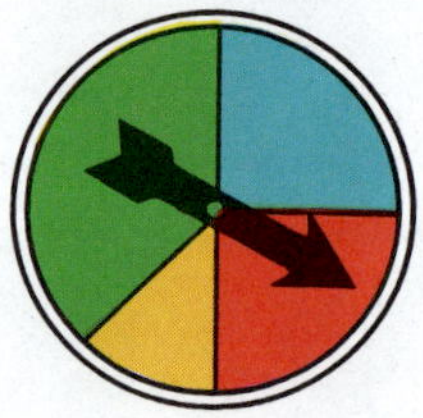

EXERCISES

First tell how many possible outcomes there are for tossing each block. Then tell whether or not the outcomes are equally likely.

1.

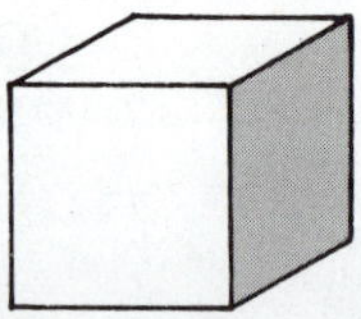

cube

2.

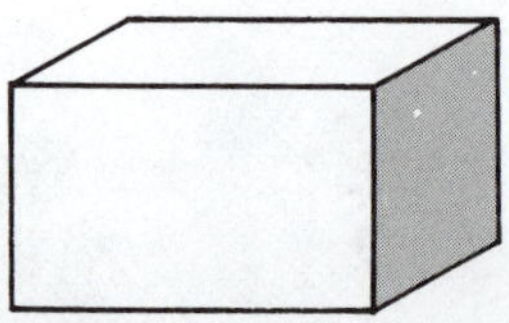

rectangular prism

3.

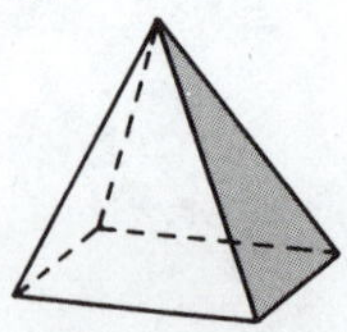

pyramid

4. What are the possible outcomes for this spinner?

   a. Which color would you expect to get most often?

   b. Which color would you expect to get least often?

5. a. Make a chart like this:

Red	*Yellow*	*Blue*

   b. Bend a paper clip like this:

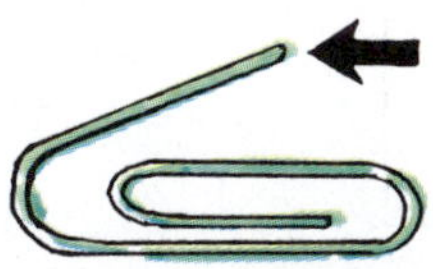

   Drop it 60 times on the square. (If it lands outside the square, drop it again.) The tip of the clip indicates the color to record.

   c. Record your results in the chart.

   d. What outcome occurred most often?

   e. What fraction of the outcomes were red? yellow? blue?

   f. What would you guess about the shaded areas?

A basic counting principle

Tree Diagram of Outcomes

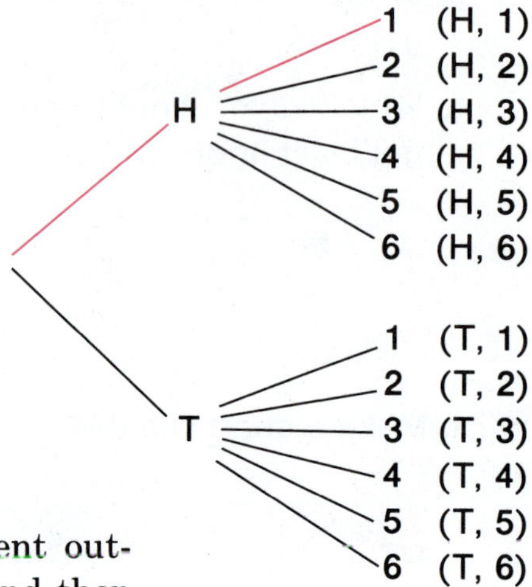

A tree diagram may be used to show the possible outcomes of an event. Here is an example.

Event:

First tossing a coin
and then
tossing a number cube.

Notice that each "branch" of the tree gives a different outcome. The red branch stands for first tossing heads and then tossing a 1.

The tree diagram shows that if a coin is tossed and then a cube is tossed, there are 2×6 (or 12) possible outcomes. This is an example of a basic counting principle.

If a first event has m outcomes, and if a second event has n outcomes, then the first event followed by the second event has $m \times n$ possible outcomes.

EXERCISES

Copy and complete the tree diagram. Then tell how many possible outcomes.

1. *Event:*
Toss a coin
and then spin this spinner.

Tree Diagram of Outcomes

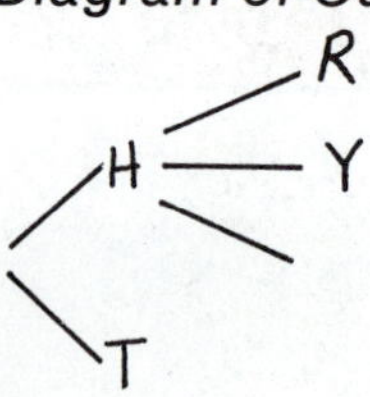

2. *Event:*
Spin this spinner
and then spin this spinner.

Tree Diagram of Outcomes

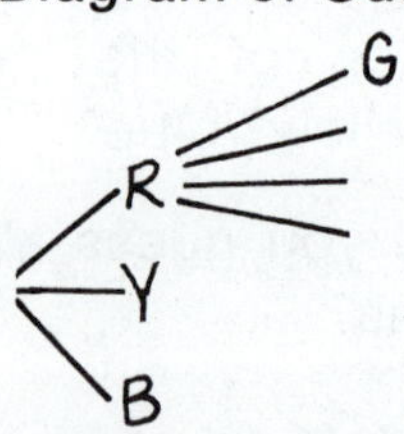

How many possible outcomes?

3. *Event:*
   Toss a dime
   and then toss a nickel.

4. *Event:*
   Toss a red number cube
   and then toss a green number cube.

5. Think about building a 2-digit number by picking one of these cards and then another.

   4 8 3 5 2

   a. How many choices would you have for the first digit?
   b. How many choices would you have for the second digit?
   c. How many 2-digit numbers could you build?

★ 6. The combination for this lock is 35-42-8. (Turn the dial right to 35, left to 42, and right to 8.) How many of these locks could be made without using the same combination twice?

★ 7. If you forgot your combination, how long would it take to try every possible combination? Assume that it takes 10 seconds to try each combination.

KEEPING SKILLS SHARP

Solve each proportion.

1. $\frac{n}{9} = \frac{3}{8}$
2. $\frac{8}{n} = \frac{5}{2}$
3. $\frac{n}{7} = \frac{1}{5}$
4. $\frac{3}{n} = \frac{4}{9}$
5. $\frac{n}{15} = \frac{6}{5}$
6. $\frac{7}{8} = \frac{n}{16}$
7. $\frac{9}{5} = \frac{n}{19}$
8. $\frac{n}{7} = \frac{3}{5}$
9. $\frac{12}{n} = \frac{2}{9}$
10. $\frac{11}{n} = \frac{1}{6}$
11. $\frac{n}{7} = \frac{0}{3}$
12. $\frac{8}{3} = \frac{11}{n}$
13. $\frac{6}{n} = \frac{2}{7}$
14. $\frac{n}{8} = \frac{7}{10}$
15. $\frac{7}{8} = \frac{n}{5}$
16. $\frac{n}{4} = \frac{1}{6}$
17. $\frac{5}{6} = \frac{n}{3}$

Probability

When a number cube is tossed, there are these 6 equally likely outcomes.

The **probability** of each outcome is $\frac{1}{6}$.

For example, when the cube is tossed, the probability of its landing "1 up" is $\frac{1}{6}$. We write

$P(1) = \frac{1}{6}$ (Read as "The probability of 1 equals $\frac{1}{6}$.")

What is $P(6)$? What is $P(\text{not } 6)$?

> If an event has n **equally likely** outcomes, then the probability of one of the outcomes is $\frac{1}{n}$.

Notice the words *equally likely* in the sentence above. The outcomes of spinning this spinner are **not equally likely.** Study these probabilities:

$P(\text{red}) = \frac{1}{2}$

$P(\text{green}) = \frac{1}{4}$

$P(\text{yellow}) = \frac{1}{4}$

The probability of an impossible outcome is 0.
For example, $P(\text{blue}) = 0$.

EXERCISES
Complete.

1. *Event:*
Spin this spinner.

a. $P(\text{red}) = ?$
b. $P(\text{yellow}) = ?$
c. $P(\text{green}) = ?$
d. $P(\text{blue}) = ?$
e. $P(\text{not blue}) = ?$

2. *Event:*
Spin this spinner.

a. $P(\text{red}) = ?$
b. $P(\text{yellow}) = ?$
c. $P(\text{blue}) = ?$
d. $P(\text{green}) = ?$
e. $P(\text{purple}) = ?$

3. *Event:*

Reach in a bag without looking and pick one of these marbles.

$P(\text{green}) = \frac{2}{7}$

a. $P(\text{yellow}) = ?$ **b.** $P(\text{blue}) = ?$

c. $P(\text{red}) = ?$ **d.** $P(\text{not red}) = ?$

4. *Event:*

Reach in a bag and pick one of these marbles.

a. $P(\text{red}) = ?$ **b.** $P(\text{blue}) = ?$

c. $P(\text{yellow}) = ?$ **d.** $P(\text{green}) = ?$

e. $P(\text{not green}) = ?$ **f.** $P(\text{purple}) = ?$

5. If you toss a number cube, what is the probability of getting

a. 6? **b.** an even number? **c.** a number less than 5?

6. Suppose that these cards were thoroughly shuffled and that you picked a card without looking.

1	2	3	4	5	6	7	8	9	10	11	12

What is the probability of picking

a. 4? **b.** an odd number? **c.** a prime number?
d. a multiple of 5? **e.** a composite number? **f.** a number less than 16?

PROJECT

1. Toss two different-colored number cubes at least 50 times and record each outcome with a number pair. For example, the outcome shown can be recorded by using the ordered pair (1, 5).

2. How many times did you get a sum of 7? of 5? of 12? What other sums did you get and how many times?

3. List all possible outcomes of tossing two cubes as shown in step 1.

4. Give the probability of tossing a sum of 7; of 5; of 12.

Probability (independent events)

The tree diagram shows all the possible outcomes of first tossing a number cube and then tossing a coin.

First Event:
Toss this cube.

Second Event:
Toss a coin.

Outcomes of First Event

Outcomes of Second Event

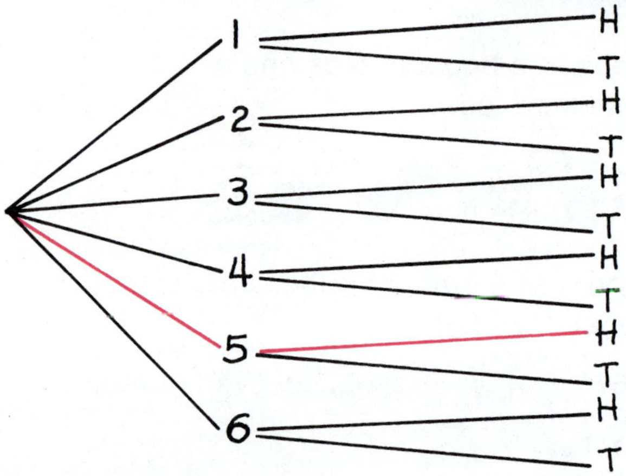

The red branch shows the outcome of first tossing a 5 with the cube and then tossing a head with the coin. Notice that the probability of tossing a 5 is $\frac{1}{6}$, the probability of tossing heads is $\frac{1}{2}$, and the probability of first tossing a 5 **and then** tossing heads is $\frac{1}{6} \times \frac{1}{2}$, or $\frac{1}{12}$. $P(5 \text{ and then H}) = \frac{1}{12}$. This is an example of the following probability principle:

> If the probability of an outcome of a first event is $\frac{a}{m}$, and if the probability of an outcome of a second event is $\frac{b}{n}$, then the probability of the first outcome followed by the second outcome is $\frac{a}{m} \cdot \frac{b}{n}$.

EXERCISES

Give each probability in lowest terms.

1. *First Event:* Spin this spinner.

Second Event: Toss a coin.

a. P(red and then heads) = ?

b. P(yellow and then tails) = ?

c. P(blue and then heads) = ?

2. *First Event:* Spin this spinner.

Second Event: Toss this cube.

a. P(red and then 1) = ?

b. P(green and then 4) = ?

c. P(blue and then 6) = ?

d. P(blue and then an even number) = ?

3. *First Event:* Toss this cube.

Second Event: Toss the cube again.

4 and then 3

a. P(4, 3) = ?

b. P(5, 1) = ?

c. P(6, even number) = ?

d. P(odd number, even number) = ?

4. *First Event:* Toss a coin.

Second Event: Toss the coin again.

Possible outcomes:

(H, H), (T, H)
(H, T), (T, T)

a. What is the probability of getting both heads?
b. What is the probability of getting one head and one tail?

★ 5. Imagine tossing a penny 10 times. What is the probability that you would toss heads all 10 times?

PROJECT

1. Make a spinner like this:

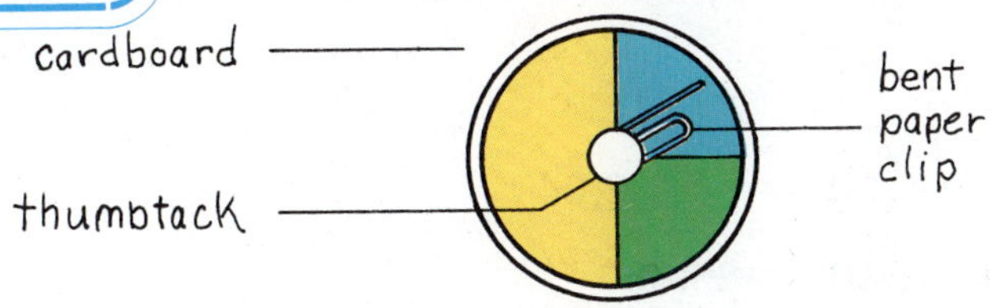

2. Spin the spinner twice and record the outcome by using notation like this:
yellow and then green
(Y, G)
Repeat 50 times.

3. Which outcome did you get the most? Which outcome did you get the least? Did this turn out as you expected?

4. Give the probability of each possible outcome.

Sampling

Suppose that you want to find out how many people in your city can swim. Instead of asking each person, you could ask only some of the people (a **sample**) and then use what you found in your sample to predict how many people in your city can swim. When taking a sample, you should be sure that the sample is not too small and that the sample is not **biased.**

The second sample is biased because most of the people who are in a swimming pool know how to swim.

Suppose that, using an unbiased sample, you found that 140 out of 200 people can swim. You could then use your sample to predict how many people in your city can swim. To do this, you could solve a proportion.

number of swimmers in sample → 140 ; number of people in sample → 200 ; number of swimmers in city → n ; number of people in city → 12,240

$$\frac{140}{200} = \frac{n}{12{,}240}$$

$$\frac{7}{10} = \frac{n}{12{,}240}$$

$$10n = 7 \times 12{,}240$$

$$10n = 85{,}680$$

$$n = 8568$$

From your sample of 200, you would predict that in this city 8568 people can swim.

EXERCISES

1. Suppose that you want to determine the average height of the seventh graders in your school. Tell why one of the following samples would probably be better than the other.

2. Suppose that you wanted to find out the average distance that students travel to get to school. Where would you be more likely to get a biased sample, at the school-bus stop or inside the school? Explain your answer.

3. Below are the results of some samples taken of students at Jefferson Junior High School. The total enrollment is 720 students. Use a proportion to solve each problem.

   a. How many students try out for sports?
   Sample results:
   20 out of 80 try out for sports.

   b. How many students have their own stereo?
   Sample results:
   36 out of 60 have their own stereo.

   c. How many students get at least 8 hours of sleep at night?
   Sample results: 65 out of 90 get at least 8 hours of sleep.

   d. How many students live at least 1 mile from school?
   Sample results: 52 out of 80 live at least 1 mile from school.

Reporting data with graphs

Cascade School raised money by selling T-shirts using the school nickname, Tigers. The results of the sale are shown by the **frequency distribution** in the chart. The frequency column shows the number of students that ordered each design.

T-shirt Sales		
Design	*Tally*	*Frequency*
CJHS Tiger	𝍸 𝍸 𝍸	15
Team Tiger	𝍸 𝍸 III	13
Tigers	𝍸 𝍸	10
I'm a Tiger	𝍸 𝍸 II	12
Tiger Power	𝍸 𝍸 𝍸 III	18
#1 Tiger	𝍸 𝍸 II	12

EXERCISES

Brooke Eastman offered to serve as treasurer for the sale. She made this bar graph by using the frequency distribution.

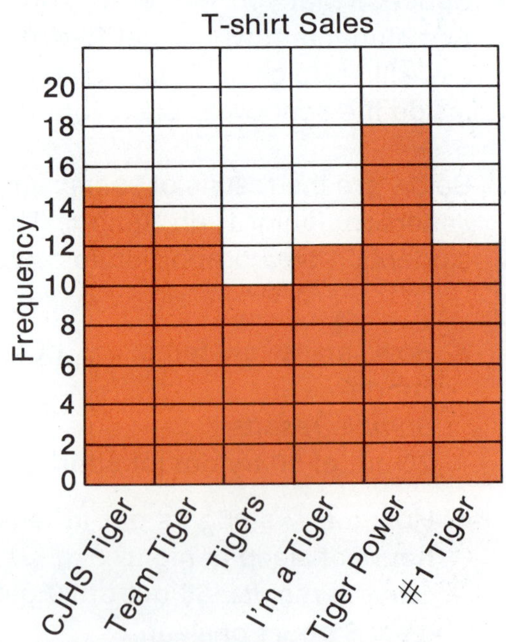

1. How many T-shirts were sold with the design of Team Tiger? I'm a Tiger? #1 Tiger?
2. How many T-shirts were sold in all?
3. What design was sold the most?
4. What design was the second-best seller?
5. What design was sold the least?
6. What design had 13 sales?
7. Did any of the designs sell the same amount? Which ones?

Peter Fairbrother kept track of his work time by making this broken-line graph.

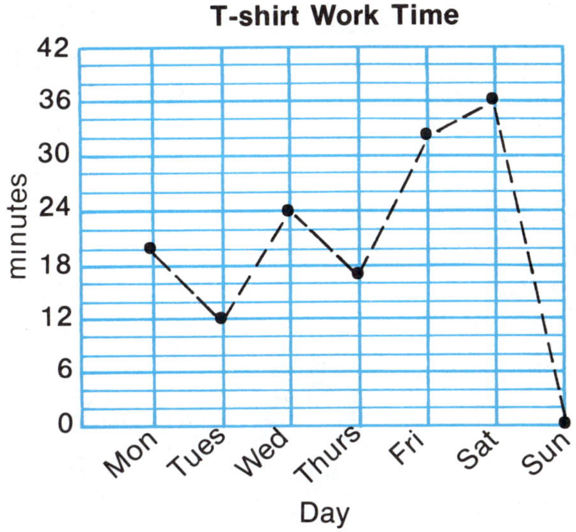

8. Estimate the number of minutes Peter worked on Tuesday. On Friday.

9. On how many days did he work more than 20 minutes?

10. About how many minutes did he work during the week?

11. How many minutes did he average per day?

Brooke made this pictograph to show how much money they took in each week.

12. What were the sales during the first week? The fourth week? The fifth week?

13. How many weeks were sales more than $30? Less than $50?

14. What was the sales total for the 6 weeks? What was the weekly average?

Brooke made a circle graph of the expenses. This graph shows percents.

15. What item represented the greatest expense?

16. How much money was spent for stencils?

17. What was the profit (sales less expenses) for the sale? (See exercise 14.)

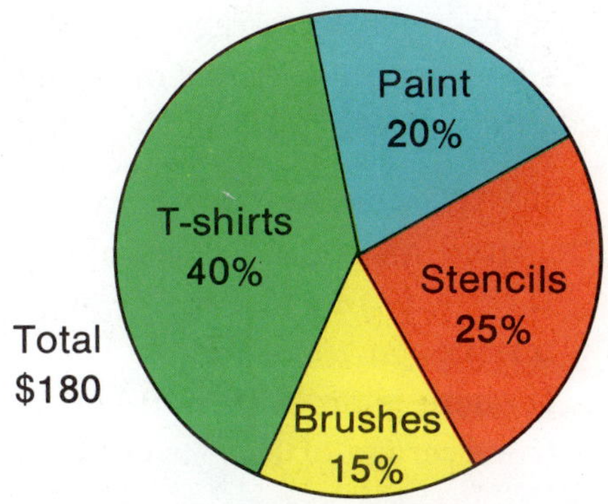

T-shirt Expenses

Range, mode, mean, and median

The table shows the heights of players on a basketball team. Now that these data have been collected, we can determine some numbers that tell us something about the data.

Player		Height (m)
Peterson		1.48
Jones		1.55
Campbell		1.59
Davis		1.63
Roberts		1.63
Garcia		1.69
Allison		1.71
Krantz	*median*	1.72
Moore		1.74
Steinburg		1.75
Mendez		1.78
Richards	*mode*	1.78
Washington		1.78
Hargrove		1.80
Cowan		1.82

range 0.34

The **range** is the difference between the least and the greatest numbers. The range, from 1.48 to 1.82, is 0.34.

The **mode** is the number that occurs most often. The mode is 1.78. It occurs 3 times.

The **mean,** or average, is the sum of all the numbers divided by the number of numbers. The mean, rounded to the nearest hundredth, is 1.70.

number of players → 15; sum of heights → 25.45

```
      1.69
 15)25.45
   -15
    10 4
   - 9 0
     1 45
   - 1 35
       10
```

Once the numbers are arranged from least to greatest, the **median** can be determined.

If there is an odd number of numbers, the median is the number in the middle.

If there is an even number of numbers, the median is the average of the two middle numbers.

The median is 1.72.

EXERCISES

1. In seven games, Krantz scored 21, 23, 19, 21, 26, 18, and 22 points. Find the range, mode, mean, and median.

2. During the first 5 games, the opposing teams scored 54, 48, 49, 56, and 50 points. Find the range, mode, mean, and median.

3. One semester Jane got the following scores on her mathematics tests: 82, 75, 93, 87, 69, 81, 83, and 84. Find the range, mode, mean, and median.

4. During the first 8 weeks of school, Alan averaged \$3.40 per week for lunches. How much did he spend for lunches during the 8 weeks?

5. Hank Aaron holds the major-league home-run record. The table shows how many home runs he hit during each year that he was in the major leagues. Find the range, mode, mean, and median.

Home Runs during Regular Season

Year	*Number*	*Year*	*Number*
1954	13	1966	44
1955	27	1967	39
1956	26	1968	29
1957	44	1969	44
1958	30	1970	38
1959	39	1971	47
1960	40	1972	34
1961	34	1973	40
1962	45	1974	20
1963	44	1975	12
1964	24	1976	10
1965	32		

PROJECT

1. Take your pulse to find out how many times your heart beats in one minute.
2. Determine the range, the mean, and the median of the heart rates of your class.

CHAPTER CHECKUP

First tell how many possible outcomes. Then tell whether or not the outcomes are equally likely. [pages 350–351]

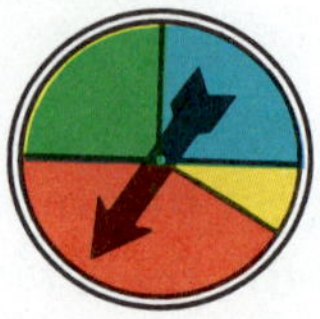

1. Toss a coin. **2.** Toss a paper cup. **3.** Spin this spinner.

Draw a tree diagram that shows the possible outcomes for these events. [pages 352–353]

4. Spin this spinner  and then toss this number cube. 

Think about placing these marbles in a bag, mixing them well, and then (without looking) picking one. Give each probability in lowest terms. [pages 354–355]

5. P(green) = ? **6.** P(blue) = ? **7.** P(yellow) = ?

8. P(red) = ? **9.** P(not red) = ? **10.** P(black) = ?

Suppose that the first event is followed by the second event. Give each probability in lowest terms. [pages 356–357]

First Event:

Spin this spinner.

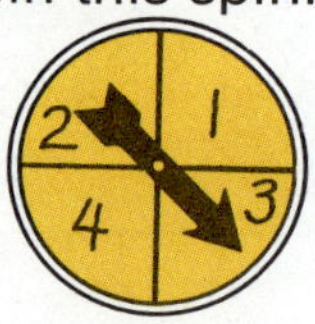

Second Event:

Then (without looking) pick a marble.

11. P(1, blue) = ?

12. P(3, green) = ?

13. P(2, red) = ?

14. P(4, not blue) = ?

[pages 362–363]

Here are the scores of a mathematics test. Give the

15. range.

16. mode.

17. mean.

18. median.

Test Scores							
100	94	88	85	83	76	72	68
100	92	86	85	80	75	70	66
96	91	86	85	78	73	70	62

CHAPTER PROJECT

Take a sample.

1. Ask some of your classmates how far from school they live. Make sure that your sample is reasonably large. Try to include classmates from all parts of your school district.

2. Find the range, the mean, and the median of your sample.

3. Complete a bar graph like the one shown.

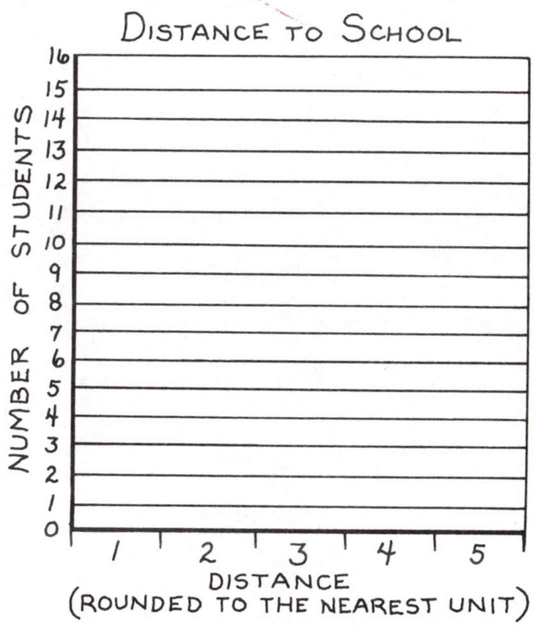

CHAPTER REVIEW

If a first event has m outcomes, and if a second event has n outcomes, then the first event followed by the second event has $m \times n$ outcomes.

First Event:
Toss this cube.

Second Event:
Toss a coin.

Tell how many outcomes there are for the

1. first event.
2. second event.
3. first event and then the second event.

If the probability of a first outcome is $\frac{a}{m}$ and if the probability of a second outcome is $\frac{b}{n}$, then the probability of the first outcome followed by the second outcome is $\frac{a}{m} \cdot \frac{b}{n}$.

Give the probability of

4. tossing a 6.
5. tossing heads.
6. tossing a 6 and then tossing heads.

The range is the difference between the least and greatest numbers.

The mean is the average.

The median is the middle number.

The mode is the most frequent number or numbers.

One week Sarah kept this record of the number of calories in her diet.
Give the

7. range.
8. mean.
9. median.
10. mode.

Daily Calories	
Sun.	1275
Mon.	1550
Tues.	1320
Wed.	1150
Thurs.	1260
Fri.	1320
Sat.	1560

CHAPTER CHALLENGE

When we wish to talk about the chances of picking a green button from the box, we can use probability. The probability of picking green is the ratio of the number of green buttons to the total number of buttons.

$$\text{probability of green} = \frac{\text{number of green}}{\text{total number}} = \frac{4}{12} = \frac{1}{3}$$

Another way of talking about the chance of getting a green button is to use **odds.** The odds in favor of picking green is the ratio of green buttons to not-green buttons.

$$\text{odds in favor of green} = \frac{\text{number of green}}{\text{number of not green}} = \frac{4}{8} = \frac{1}{2}$$

The odds against getting green is the reciprocal of the odds in favor of getting green.

$$\text{odds against green} = \frac{\text{number of not green}}{\text{number of green}} = \frac{8}{4} = \frac{2}{1}$$

Give the odds in favor of picking a

1. red button.
2. blue button.
3. yellow button.

Give the odds against picking a

4. yellow button.
5. blue button.
6. red button.

Give the odds

7. in favor of spinning yellow.
8. in favor of spinning blue.
9. against spinning red.
10. against spinning green.

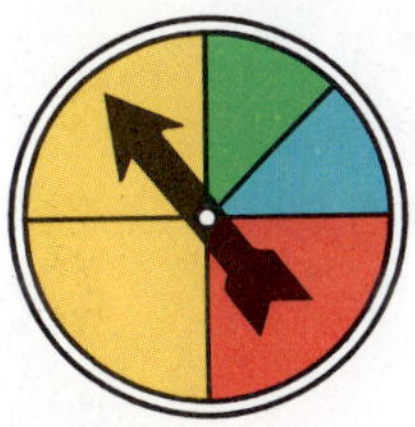

b c d 2. a b c d a b c d a b c d 5. a b c d 6. a b c d

MAJOR CHECKUP
STANDARDIZED FORMAT

b c d 8. a b c d 9. a b c d 10. a b c d 11. a b c d 12. a b c d

Choose the correct letter.

1. Round 697,298,000 to the nearest ten million.

a. 697,300,000
b. 698,000,000
c. 700,000,000
d. none of these

2. Complete.

2.3 L = ___?___

a. 0.023 mL
b. 2300 mL
c. 230 mL
d. none of these

3. The prime factorization of 48 is

a. 6×8
b. $2 \times 2 \times 3 \times 3$
c. $2 \times 2 \times 3 \times 4$
d. none of these

4. Add.

$$7\frac{5}{6} + 5\frac{1}{4}$$

a. $13\frac{1}{12}$
b. $12\frac{7}{12}$
c. $12\frac{1}{12}$
d. none of these

5. Find the product.

$6\frac{2}{3} \times 2\frac{5}{8}$

a. $17\frac{1}{2}$
b. $\frac{2}{35}$
c. $2\frac{35}{63}$
d. none of these

6. Give the perimeter.

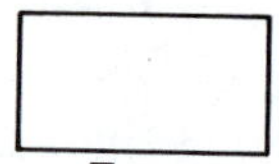
4 cm
7 cm

a. 11 cm
b. 22 cm
c. 28 cm
d. none of these

7. 25% of 192 = ___?___

a. 48
b. 768
c. $38\frac{2}{5}$
d. none of these

8. Compute the interest.

principal: $240
rate: 8% per year
time: 6 months

a. $192
b. $9.60
c. $19.20
d. none of these

9. The measure of an acute angle is

a. greater than 90°.
b. 90°.
c. less than 90°.
d. none of these

10. *P*(yellow) = ___?___

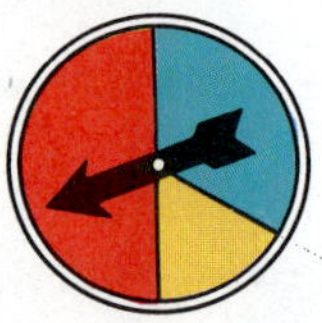

a. $\frac{1}{2}$
b. $\frac{1}{3}$
c. $\frac{1}{4}$
d. none of these

11. A gallon contains 8 pints. How many gallons contain 46 pints?

a. 6 **b.** $5\frac{3}{4}$
c. 368 **d.** none of these

12. Donna bought $2\frac{1}{2}$ dozen large eggs for $2.30. How much did 1 dozen cost?

a. $.92 **b.** $1.09
c. $5.75 **d.** none of these

Computing with Integers

13

Integers

Imagine containers that have positive or negative charges in them.

Mixtures of such charges form as many positive-negative pairs as possible. The pairs have no charge. Count the unpaired charges to find the charge in the container.

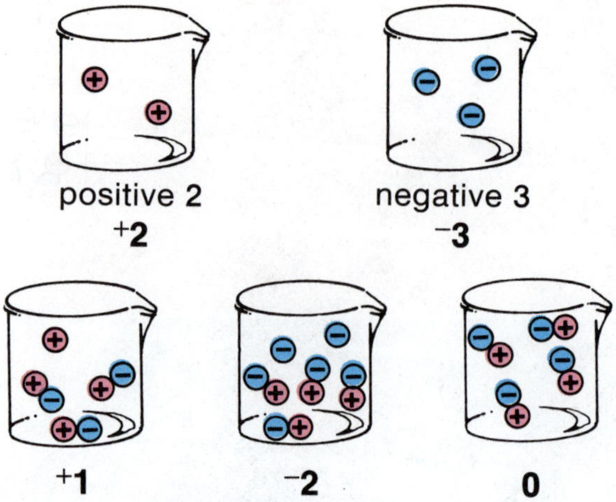

Positive numbers and **negative numbers** like these (and 0) are called **integers.**

integers: $^{-}6$, $^{+}45$, 0, $^{-}876$ *not integers:* $^{-}3.2$, $^{+}\frac{1}{2}$

The integers can be shown on the number line like this:

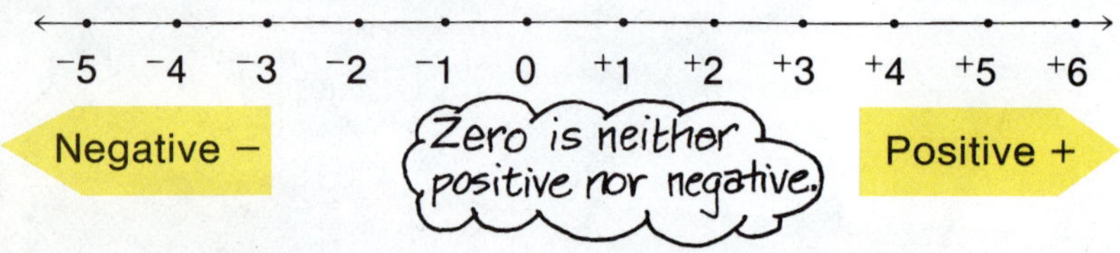

You can use a number line to compare two integers. The greater of two integers is always to the right. The number line above shows that

$^{-}4 < {}^{-}3$ $^{+}2 < {}^{+}3$ $0 < {}^{+}1$ $^{+}4 > {}^{+}3$ $^{+}1 > {}^{-}5$

(is less than) (is greater than)

Integers can be used in situations where opposites are involved.

$^{+}25$ could represent a 25° rise in temperature.
$^{-}10$ could represent a $10 decrease in the price of stock.
$^{+}15$ could represent a football team's gain of 15 yards.

EXERCISES

Use integers to represent the charges in these containers.

1. 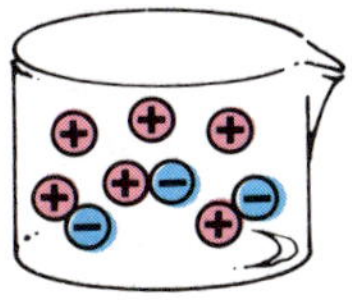

2. 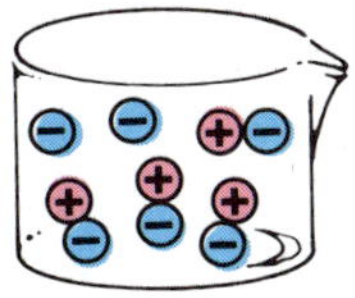

3. 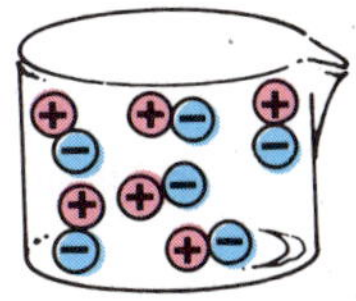

< or >?

4. $^{-}8$ ● $^{-}4$
5. $^{+}8$ ● $^{-}4$
6. $^{-}8$ ● $^{+}4$
7. $^{+}8$ ● $^{+}4$
8. $^{+}6$ ● $^{+}3$
9. $^{+}6$ ● $^{-}3$
10. $^{-}6$ ● $^{+}3$
11. $^{-}6$ ● $^{-}3$

Complete.

12. __?__ is neither positive nor negative.
13. All __?__ integers are greater than 0.
14. All __?__ integers are less than 0.
15. Zero is __?__ than any negative integer.
16. A negative integer is __?__ than any positive integer.

Solve.

17. If $^{+}8$ stands for a gain of 8 meters, what does $^{-}4$ stand for?
18. If $^{-}10$ stands for 10° below zero, what does $^{+}16$ stand for?
19. If $^{-}40$ stands for 40 meters below sea level, then what stands for 220 meters above sea level?
20. If $^{+}3$ stands for 3 centimeters above the average height, then what stands for 2 centimeters below the average height?
21. Gerry used integers to keep a record of how much she earned or spent each day. The table shows that she earned 75¢ on Monday. Did Gerry have more or less money at the end of the week than at the beginning? How much more or less?

Monday	$^{+}75$¢
Tuesday	$^{+}60$¢
Wednesday	$^{-}50$¢
Thursday	$^{+}75$¢
Friday	$^{-}80$¢
Saturday	$^{+}80$¢
Sunday	$^{+}50$¢

Adding integers

To understand how to add integers, you can think about putting charges together.

EXAMPLE 1. $^{+}5 + {}^{+}2 = ?$

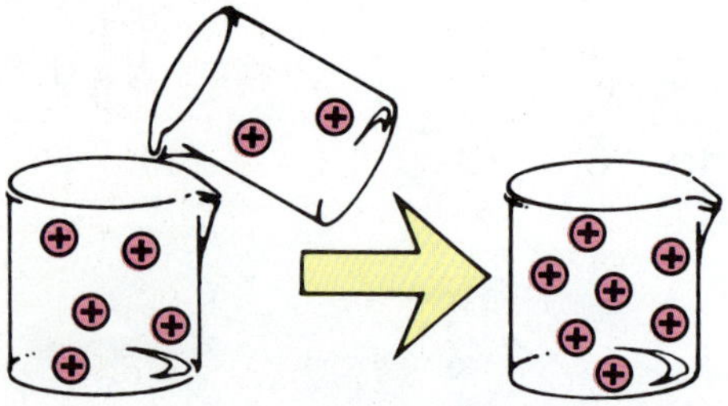

$^{+}5 + {}^{+}2 = {}^{+}7$

EXAMPLE 2. $^{-}3 + {}^{-}4 = ?$

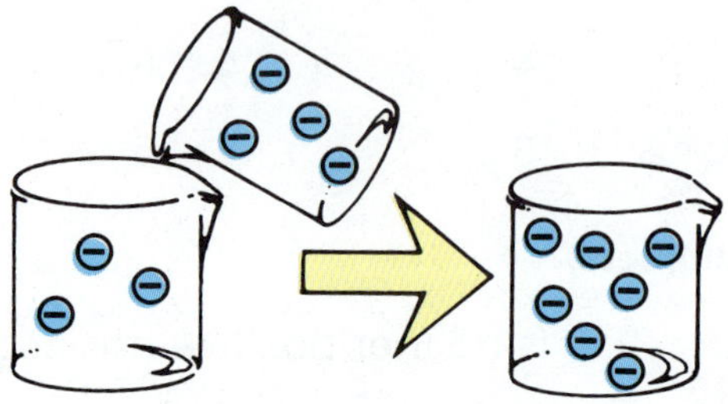

$^{-}3 + {}^{-}4 = {}^{-}7$

EXAMPLE 3. $^{+}6 + {}^{-}3 = ?$

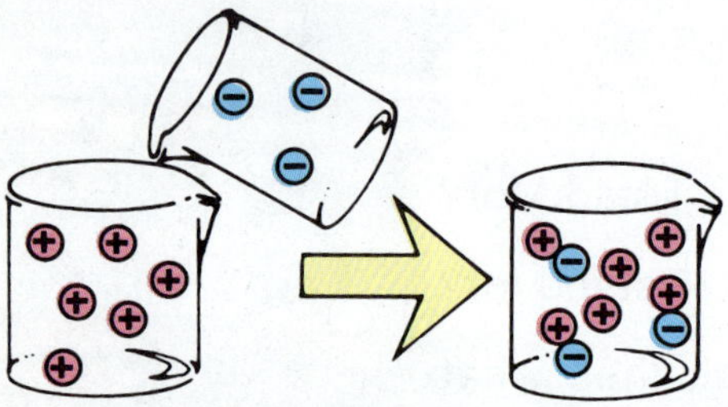

$^{+}6 + {}^{-}3 = {}^{+}3$

EXAMPLE 4. $^{-}5 + {}^{+}1 = ?$

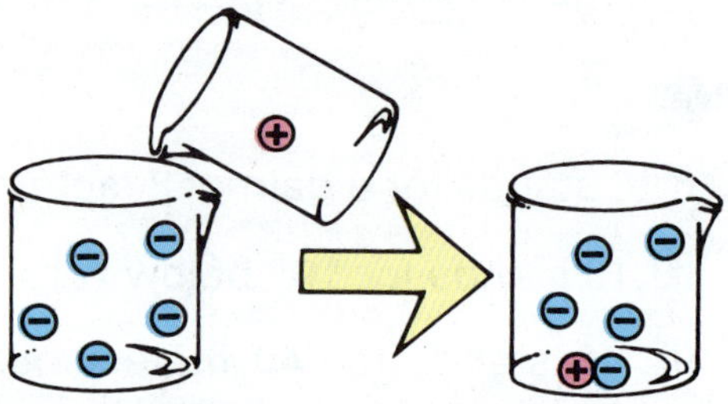

$^{-}5 + {}^{+}1 = {}^{-}4$

EXAMPLE 5. $^{+}3 + {}^{-}3 = ?$

$^{+}3 + {}^{-}3 = 0$

$^{+}3$ and $^{-}3$ are called **opposites** because their sum is 0.

$^{-}45$ is the opposite of $^{+}45$. 0 is its own opposite.

EXERCISES

Give each sum.

1. $^{+}8 + ^{+}3$	**2.** $^{+}8 + ^{-}3$	**3.** $^{-}8 + ^{+}3$	**4.** $^{-}8 + ^{-}3$
5. $^{+}6 + ^{+}9$	**6.** $^{+}6 + ^{-}9$	**7.** $^{-}6 + ^{+}9$	**8.** $^{-}6 + ^{-}9$
9. $^{+}7 + ^{+}5$	**10.** $^{+}7 + ^{-}5$	**11.** $^{-}7 + ^{+}5$	**12.** $^{-}7 + ^{-}5$
13. $^{+}9 + ^{-}9$	**14.** $^{+}4 + ^{-}10$	**15.** $^{-}4 + ^{+}10$	**16.** $^{-}6 + ^{+}11$
17. $^{+}2 + ^{-}4$	**18.** $^{-}6 + ^{-}8$	**19.** $^{+}5 + 0$	**20.** $0 + ^{-}6$
21. $^{-}9 + ^{-}2$	**22.** $^{+}6 + ^{+}9$	**23.** $^{+}10 + ^{-}4$	**24.** $^{+}12 + ^{-}3$
25. $^{-}6 + ^{-}7$	**26.** $^{-}6 + ^{+}7$	**27.** $^{+}6 + ^{-}7$	**28.** $^{+}6 + ^{+}7$
29. $^{-}24 + ^{+}39$	**30.** $^{+}23 + ^{+}36$	**31.** $^{-}53 + ^{-}26$	**32.** $^{+}74 + ^{-}60$

Solve.

★ **33.** Central High School won the city championship. The football coach used integers to record yards gained or lost each time the ball was carried. For example, he used $^{+}6$ for a gain of 6 yards.

a. How many yards did Berry gain?

b. Who netted the most yards?

c. How many yards did Martinez and Phillips net together?

Player			
Berry	*Allan*	*Martinez*	*Phillips*
$^{+}6$	$^{-}4$	$^{-}6$	$^{+}3$
$^{-}2$	$^{+}8$	$^{+}5$	0
$^{+}4$	$^{+}3$	$^{+}9$	$^{-}4$
0	$^{+}2$	$^{+}1$	$^{+}6$
$^{-}3$	$^{-}5$	$^{-}2$	$^{+}2$
$^{+}5$		$^{+}1$	$^{+}1$
$^{+}11$			$^{-}1$

Copy and complete these addition boxes.
Add across and add down.

34.

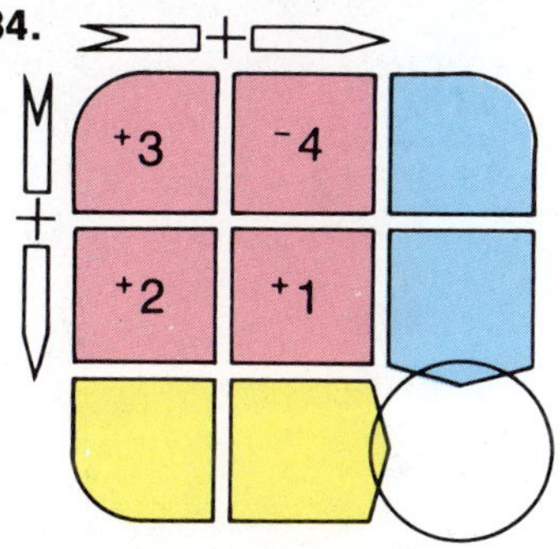

35.

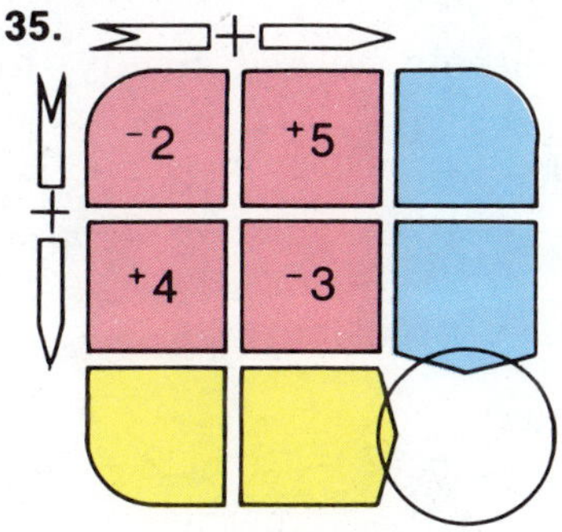

36.

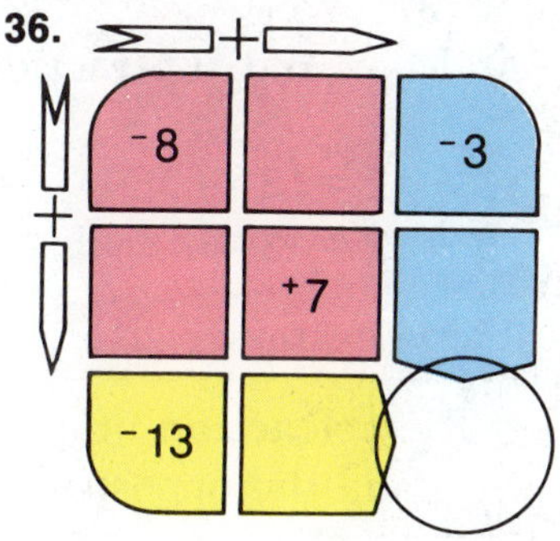

Subtracting integers

You can understand subtraction of integers by thinking about taking charges out of a container.

EXAMPLE 1. $^{+}5 - {}^{+}2 = ?$

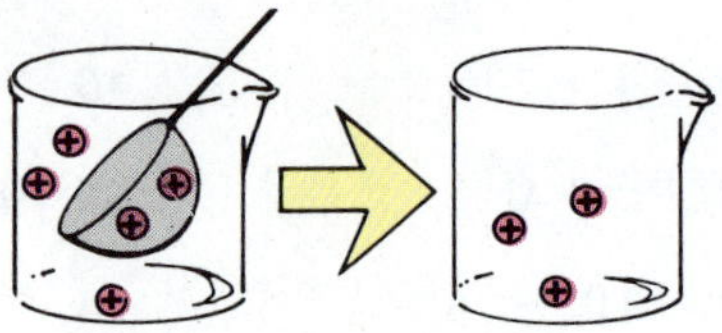

$^{+}5 - {}^{+}2 = {}^{+}3$

EXAMPLE 2. $^{-}4 - {}^{-}3 = ?$

$^{-}4 - {}^{-}3 = {}^{-}1$

EXAMPLE 3. $^{+}2 - {}^{+}4 = ?$

We do not have 4 ⊕ to take out.

Draw at least two ⊕⊖ pairs.

$^{+}2 - {}^{+}4 = {}^{-}2$

EXAMPLE 4. $^{+}2 - {}^{-}3 = ?$

We do not have 3 ⊖ to take out.

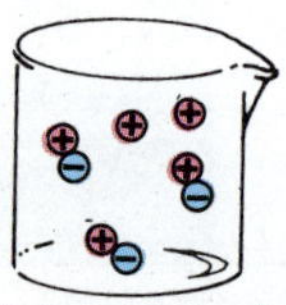

Draw at least three ⊕⊖ pairs.

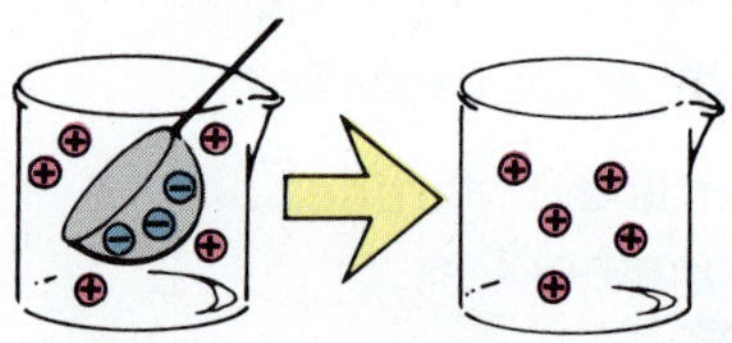

$^{+}2 - {}^{-}3 = {}^{+}5$

Addition and subtraction are closely related. Study these examples.

$^{+}5 - {}^{+}2 = {}^{+}3$	$^{-}4 - {}^{-}3 = {}^{-}1$	$^{+}2 - {}^{+}4 = {}^{-}2$	$^{+}2 - {}^{-}3 = {}^{+}5$
$^{+}5 + {}^{-}2 = {}^{+}3$	$^{-}4 + {}^{+}3 = {}^{-}1$	$^{+}2 + {}^{-}4 = {}^{-}2$	$^{+}2 + {}^{+}3 = {}^{+}5$

These examples give a way to subtract.

> To subtract an integer, add the opposite.

$^{-}3 - {}^{+}2 = {}^{-}3 + {}^{-}2 = {}^{-}5$

EXERCISES

Complete.

1. To subtract $^{+}2$, add ___?___.
2. To subtract $^{-}5$, add ___?___.
3. To subtract an integer, add the ___?___ of the integer.

Give each difference.

$^{+}8 + {}^{-}3 = {}^{+}5$

4. $^{+}8 - {}^{+}3$

$^{-}5 + {}^{+}2 = {}^{-}3$

5. $^{-}5 - {}^{-}2$

6. $^{-}3 - {}^{+}9$
7. $^{+}12 - {}^{-}3$
8. $^{-}5 - {}^{-}7$
9. $^{+}7 - 0$
10. $^{-}8 - {}^{+}3$
11. $^{-}2 - {}^{-}3$
12. $^{+}3 - {}^{-}8$
13. $^{-}5 - {}^{-}6$
14. $^{-}10 - {}^{+}8$
15. $^{-}3 - {}^{-}8$
16. $^{+}3 - {}^{+}6$
17. $^{-}9 - {}^{+}7$
18. $0 - {}^{-}5$
19. $^{+}15 - {}^{-}4$
20. $^{-}4 - {}^{+}6$
21. $^{-}8 - {}^{-}7$
22. $^{+}12 - 0$
23. $^{+}8 - {}^{-}12$
24. $^{-}3 - {}^{+}13$
25. $^{+}15 - {}^{-}6$
26. $0 - {}^{-}4$
27. $^{+}13 - {}^{-}3$
28. $^{-}10 - {}^{-}12$
29. $^{+}18 - {}^{+}13$
30. $^{-}35 - {}^{-}35$
31. $^{+}35 - {}^{-}35$
32. $^{-}19 - {}^{+}6$
33. $0 - {}^{-}6$
34. $0 - {}^{+}19$
35. $^{+}23 - {}^{-}17$
36. $^{-}26 - {}^{-}35$
37. $^{+}38 - {}^{-}59$
38. $0 - 0$
39. $^{+}16 - {}^{+}4$
40. $^{+}13 - {}^{-}14$
41. $^{-}8 - {}^{+}3$
42. $^{-}18 - {}^{+}13$
43. $^{+}18 - {}^{-}13$
44. $^{-}18 - {}^{-}13$

45. $1 - 2 + 3 - 4 + 5 - 6 + 7 - \cdots + 1001 = ?$

KEEPING SKILLS SHARP

Change to a percent.

1. $\frac{1}{2}$
2. $\frac{1}{4}$
3. $\frac{3}{4}$
4. $\frac{3}{5}$
5. $\frac{1}{3}$
6. $\frac{2}{3}$
7. $\frac{1}{8}$
8. $\frac{9}{8}$
9. $\frac{5}{6}$
10. $\frac{11}{6}$
11. $\frac{1}{9}$
12. $\frac{9}{16}$
13. $\frac{8}{5}$
14. $\frac{7}{9}$
15. $\frac{5}{8}$
16. $\frac{7}{6}$
17. $\frac{4}{5}$
18. $\frac{3}{10}$
19. $\frac{11}{9}$
20. $\frac{7}{8}$
21. $\frac{3}{2}$
22. $\frac{4}{15}$
23. $\frac{8}{9}$
24. $\frac{5}{12}$
25. $\frac{2}{5}$
26. $\frac{3}{8}$
27. $\frac{4}{9}$
28. $\frac{2}{9}$
29. $\frac{7}{15}$
30. $\frac{11}{20}$

Practice exercises

Give each sum.

1. $^{+}2 + {}^{+}6$
2. $^{-}5 + {}^{-}9$
3. $^{+}3 + {}^{-}7$
4. $0 + {}^{-}9$
5. $^{-}7 + {}^{+}8$
6. $^{+}3 + {}^{+}5$
7. $^{+}4 + {}^{-}5$
8. $^{-}9 + {}^{-}5$
9. $^{-}5 + {}^{-}8$
10. $^{-}7 + {}^{-}9$
11. $^{-}2 + {}^{+}8$
12. $^{+}6 + {}^{+}4$
13. $0 + {}^{+}8$
14. $^{-}8 + {}^{+}5$
15. $^{-}6 + {}^{-}9$
16. $^{-}8 + {}^{+}8$
17. $^{+}8 + {}^{+}9$
18. $^{+}3 + {}^{-}9$
19. $^{+}7 + {}^{+}4$
20. $0 + 0$
21. $^{+}9 + {}^{-}9$
22. $^{-}6 + {}^{+}8$
23. $^{-}9 + {}^{-}8$
24. $^{+}8 + {}^{+}8$
25. $^{+}10 + {}^{+}7$
26. $^{+}15 + {}^{-}9$
27. $^{+}13 + {}^{-}10$
28. $^{-}9 + {}^{-}15$
29. $^{-}23 + 0$
30. $^{+}28 + {}^{+}19$
31. $^{+}21 + {}^{-}21$
32. $^{-}24 + {}^{-}16$

Give each difference.

33. $^{+}9 - {}^{+}2$
34. $^{-}5 - 0$
35. $^{+}8 - {}^{+}5$
36. $^{-}6 - {}^{+}6$
37. $^{-}6 - {}^{-}7$
38. $^{+}6 - {}^{-}5$
39. $^{-}8 - {}^{-}7$
40. $^{-}5 - {}^{+}6$
41. $^{-}9 - {}^{+}9$
42. $^{+}3 - {}^{+}9$
43. $^{-}4 - {}^{-}9$
44. $^{+}8 - {}^{+}4$
45. $^{+}8 - {}^{-}3$
46. $^{-}9 - {}^{-}8$
47. $^{+}2 - {}^{-}9$
48. $0 - {}^{+}9$
49. $^{-}9 - {}^{+}6$
50. $^{+}3 - {}^{-}8$
51. $^{+}8 - {}^{+}8$
52. $^{-}6 - {}^{-}6$
53. $^{+}5 - {}^{-}7$
54. $^{+}6 - {}^{+}8$
55. $^{+}7 - {}^{-}6$
56. $^{-}5 - {}^{+}9$
57. $^{+}15 - {}^{-}6$
58. $^{+}18 - {}^{-}9$
59. $^{-}17 - {}^{-}9$
60. $^{+}15 - {}^{+}6$
61. $^{+}23 - {}^{+}5$
62. $^{-}26 - {}^{+}24$
63. $^{-}25 - {}^{+}28$
64. $^{+}28 - {}^{-}21$

Problem solving

Solve.

1. Which city on the map had a high temperature of 9°C and a low temperature of ⁻3°C?

2. Which city had a difference of 11° between its low and high temperatures?

3. Which city had a high temperature that was 16° more than Chicago's high?

4. Which city had a low temperature that was 32° less than Miami's low?

5. At 6:00 A.M. the temperature in Denver was ⁻1°C. By noon it had risen 8°. By sunset it had fallen 6° from the noon temperature. What was the temperature at sunset?

6. In Detroit the temperature was 5°C at noon. It rose 4° by 3:00 P.M. Then it dropped 11° by midnight. What was the temperature at midnight?

7. The temperature in New York rose 2° per hour from 9:00 A.M. until noon. At 9:00 A.M. the temperature was ⁻1°C. What was the temperature at noon?

8. The temperature in Kansas City dropped 1° per hour from 7:00 P.M. until midnight. At 9:00 P.M. the temperature was 2°C. What was the temperature at midnight?

Multiplying integers

To multiply integers, you can think of "putting charges in" as positive and "taking charges out" as negative.

EXAMPLE 1. $^{+}2 \times {}^{+}3 = ?$

Put in two sets of $^{+}3$ charges.

$^{+}2 \times {}^{+}3 = {}^{+}6$

EXAMPLE 2. $^{+}2 \times {}^{-}3 = ?$

Put in two sets of $^{-}3$ charges.

$^{+}2 \times {}^{-}3 = {}^{-}6$

EXAMPLE 3. $^{-}2 \times {}^{+}3 = ?$

We need to take out two sets of $^{+}3$ charges. So, draw at least 6 ⊕⊖ pairs.

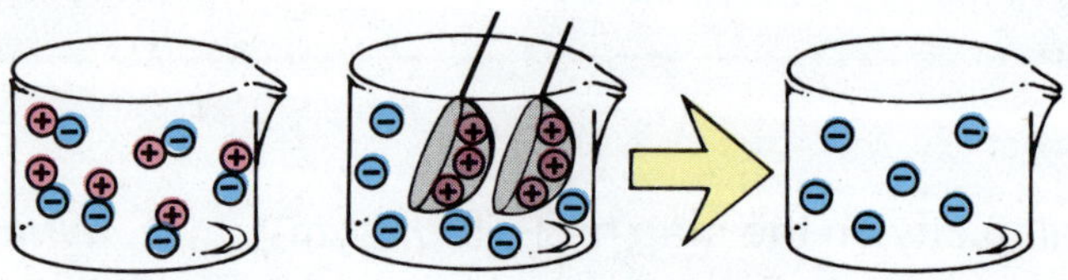

$^{-}2 \times {}^{+}3 = {}^{-}6$

EXAMPLE 4. $^{-}2 \times {}^{-}3 = ?$

We need to take out two sets of $^{-}3$ charges. So, draw at least 6 ⊕⊖ pairs.

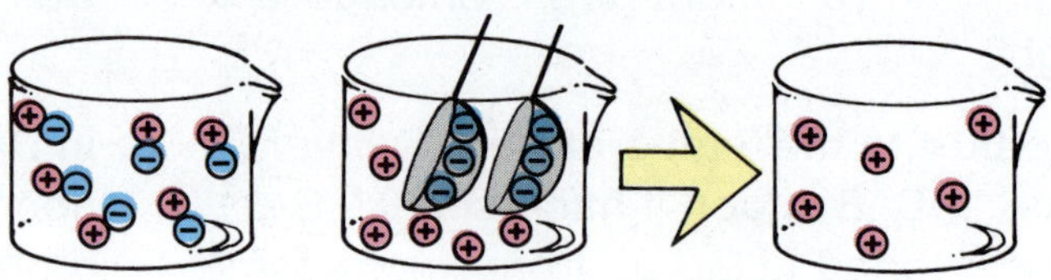

$^{-}2 \times {}^{-}3 = {}^{+}6$

These examples give us a way to multiply.

> The product of two integers with the *same* sign is *positive*.
> The product of two integers with *different* signs is *negative*.
> The product of any integer and 0 is 0.

Complete.
Positive, negative, or zero?

1. The product of two positive integers is a ___?___ integer.

2. The product of two negative integers is a ___?___ integer.

3. The product of a positive integer and a negative integer is a ___?___ integer.

4. The product of an integer and 0 is ___?___.

Give each product.

5. $^{+}3 \times {}^{+}4$	6. $^{+}3 \times {}^{-}4$	7. $^{-}3 \times {}^{+}4$	8. $^{-}3 \times {}^{-}4$
9. $^{+}5 \times {}^{+}6$	10. $^{+}5 \times {}^{-}6$	11. $^{-}5 \times {}^{+}6$	12. $^{-}5 \times {}^{-}6$
13. $^{+}7 \times {}^{+}5$	14. $^{+}7 \times {}^{-}5$	15. $^{-}7 \times {}^{+}5$	16. $^{-}7 \times {}^{-}5$
17. $^{+}8 \times {}^{+}6$	18. $^{+}8 \times {}^{-}6$	19. $^{-}8 \times {}^{+}6$	20. $^{-}8 \times {}^{-}6$
21. $^{+}9 \times {}^{+}6$	22. $^{+}9 \times {}^{-}6$	23. $^{-}9 \times {}^{+}6$	24. $^{-}9 \times {}^{-}6$
25. $^{+}9 \times {}^{+}9$	26. $^{-}7 \times {}^{+}7$	27. $^{-}6 \times 0$	28. $^{-}7 \times {}^{-}8$
29. 0×0	30. $0 \times {}^{+}5$	31. $^{-}8 \times {}^{+}8$	32. $^{+}8 \times {}^{-}9$
33. $^{-}7 \times {}^{-}9$	34. $^{+}9 \times {}^{+}8$	35. $^{-}6 \times {}^{-}7$	36. $^{+}3 \times {}^{-}9$
37. $^{-}12 \times {}^{+}5$	38. $^{-}16 \times {}^{+}3$	39. $^{+}9 \times {}^{+}15$	40. $^{-}16 \times {}^{+}4$
41. $^{+}24 \times {}^{-}3$	42. $^{+}12 \times {}^{+}18$	43. $^{+}15 \times {}^{-}15$	44. $^{-}19 \times 0$

Is the missing factor positive, negative, or zero?

45. $? \times {}^{-}3 = {}^{-}24$	46. $^{+}6 \times ? = {}^{+}18$	47. $? \times {}^{-}8 = {}^{-}72$
48. $? \times {}^{+}7 = {}^{+}35$	49. $^{-}3 \times ? = {}^{+}3$	50. $? \times {}^{+}3 = 0$

Give the missing factor.

51. $? \times {}^{-}3 = {}^{-}24$	52. $^{+}6 \times ? = {}^{+}18$	53. $? \times {}^{-}8 = {}^{-}72$
54. $? \times {}^{+}7 = {}^{+}35$	55. $^{-}3 \times ? = {}^{+}3$	56. $? \times {}^{+}3 = 0$

Dividing integers

You can find the quotient of two integers by finding a missing factor. Remember that multiplication and division are related.

EXAMPLE 1. $^{+}24 \div {}^{+}8 = ?$

$^{+}8 \times ? = {}^{+}24$ $\quad$ $^{+}8 \times {}^{+}3 = 24$ $\qquad$ $^{+}24 \div {}^{+}8 = {}^{+}3$

EXAMPLE 2. $^{-}24 \div {}^{+}8 = ?$

$^{+}8 \times ? = {}^{-}24$ $\quad$ $^{+}8 \times {}^{-}3 = {}^{-}24$ $\qquad$ $^{-}24 \div {}^{+}8 = {}^{-}3$

EXAMPLE 3. $^{-}24 \div {}^{-}8 = ?$

$^{-}8 \times ? = {}^{-}24$ $\quad$ $^{-}8 \times {}^{+}3 = {}^{-}24$ $\qquad$ $^{-}24 \div {}^{-}8 = {}^{+}3$

EXAMPLE 4. $^{+}24 \div {}^{-}8 = ?$

$^{-}8 \times ? = {}^{+}24$ $\quad$ $^{-}8 \times {}^{-}3 = {}^{+}24$ $\qquad$ $^{+}24 \div {}^{-}8 = {}^{-}3$

These examples give us a way to divide.

> The quotient of two integers with the *same* sign is *positive*.
> The quotient of two integers with *different* signs is *negative*.
> The quotient of 0 divided by any nonzero integer is 0.

EXERCISES

Positive, negative, or zero?

1. If you divide a positive integer by a positive integer the quotient is ___?___.
2. If you divide a negative integer by a negative integer the quotient is ___?___.

3. The quotient of a positive integer divided by a negative integer is __?__.

4. The quotient of a negative integer divided by a positive integer is __?__.

5. The quotient of 0 divided by any integer is __?__.

Give each quotient.

6. $^{+}12 \div {}^{+}6$	7. $^{-}16 \div {}^{-}2$	8. $^{-}18 \div {}^{+}9$
9. $^{+}12 \div {}^{-}4$	10. $^{-}20 \div {}^{-}4$	11. $^{+}21 \div {}^{+}7$
12. $^{-}40 \div {}^{-}8$	13. $^{+}24 \div {}^{-}3$	14. $^{-}18 \div {}^{+}3$
15. $^{-}28 \div {}^{+}7$	16. $^{+}64 \div {}^{+}8$	17. $^{+}42 \div {}^{-}7$
18. $^{-}36 \div {}^{-}9$	19. $^{+}48 \div {}^{-}6$	20. $0 \div {}^{-}6$
21. $^{-}63 \div {}^{-}9$	22. $^{+}36 \div {}^{+}4$	23. $^{-}54 \div {}^{-}9$
24. $0 \div {}^{-}8$	25. $^{+}72 \div {}^{-}8$	26. $^{-}45 \div {}^{+}5$
27. $^{-}32 \div {}^{+}2$	28. $^{+}45 \div {}^{-}3$	29. $^{+}84 \div {}^{+}4$
30. $^{-}52 \div {}^{-}4$	31. $^{+}95 \div {}^{+}5$	32. $^{-}72 \div {}^{-}6$
33. $^{-}96 \div {}^{+}24$	34. $^{-}98 \div {}^{-}14$	35. $^{+}65 \div {}^{-}13$

$<$, $=$, or $>$?

36. $^{+}3 + {}^{-}3$ ● $^{+}3$	37. $^{-}6 - {}^{+}2$ ● 0
38. $^{-}2 \times {}^{-}1$ ● $^{-}4$	39. $^{-}8 \div {}^{+}2$ ● $^{-}4$
40. $^{-}4 \times {}^{+}2$ ● $^{+}8$	41. $^{-}5 - {}^{-}6$ ● $^{-}11$
★ 42. $^{+}2 \times {}^{-}3$ ● $^{+}2 + {}^{-}3$	★ 43. $^{-}12 \times {}^{+}2$ ● $^{-}12 \div {}^{+}2$
★ 44. $0 \times {}^{+}6$ ● $^{-}4 + {}^{+}5$	★ 45. $^{-}6 - {}^{-}4$ ● $^{-}6 + {}^{+}4$
★ 46. $^{-}6 \div {}^{-}2$ ● $^{-}6 + {}^{-}2$	★ 47. $^{+}8 + {}^{-}4$ ● $^{+}8 \div {}^{-}4$

48. True or false?

$$1 \div 2 \times 3 \div 4 \times 5 \div 6 \times 7 \div 8 \times \cdots \div 50 > 0.1$$

KEEPING SKILLS SHARP

Solve.

1. 18% of 50 = n
2. 46% of 150 = n
3. 36% of 25 = n
4. $37\frac{1}{2}$% of 96 = n
5. $66\frac{2}{3}$% of 81 = n
6. 110% of 80 = n
7. 16% of 50 = n
8. $83\frac{1}{3}$% of 300 = n
9. 1% of 1000 = n
10. 200% of 25 = n

11. 15% of n = 15
12. 23% of n = 46
13. 20% of n = 16
14. $33\frac{1}{3}$% of n = 18
15. $62\frac{1}{2}$% of n = 55
16. $83\frac{1}{3}$% of n = 95
17. 40% of n = 28
18. 75% of n = 93
19. $16\frac{2}{3}$% of n = 13
20. 128% of n = 128

Practice exercises

Give each product.

1. $^{+}9 \times {}^{+}5$
2. $^{-}9 \times {}^{+}6$
3. $^{-}7 \times {}^{-}6$
4. $^{+}6 \times {}^{-}5$
5. $^{-}7 \times {}^{-}8$
6. $^{+}4 \times {}^{-}8$
7. $^{+}7 \times {}^{-}7$
8. $^{-}8 \times {}^{-}9$
9. $^{+}9 \times {}^{-}4$
10. $^{+}8 \times {}^{+}8$
11. $^{-}3 \times {}^{+}9$
12. $0 \times {}^{-}7$
13. $^{-}6 \times {}^{-}8$
14. 0×0
15. $^{-}9 \times {}^{+}8$
16. $^{+}4 \times {}^{+}7$
17. $^{-}9 \times {}^{+}9$
18. $^{+}5 \times {}^{+}5$
19. $^{-}6 \times 0$
20. $^{-}8 \times {}^{+}4$
21. $^{+}10 \times {}^{-}5$
22. $^{-}11 \times {}^{-}3$
23. $^{-}12 \times {}^{-}2$
24. $^{+}5 \times {}^{-}11$
25. $^{-}13 \times {}^{+}4$
26. $^{+}6 \times {}^{+}15$
27. $^{+}19 \times 0$
28. $^{-}15 \times {}^{+}12$
29. $^{+}16 \times {}^{-}16$
30. $^{-}18 \times {}^{+}20$
31. $^{+}23 \times {}^{+}13$
32. $0 \times {}^{-}22$

Give each quotient.

33. $^{+}21 \div {}^{+}7$
34. $^{+}24 \div {}^{-}3$
35. $^{-}45 \div {}^{+}5$
36. $^{-}25 \div {}^{-}5$
37. $^{-}72 \div {}^{-}9$
38. $^{+}54 \div {}^{+}6$
39. $^{-}28 \div {}^{+}4$
40. $^{+}48 \div {}^{-}6$
41. $^{+}30 \div {}^{-}5$
42. $^{-}36 \div {}^{+}6$
43. $^{-}72 \div {}^{-}8$
44. $^{+}32 \div {}^{+}8$
45. $^{-}63 \div {}^{+}7$
46. $^{+}35 \div {}^{+}7$
47. $^{-}36 \div {}^{-}9$
48. $^{+}56 \div {}^{-}8$
49. $^{+}42 \div {}^{-}6$
50. $^{-}81 \div {}^{+}9$
51. $^{+}49 \div {}^{+}7$
52. $^{-}42 \div {}^{-}7$
53. $^{-}24 \div {}^{+}2$
54. $0 \div {}^{-}8$
55. $^{+}45 \div {}^{+}5$
56. $^{+}42 \div {}^{-}21$
57. $0 \div {}^{+}25$
58. $^{+}36 \div {}^{+}3$
59. $^{+}72 \div {}^{-}18$
60. $^{-}75 \div {}^{+}25$
61. $^{+}45 \div {}^{-}15$
62. $^{-}60 \div {}^{-}20$
63. $^{-}96 \div {}^{+}24$
64. $^{-}96 \div {}^{+}12$

Mathematics in careers

Property such as houses, buildings, and land is called *real estate*. Real estate is generally sold by a real-estate agent. For selling the property, the agent receives a commission, which is generally a percentage of the selling price.

Often several real-estate agents in the same area belong to a multiple listing service. When such an agent lists (signs an agreement to try to sell) some real estate, he informs the other agents who belong to the multiple listing service. Any of these agents can then sell the property. If an agent other than the one who listed the property sells it, the commission is divided between the agent who listed the property and the one who sold it.

EXERCISES

During one week, the R. E. Johnson Real Estate Company received commissions from the sale of the properties listed below. The total commission paid was 6% of the selling price. How much commission did the R. E. Johnson Real Estate Company receive during this week?

1. Listed by R. E. Johnson
   Sold by R. E. Johnson

   Well-maintained, professionally landscaped 3-bedroom ranch home, near schools, with living room, kitchen, utility room, and large bath. Attic with pull-down stairs, patio, attached garage. Has central air conditioning, carpet and tile floors, and maintenance-free aluminum siding. This home is a real buy at $69,900.

2. Listed by A. J. Thomas
   Sold by R. E. Johnson

   Large, exceptionally nice bi-level in Kenwood School area, with 3 large bedrooms and bath up. Lower level has big L-shaped family room, 4th bedroom, large bath, utility area. Storm windows, carpet and tile floors, redwood sun deck off kitchen. A beautiful white aluminum-sided home in excellent condition. Will consider trade. $84,900.

3. Listed by R. E. Johnson
   Sold by C. R. Davidson

   Beautifully cared for older air-conditioned brick-on-tile home in established area. 4 bedrooms and lovely bath on 2nd, study and family room on first. Brick fireplace in carpeted and draped living room. Beamed ceiling in magnificent dining room. recreation room, work room, etc, in basement. Attic, 2-car garage. $89,900.

4. Listed by R. E. Johnson
   Sold by R. E. Johnson

   12-unit brick apartment building in good location. 8 units furnished and 4 unfurnished. Furnishings include washers and dryers, disposals in most. This apartment building has a net income of $60,000 per year. Call for more details. $790,000.

Graphing ordered pairs

Here is a **coordinate plane.** The red horizontal number line is the **first coordinate axis.** The red vertical number line is the **second coordinate axis.** The point where the axes intersect is called the **origin.**

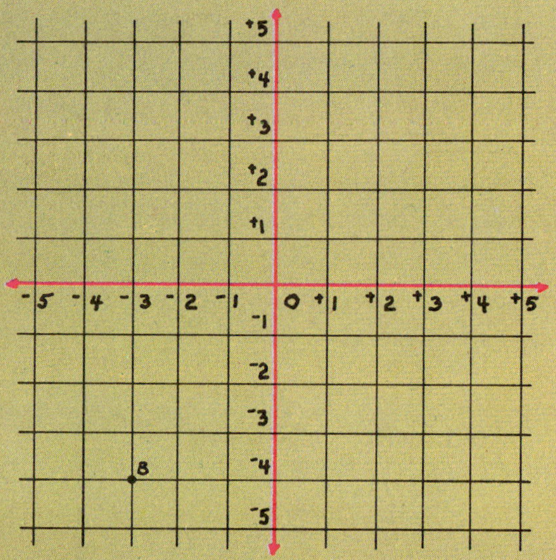

Ordered pairs of numbers may be graphed on a coordinate plane. Here is how to graph the ordered pair $(^-3, ^-4)$.

Step 1. Start at the origin. Move along the first coordinate axis to $^-3$.

Step 2. Then move parallel to the second coordinate axis to $^-4$.

The ordered pair $(^-3, ^-4)$ gives the **coordinates** of point *B*. Point *B* is the **graph** of the ordered pair $(^-3, ^-4)$.

EXERCISES

Give the coordinates of each point.

1. *A*	**2.** *B*	**3.** *C*
4. *D*	**5.** *E*	**6.** *F*
7. *G*	**8.** *H*	**9.** *I*
10. *J*	**11.** *K*	**12.** *L*
13. *M*	**14.** *N*	**15.** *P*

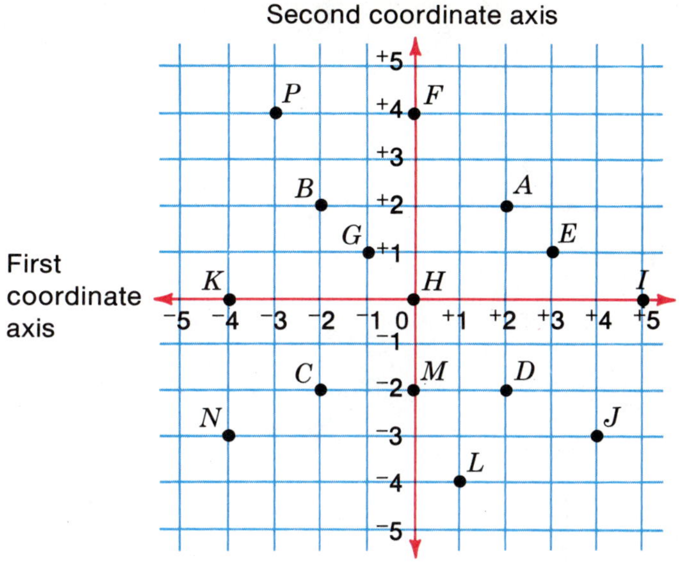

Graph the following ordered pairs.
Label each point with its coordinates.

16. $(^+3, ^+2)$	**17.** $(^+3, ^-2)$	**18.** $(^-3, ^+2)$	**19.** $(^-3, ^-2)$
20. $(0, 0)$	**21.** $(^-4, 0)$	**22.** $(^-4, ^-1)$	**23.** $(^-4, ^+1)$
24. $(^+4, ^-1)$	**25.** $(^+4, ^+1)$	**26.** $(0, ^-4)$	**27.** $(^+1, ^-2)$
28. $(^-3, ^+3)$	**29.** $(^-2, ^-2)$	**30.** $(^+3, 0)$	**31.** $(^+4, ^-3)$

32. Copy the triangle on graph paper and give the coordinates of each vertex.

33. Add $^+3$ to the first number of each ordered pair. Graph the new ordered pairs. Connect the points to make a new triangle.

34. Multiply the second number of each original pair by $^-1$. Graph the ordered pairs and draw the triangle.

35. Draw the triangle that you get by multiplying all the original numbers by $^-2$.

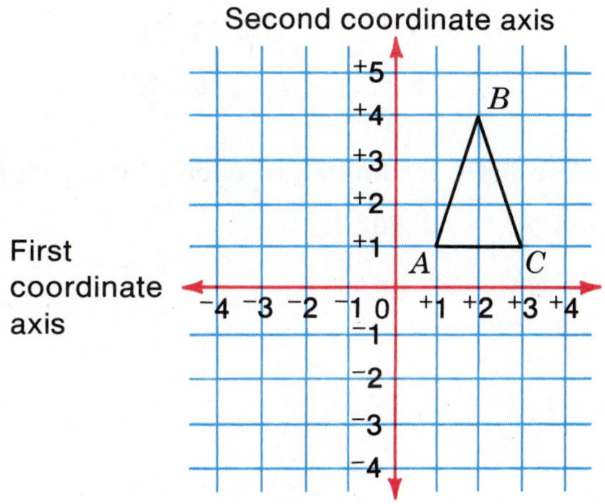

CHAPTER CHECKUP

Give each sum. [pages 372–373, 376]

1. $^{+}6 + ^{+}3$
2. $^{-}5 + ^{+}5$
3. $^{-}3 + ^{-}9$
4. $^{-}6 + ^{+}3$
5. $^{+}5 + ^{-}5$
6. $^{-}1 + ^{-}6$
7. $0 + ^{-}2$
8. $^{+}9 + ^{+}7$
9. $^{+}10 + ^{-}3$
10. $^{+}8 + ^{+}9$
11. $^{-}4 + ^{+}10$
12. $^{-}5 + ^{-}8$

Give each difference. [pages 374–376]

13. $^{-}5 - ^{+}3$
14. $^{-}2 - ^{-}1$
15. $^{+}3 - ^{+}5$
16. $^{-}4 - ^{+}5$
17. $^{+}6 - ^{+}6$
18. $^{-}6 - ^{+}2$
19. $^{+}8 - ^{-}8$
20. $^{-}3 - ^{-}6$
21. $^{-}9 - ^{-}2$
22. $^{+}6 - 0$
23. $^{-}8 - ^{+}3$
24. $^{+}7 - ^{-}9$

Give each product. [pages 378–379, 382]

25. $^{-}6 \times ^{-}4$
26. $^{+}5 \times ^{+}5$
27. $^{-}5 \times ^{+}7$
28. $^{+}10 \times 0$
29. $^{+}8 \times ^{-}6$
30. $^{+}8 \times ^{+}7$
31. $^{-}4 \times 0$
32. $^{-}6 \times ^{-}9$
33. $0 \times ^{-}3$
34. $^{-}8 \times ^{-}9$
35. $^{-}9 \times ^{+}8$
36. $^{+}6 \times ^{+}9$

Give each quotient. [pages 380–382]

37. $^{+}45 \div ^{-}9$
38. $^{-}49 \div ^{+}7$
39. $^{-}40 \div ^{+}8$
40. $^{+}64 \div ^{-}8$
41. $^{+}36 \div ^{+}4$
42. $^{+}56 \div ^{+}7$
43. $0 \div ^{-}9$
44. $^{-}40 \div ^{-}8$
45. $^{-}72 \div ^{+}8$
46. $^{-}81 \div ^{-}9$
47. $^{-}48 \div ^{-}6$
48. $^{+}42 \div ^{+}7$

Give the coordinates of each point. [pages 384–385]

49. A
50. B
51. C
52. D
53. E
54. F
55. G
56. H

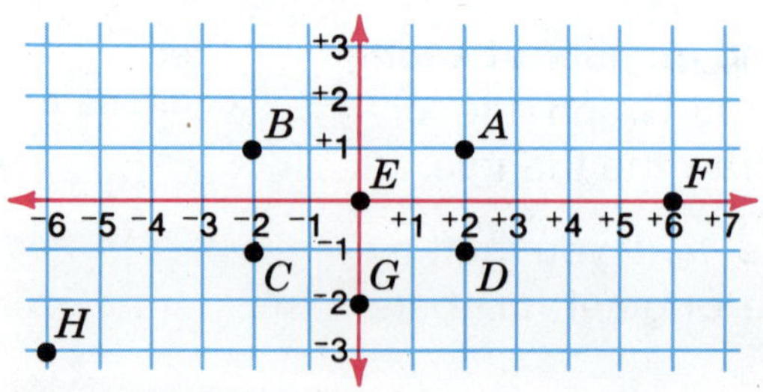

CHAPTER PROJECT

1. You have located points on a coordinate plane by referring to two lines, or axes. Circles around the earth called lines of latitude and longitude are used to locate places on the earth's surface. In an encyclopedia, see what you can learn about latitude and longitude.

2. Use a map in an atlas to find the latitude and longitude of the place where you live. What are the latitude and longitude of the point on the side of the earth opposite the place where you live?

3. The earth is divided into 24 time zones based on the lines of longitude. In an encyclopedia, see what you can learn about time zones. Which time zone do you live in?

CHAPTER REVIEW

To subtract an integer, add its opposite.

$${}^-4 - {}^-5 = {}^-4 + {}^+5$$
$$= {}^+1$$

The product of two integers with the same sign is positive.

$${}^-8 \times {}^-3 = {}^+24$$

The product of two integers with different signs is negative.

$${}^+8 \times {}^-2 = {}^-16$$

The quotient of two integers with the same sign is positive.

$${}^+12 \div {}^+4 = {}^+3$$

The quotient of two integers with different signs is negative.

$${}^-18 \div {}^+9 = {}^-2$$

Give each sum.

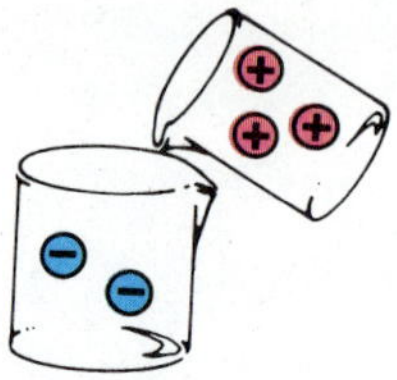

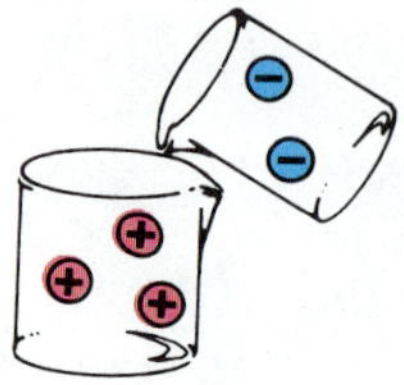

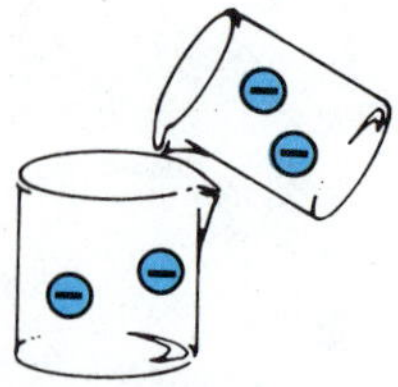

1. ${}^-2 + {}^+3$
2. ${}^+3 + {}^-2$
3. ${}^-2 + {}^-2$
4. ${}^+5 + {}^-4$
5. ${}^-5 + {}^+4$
6. ${}^-8 + {}^-3$
7. ${}^+6 + {}^+2$
8. ${}^-7 + 0$
9. $0 + {}^+6$

Give each difference.

10. ${}^-8 - {}^-2$
11. ${}^+7 - {}^+2$
12. ${}^+5 - 0$
13. ${}^+9 - {}^-4$
14. ${}^-7 - {}^-3$
15. ${}^-10 - {}^+8$
16. ${}^+15 - {}^+7$
17. ${}^+16 - {}^-8$
18. ${}^-12 - {}^+4$

Give each product.

19. ${}^+9 \times {}^+8$
20. ${}^-6 \times {}^-7$
21. ${}^-6 \times {}^+8$
22. $0 \times {}^-7$
23. ${}^+7 \times {}^+9$
24. ${}^+5 \times {}^-9$
25. ${}^-6 \times {}^+9$
26. 0×0
27. ${}^-8 \times {}^-7$

Give each quotient.

28. ${}^+63 \div {}^+9$
29. ${}^-56 \div {}^+8$
30. ${}^-54 \div {}^-6$
31. ${}^-42 \div {}^-6$
32. ${}^+48 \div {}^+6$
33. ${}^-36 \div {}^+4$
34. ${}^+32 \div {}^-4$
35. $0 \div {}^-8$
36. ${}^-27 \div {}^-9$

CHAPTER CHALLENGE

To get the **output number,** this function machine adds $^{-}3$ to the **input number.**

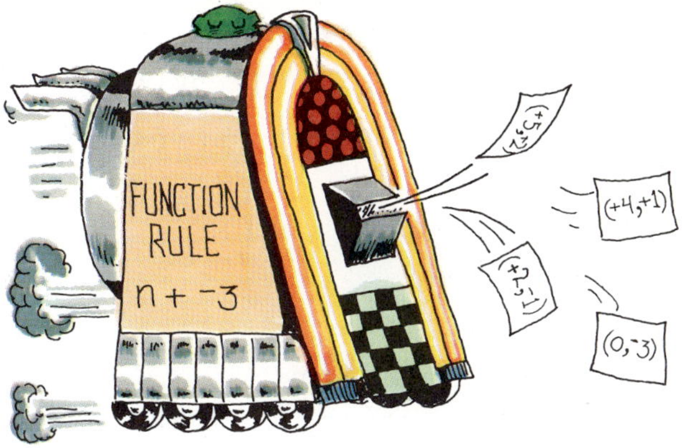

We can describe the function by

$$f: n \rightarrow \underbrace{n + {}^{-}3}$$

input number (arrow to n) — output number (under $n + {}^{-}3$)

Read as "f takes n to $n + {}^{-}3$."

To graph a function, we graph the ordered pairs of the function. Here we have graphed some ordered pairs of the function

$$f: n \rightarrow n + {}^{-}3$$

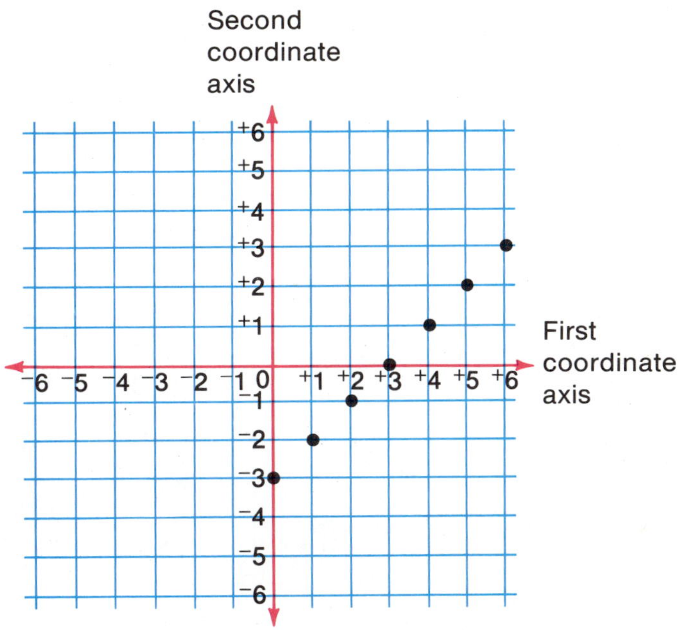

What do you notice about the points of the graph? What are some other points that belong to the graph of this function?

1. Copy and complete these ordered pairs of the function $f: n \rightarrow {}^{-}3n$.

a. $({}^{-}3, ?)$ **b.** $({}^{+}1, ?)$ **c.** $(0, ?)$

d. $({}^{-}4, ?)$ **e.** $({}^{-}2, ?)$ **f.** $({}^{+}3, ?)$

g. $({}^{+}2, ?)$ **h.** $({}^{+}4, ?)$ **i.** $({}^{-}1, ?)$

j. Graph these ordered pairs.

2. Graph some ordered pairs of these functions.

a. $g: n \rightarrow {}^{+}2n$ **b.** $h: n \rightarrow n - {}^{-}4$

c. $j: n \rightarrow n^2$ **d.** $f: n \rightarrow n^2 + 1$

MAJOR CHECKUP
STANDARDIZED FORMAT

Choose the correct letter.

1. $3^4 =$ ___?___

a. 12
b. 64
c. 81
d. none of these

2. Divide.

$0.016\overline{)75.68}$

a. 47.3
b. 4.73
c. 473
d. none of these

3. Subtract.

$$\begin{array}{r} 12 \\ -5\frac{3}{4} \\ \hline \end{array}$$

a. $7\frac{3}{4}$
b. $7\frac{1}{4}$
c. $6\frac{1}{4}$
d. none of these

4. Find the product.

$\frac{3}{8} \times \frac{2}{3}$

a. $\frac{1}{4}$
b. $\frac{9}{16}$
c. 4
d. none of these

5. 40% of ___?___ = 60

a. 24
b. 150
c. 240
d. none of these

6. A bicycle that regularly sells for \$80 was put on sale for \$64. The percent of discount was

a. 25%
b. 20%
c. 16%
d. none of these

7. How many lines of symmetry does a parallelogram have?

a. 1
b. 2
c. 4
d. none of these

8. Imagine tossing a dime and then tossing a nickel. The probability of getting both tails is

a. $\frac{1}{2}$ **b.** $\frac{3}{4}$
c. $\frac{1}{4}$ **d.** none of these

9. Add.

$^{+}7 + {}^{-}12$

a. $^{-}5$
b. $^{+}5$
c. $^{-}19$
d. none of these

10. Multiply.

$^{-}6 \times {}^{-}4$

a. $^{-}10$
b. $^{-}24$
c. $^{+}24$
d. none of these

11. Rolls are \$1.60 per dozen. What is the cost of 9 rolls?

a. \$1.20 **b.** \$1.26
c. \$14.40 **d.** none of these

12. A closet is 1 m wide, 0.8 m deep, and 2.2 m high. Which volume is closest to the volume of the closet?

a. 1 m^3 **b.** 2 m^3
c. 3 m^3 **d.** 4 m^3

Skill Test

1	Add two or more numbers	475 +312	296 +352	596 628 +327	3758 5863 +9479
2	Subtract any two numbers	352 −121	462 −139	526 −347	6000 −2382
3	Multiply by a 1-digit number	241 ×2	28 ×3	516 ×6	2371 ×8
4	Multiply by 10 or 100	67 ×10	38 ×100	576 ×10	3241 ×100
5	Multiply by a 2-digit number	431 ×26	358 ×34	544 ×61	8075 ×85
6	Multiply by a 3-digit number	543 ×122	308 ×496	615 ×204	8312 ×426
7	Divide by a 1-digit number	3)252	6)594	8)462	5)4295
8	Divide by a 2-digit number	23)713	45)1080	93)5762	42)8320
9	Divide by a 3-digit number	200)4600	120)960	431)13392	486)2483
10	Add any two decimals	3.51 + 2.81 47.5 + 2.68		67.5 + 9.8 4.832 + 0.96	

11	Subtract any two decimals	$0.483 - 0.265$ $36.4 - 2.7$		$3.52 - 1.7$ $71.6 - 4.83$	
12	Multiply any two decimals	5.3×2.7	0.41×0.08	0.62×0.37	1.46×3.8
13	Divide a decimal by a whole number	$3\overline{)85.2}$	$5\overline{)6.25}$	$56\overline{)71.68}$	$50\overline{)4.6}$
14	Divide any two decimals	$0.9\overline{)1.08}$	$0.06\overline{)0.192}$	$2.3\overline{)1.334}$	$0.46\overline{)142.6}$
15	Divide and round to the nearest tenth	$0.6\overline{)5.23}$	$0.9\overline{)1}$	$1.6\overline{)85.2}$	$0.45\overline{)17.6}$
16	Find an equivalent fraction	$\frac{2}{3} = \frac{?}{12}$	$\frac{5}{8} = \frac{?}{64}$	$\frac{5}{6} = \frac{?}{30}$	$\frac{5}{4} = \frac{?}{32}$
17	Write a fraction in lowest terms	$\frac{6}{12}$	$\frac{18}{24}$	$\frac{15}{35}$	$\frac{45}{63}$
18	Compare fractions	$<$ or $>$? $\frac{3}{7}$ ● $\frac{4}{7}$	$\frac{2}{3}$ ● $\frac{3}{5}$	$\frac{4}{8}$ ● $\frac{4}{7}$	$\frac{4}{7}$ ● $\frac{5}{8}$
19	Add fractions	$\frac{1}{8} + \frac{3}{4}$	$\frac{1}{6} + \frac{1}{3}$	$\frac{2}{3} + \frac{2}{5}$	$\frac{3}{4} + \frac{2}{3}$
20	Subtract fractions	$\frac{3}{4} - \frac{1}{8}$	$\frac{1}{3} - \frac{1}{6}$	$\frac{2}{3} - \frac{2}{5}$	$\frac{3}{4} - \frac{2}{3}$

21	Change a mixed number to a fraction	$2\frac{1}{2} = \underline{?}$	$3\frac{2}{3} = \underline{?}$	$5\frac{3}{4} = \underline{?}$	$4\frac{7}{8} = \underline{?}$
22	Change a fraction to a mixed number	$\frac{7}{2} = \underline{?}$	$\frac{8}{3} = \underline{?}$	$\frac{15}{4} = \underline{?}$	$\frac{23}{8} = \underline{?}$
23	Add mixed numbers	$3\frac{5}{7} + 2\frac{3}{7}$	$4\frac{2}{3} + 1\frac{1}{3}$	$12\frac{5}{6} + 4\frac{3}{8}$	$5\frac{2}{3} + 4\frac{3}{4}$
24	Subtract mixed numbers	$5\frac{2}{7} - 1\frac{1}{7}$	$6\frac{2}{9} - 3\frac{5}{9}$	$6\frac{3}{4} - 4\frac{5}{6}$	$8\frac{2}{3} - 3\frac{3}{4}$
25	Multiply fractions	$\frac{2}{3} \times \frac{4}{7}$	$\frac{3}{5} \times \frac{5}{8}$	$\frac{6}{7} \times \frac{5}{3}$	$\frac{12}{15} \times \frac{5}{4}$
26	Find a fraction of a number	$\frac{1}{2}$ of 18	$\frac{2}{3}$ of 21	$\frac{4}{3}$ of 12	$\frac{5}{6}$ of 18
27	Divide fractions	$\frac{3}{4} \div \frac{5}{2}$	$\frac{3}{4} \div \frac{3}{8}$	$\frac{2}{3} \div \frac{3}{4}$	$\frac{5}{7} \div \frac{3}{7}$
28	Simplify fractions	$\frac{\frac{2}{3}}{\frac{2}{5}}$	$\frac{\frac{3}{4}}{\frac{3}{8}}$	$\frac{\frac{2}{3}}{\frac{3}{4}}$	$\frac{\frac{5}{9}}{\frac{2}{3}}$
29	Change a decimal to a fraction or mixed number	$0.65 = \underline{?}$ $1.375 = \underline{?}$		$0.5 = \underline{?}$ $2.05 = \underline{?}$	

30	Change a fraction or mixed number to a decimal	$\frac{2}{5} = \underline{?}$	$\frac{3}{2} = \underline{?}$	$2\frac{5}{8} = \underline{?}$	$1\frac{3}{4} = \underline{?}$
31	Change a fraction to a mixed decimal	$\frac{1}{3} = \underline{?}$	$\frac{5}{3} = \underline{?}$	$\frac{5}{6} = \underline{?}$	$\frac{4}{9} = \underline{?}$
32	Multiply mixed numbers	$3\frac{1}{2} \times 2$ $2\frac{1}{3} \times 3\frac{2}{3}$		$1\frac{1}{4} \times 1\frac{1}{2}$ $1\frac{2}{3} \times 2\frac{3}{4}$	
33	Divide mixed numbers	$3\frac{1}{2} \div 2$ $2\frac{1}{3} \div 1\frac{1}{2}$		$2\frac{3}{4} \div 1\frac{1}{3}$ $4\frac{2}{3} \div 2\frac{3}{8}$	
34	Solve a proportion	$\frac{5}{3} = \frac{n}{4}$ $\frac{4}{5} = \frac{6}{n}$		$\frac{2}{n} = \frac{3}{8}$ $\frac{n}{5} = \frac{6}{7}$	
35	Change a percent to a fraction	$60\% = \underline{?}$	$5\% = \underline{?}$	$25\% = \underline{?}$	$90\% = \underline{?}$
36	Change a decimal to a percent	$0.1 = \underline{?}$	$0.15 = \underline{?}$	$0.236 = \underline{?}$	$1.7 = \underline{?}$
37	Change a percent to a decimal	$18\% = \underline{?}$ $12.2\% = \underline{?}$		$65.3\% = \underline{?}$ $8.6\% = \underline{?}$	
38	Change a fraction to a percent	$\frac{7}{10} = \underline{?}$	$\frac{9}{4} = \underline{?}$	$\frac{5}{9} = \underline{?}$	$\frac{7}{6} = \underline{?}$

39	Find a percent of a number	25% of $28 = n$		35% of $50 = n$	
		$33\frac{1}{3}\%$ of $51 = n$		$83\frac{1}{3}\%$ of $72 = n$	
40	Find the number when a percent is given	10% of $n = 6$		75% of $n = 45$	
		$12\frac{1}{2}\%$ of $n = 16$		32% of $n = 8$	
41	Compare integers	$<$ **or** $>$**?**			
		${}^{+}3$ ● ${}^{-}2$	0 ● ${}^{-}5$	${}^{+}6$ ● ${}^{-}8$	${}^{-}5$ ● ${}^{-}6$
42	Add two integers	${}^{+}8 + {}^{+}6$	${}^{-}4 + {}^{+}7$	${}^{+}9 + {}^{-}3$	${}^{-}5 + {}^{-}7$
43	Subtract two integers	${}^{+}5 - {}^{+}2$	${}^{+}6 - {}^{-}4$	${}^{-}5 - {}^{-}2$	${}^{-}7 - {}^{+}3$
44	Multiply two integers	${}^{+}8 \times {}^{-}3$	${}^{+}2 \times {}^{+}7$	${}^{-}6 \times {}^{-}4$	${}^{-}3 \times {}^{+}7$
45	Divide two integers	${}^{-}8 \div {}^{+}2$	${}^{+}18 \div {}^{-}6$	${}^{+}9 \div {}^{+}3$	${}^{-}28 \div {}^{-}7$

Extra Practice

Set 1 Add.

1. 43 +26	2. 70 +28	3. 32 +32	4. 41 +35	5. 51 +20	6. 63 +31
7. 62 +33	8. 33 +62	9. 81 +12	10. 54 +62	11. 73 +55	12. 86 +21
13. 41 53 +12	14. 30 25 +84	15. 43 61 +50	16. 52 33 +24	17. 44 20 +64	18. 35 72 +22

Set 2 Add.

1. 78 +29	2. 36 +38	3. 19 +45	4. 45 +63	5. 39 +82	6. 46 +29
7. 54 +38	8. 94 +65	9. 58 +58	10. 28 +36	11. 63 +78	12. 57 +95
13. 75 67 +45	14. 83 78 +41	15. 89 21 +57	16. 79 32 +46	17. 94 56 +37	18. 68 75 +98

Set 3 Subtract.

1. 65 −23	2. 38 −15	3. 65 −22	4. 56 −14	5. 49 −23	6. 78 −35
7. 39 −15	8. 87 −24	9. 79 −32	10. 88 −43	11. 55 −40	12. 86 −32
13. 148 −82	14. 184 −91	15. 177 −93	16. 166 −84	17. 156 −92	18. 135 −83

Set 4 Subtract.

1. $\begin{array}{r}82\\-23\\\hline\end{array}$	2. $\begin{array}{r}75\\-36\\\hline\end{array}$	3. $\begin{array}{r}64\\-39\\\hline\end{array}$	4. $\begin{array}{r}53\\-16\\\hline\end{array}$	5. $\begin{array}{r}45\\-36\\\hline\end{array}$	6. $\begin{array}{r}72\\-58\\\hline\end{array}$
7. $\begin{array}{r}76\\-39\\\hline\end{array}$	8. $\begin{array}{r}82\\-46\\\hline\end{array}$	9. $\begin{array}{r}60\\-58\\\hline\end{array}$	10. $\begin{array}{r}75\\-29\\\hline\end{array}$	11. $\begin{array}{r}42\\-18\\\hline\end{array}$	12. $\begin{array}{r}93\\-47\\\hline\end{array}$
13. $\begin{array}{r}164\\-95\\\hline\end{array}$	14. $\begin{array}{r}143\\-75\\\hline\end{array}$	15. $\begin{array}{r}132\\-58\\\hline\end{array}$	16. $\begin{array}{r}120\\-46\\\hline\end{array}$	17. $\begin{array}{r}165\\-97\\\hline\end{array}$	18. $\begin{array}{r}174\\-86\\\hline\end{array}$

Set 5 Multiply.

1. $\begin{array}{r}23\\\times 2\\\hline\end{array}$	2. $\begin{array}{r}34\\\times 2\\\hline\end{array}$	3. $\begin{array}{r}21\\\times 4\\\hline\end{array}$	4. $\begin{array}{r}23\\\times 3\\\hline\end{array}$	5. $\begin{array}{r}22\\\times 3\\\hline\end{array}$	6. $\begin{array}{r}20\\\times 4\\\hline\end{array}$
7. $\begin{array}{r}21\\\times 6\\\hline\end{array}$	8. $\begin{array}{r}42\\\times 3\\\hline\end{array}$	9. $\begin{array}{r}30\\\times 5\\\hline\end{array}$	10. $\begin{array}{r}42\\\times 4\\\hline\end{array}$	11. $\begin{array}{r}82\\\times 2\\\hline\end{array}$	12. $\begin{array}{r}61\\\times 8\\\hline\end{array}$
13. $\begin{array}{r}84\\\times 2\\\hline\end{array}$	14. $\begin{array}{r}30\\\times 6\\\hline\end{array}$	15. $\begin{array}{r}51\\\times 6\\\hline\end{array}$	16. $\begin{array}{r}41\\\times 5\\\hline\end{array}$	17. $\begin{array}{r}64\\\times 2\\\hline\end{array}$	18. $\begin{array}{r}70\\\times 7\\\hline\end{array}$

Set 6 Multiply.

1. $\begin{array}{r}29\\\times 2\\\hline\end{array}$	2. $\begin{array}{r}46\\\times 4\\\hline\end{array}$	3. $\begin{array}{r}38\\\times 3\\\hline\end{array}$	4. $\begin{array}{r}54\\\times 5\\\hline\end{array}$	5. $\begin{array}{r}63\\\times 7\\\hline\end{array}$	6. $\begin{array}{r}75\\\times 6\\\hline\end{array}$
7. $\begin{array}{r}39\\\times 8\\\hline\end{array}$	8. $\begin{array}{r}46\\\times 6\\\hline\end{array}$	9. $\begin{array}{r}74\\\times 3\\\hline\end{array}$	10. $\begin{array}{r}65\\\times 5\\\hline\end{array}$	11. $\begin{array}{r}84\\\times 4\\\hline\end{array}$	12. $\begin{array}{r}95\\\times 7\\\hline\end{array}$
13. $\begin{array}{r}65\\\times 6\\\hline\end{array}$	14. $\begin{array}{r}58\\\times 8\\\hline\end{array}$	15. $\begin{array}{r}73\\\times 5\\\hline\end{array}$	16. $\begin{array}{r}92\\\times 9\\\hline\end{array}$	17. $\begin{array}{r}83\\\times 7\\\hline\end{array}$	18. $\begin{array}{r}69\\\times 4\\\hline\end{array}$

Set 7 Divide.

1. $4\overline{)33}$	2. $3\overline{)17}$	3. $2\overline{)19}$	4. $5\overline{)33}$	5. $8\overline{)47}$	6. $7\overline{)23}$
7. $6\overline{)50}$	8. $9\overline{)37}$	9. $3\overline{)25}$	10. $5\overline{)43}$	11. $6\overline{)40}$	12. $8\overline{)44}$
13. $9\overline{)70}$	14. $7\overline{)29}$	15. $4\overline{)38}$	16. $6\overline{)46}$	17. $8\overline{)53}$	18. $7\overline{)39}$

Set 8 Divide.

1. $3\overline{)69}$ 2. $2\overline{)64}$ 3. $4\overline{)88}$ 4. $3\overline{)66}$ 5. $2\overline{)86}$ 6. $3\overline{)93}$

7. $5\overline{)105}$ 8. $8\overline{)168}$ 9. $7\overline{)217}$ 10. $6\overline{)120}$ 11. $5\overline{)155}$ 12. $9\overline{)279}$

13. $6\overline{)426}$ 14. $8\overline{)488}$ 15. $7\overline{)287}$ 16. $9\overline{)369}$ 17. $5\overline{)350}$ 18. $7\overline{)427}$

Set 9 Divide.

1. $3\overline{)48}$ 2. $4\overline{)56}$ 3. $2\overline{)96}$ 4. $5\overline{)70}$ 5. $7\overline{)84}$ 6. $3\overline{)51}$

7. $7\overline{)98}$ 8. $3\overline{)87}$ 9. $6\overline{)84}$ 10. $5\overline{)75}$ 11. $6\overline{)72}$ 12. $6\overline{)96}$

13. $4\overline{)76}$ 14. $4\overline{)92}$ 15. $8\overline{)96}$ 16. $3\overline{)84}$ 17. $2\overline{)58}$ 18. $5\overline{)60}$

Set 10. Give the standard numeral.

1. eight thousand
2. five thousand one hundred twenty
3. forty-two thousand, nine hundred six
4. nineteen thousand, seven hundred fifty-three
5. three hundred twenty-eight thousand, six hundred seventy-six
6. five million, two hundred twelve thousand, four hundred thirty-seven
7. two hundred forty-two million, six thousand, two hundred eighty-three
8. five billion, two hundred twenty million, two hundred eleven thousand, sixty-six
9. twenty-four billion, nine hundred forty million, five thousand eighty-two
10. seven trillion, one hundred thirty-four million, two hundred forty thousand, twenty

Set 11 Round each number to the nearest hundred, thousand, and million.

1. 6,834,527 2. 9,498,937 3. 9,672,403 4. 12,291,742

5. 233,035,555 6. 456,621,384 7. 789,125,246 8. 919,291,350

Set 12 < or >?

1. 4689 ● 4731
2. 6390 ● 6382
3. 4934 ● 4898
4. 26,428 ● 27,222
5. 8999 ● 10,000
6. 72,958 ● 73,094
7. 85,395 ● 83,595
8. 100,000 ● 99,999
9. 168,749 ● 168,579
10. 298,431 ● 298,341
11. 835,822 ● 843,262
12. 987,342 ● 978,342

Set 13 Add.

1. 329 + 437
2. 509 + 388
3. 625 + 158
4. 978 + 388
5. 659 + 594
6. 746 + 896
7. 3565 + 828
8. 4836 + 565
9. 9024 + 889
10. 7477 + 3593
11. 6185 + 6189
12. 4778 + 8696
13. 384, 292, + 743
14. 567, 308, + 826
15. 295, 762, + 547
16. 1835, 6916, + 2159
17. 7068, 9383, + 4555
18. 5347, 8274, + 3339

Set 14 Subtract.

1. 863 − 235
2. 752 − 259
3. 694 − 346
4. 538 − 162
5. 453 − 360
6. 725 − 581
7. 3210 − 2483
8. 5632 − 1955
9. 9322 − 3996
10. 6033 − 1588
11. 5014 − 2939
12. 7003 − 4668
13. 5032 − 1486
14. 6233 − 2485
15. 9157 − 3469
16. 7024 − 3299
17. 7335 − 2467
18. 5001 − 2438

Set 15 Multiply.

1. 23 × 3
2. 41 × 2
3. 52 × 3
4. 64 × 4
5. 83 × 5
6. 90 × 4
7. 142 × 5
8. 238 × 9
9. 196 × 8
10. 370 × 7
11. 415 × 6
12. 467 × 5
13. 4916 × 7
14. 5328 × 2
15. 6742 × 4
16. 2864 × 5
17. 8955 × 8
18. 3780 × 9

Set 16 **Give each product.**

1. a. 24×10
   b. 24×100
   c. 24×1000
2. a. 57×10
   b. 57×100
   c. 57×1000
3. a. 80×10
   b. 80×100
   c. 80×1000
4. a. 142×10
   b. 142×100
   c. 142×1000
5. a. 200×10
   b. 200×100
   c. 200×1000
6. a. 1242×10
   b. 1242×100
   c. 1242×1000
7. a. 3450×10
   b. 3450×100
   c. 3450×1000
8. a. 8100×10
   b. 8100×100
   c. 8100×1000

Set 17 **Multiply.**

1. 352×24
2. 592×91
3. 766×48
4. 691×63
5. 829×83
6. 527×38
7. 635×64
8. 863×96
9. 478×27
10. 409×75
11. 742×41
12. 312×53
13. 820×76
14. 371×40
15. 602×16
16. 547×37
17. 608×50
18. 417×93

Set 18 **Multiply.**

1. 527×123
2. 658×243
3. 492×174
4. 581×325
5. 653×208
6. 842×258
7. 952×333
8. 378×156
9. 834×240
10. 815×524
11. 706×430
12. 674×462
13. 640×306
14. 725×301
15. 534×320
16. 906×254
17. 829×306
18. 935×256

Set 19 **Give the standard numeral.**

1. 5^2 2. 2^4 3. 4^3 4. 3^5 5. 1^4 6. 10^2 7. 3^3 8. 5^4 9. 10^3

10. 2^6 11. 8^3 12. 6^2 13. 1^7 14. 2^5 15. 6^3 16. 10^4 17. 7^2 18. 9^3

Set 20 **Divide.**

1. $5\overline{)706}$ 2. $8\overline{)359}$ 3. $5\overline{)592}$ 4. $7\overline{)488}$ 5. $4\overline{)969}$

6. $9\overline{)5218}$ 7. $6\overline{)3961}$ 8. $5\overline{)7438}$ 9. $3\overline{)9064}$ 10. $9\overline{)8213}$

11. $2\overline{)62{,}943}$ 12. $7\overline{)71{,}826}$ 13. $8\overline{)54{,}391}$ 14. $6\overline{)70{,}658}$ 15. $6\overline{)42{,}653}$

Set 21 **Divide.**

1. $18\overline{)7364}$ 2. $34\overline{)5294}$ 3. $57\overline{)1783}$ 4. $24\overline{)9642}$ 5. $42\overline{)7538}$

6. $78\overline{)3592}$ 7. $95\overline{)6000}$ 8. $45\overline{)5783}$ 9. $72\overline{)7044}$ 10. $80\overline{)6903}$

11. $45\overline{)38{,}216}$ 12. $83\overline{)53{,}416}$ 13. $16\overline{)29{,}784}$ 14. $64\overline{)36{,}952}$ 15. $38\overline{)74{,}615}$

Set 22 **Divide.**

1. $125\overline{)3956}$ 2. $203\overline{)7834}$ 3. $138\overline{)5916}$ 4. $351\overline{)7538}$ 5. $150\overline{)2964}$

6. $522\overline{)9678}$ 7. $358\overline{)6398}$ 8. $212\overline{)9000}$ 9. $534\overline{)9008}$ 10. $242\overline{)8634}$

11. $156\overline{)29{,}783}$ 12. $618\overline{)52{,}006}$ 13. $425\overline{)39{,}481}$ 14. $625\overline{)62{,}934}$ 15. $481\overline{)49{,}538}$

Set 23 $<$ **or** $>$**?**

1. 0.5 ● 0.4 2. 0.09 ● 0.08 3. 0.006 ● 0.005 4. 0.5 ● 0.05

5. 0.003 ● 0.02 6. 0.09 ● 0.1 7. 0.2 ● 0.008 8. 0.5 ● 0.004

9. 0.75 ● 0.76 10. 9.24 ● 9.42 11. 7.85 ● 7.58 12. 18.79 ● 18.8

13. 0.599 ● 1 14. 2 ● 1.897 15. 9.534 ● 9.0541 16. 6.375 ● 63.75

17. 56.37 ● 5.637 18. 5 ● 5.999 19. 8.001 ● 8.01 20. 19.02 ● 19.019

Set 24 **Round each number to the nearest tenth, hundredth, and thousandth.**

1. 16.1158 2. 43.2565 3. 82.3423 4. 65.8166 5. 58.3454

6. 9.9998 7. 15.4083 8. 29.9602 9. 70.1528 10. 37.0765

Set 25 Add.

1. $\begin{array}{r} 58.6 \\ +68.8 \\ \hline \end{array}$ 2. $\begin{array}{r} 4.63 \\ +0.34 \\ \hline \end{array}$ 3. $\begin{array}{r} 0.644 \\ +0.378 \\ \hline \end{array}$ 4. $\begin{array}{r} 91.2 \\ +25.9 \\ \hline \end{array}$ 5. $\begin{array}{r} 8.92 \\ +6.16 \\ \hline \end{array}$ 6. $\begin{array}{r} 5.39 \\ +0.53 \\ \hline \end{array}$

7. 28.4 + 73.6 8. 22.4 + 4.78 9. 1.86 + 43 10. 4.16 + 0.379

11. 27.6 + 9.28 12. 7.88 + 0.594 13. 371 + 0.853 14. 8.62 + 5.95

15. 0.174 + 6.76 16. 99.7 + 4.89 17. 0.145 + 5.57 18. 0.0683 + 0.646

Set 26 Subtract.

1. $\begin{array}{r} 95.6 \\ -83.1 \\ \hline \end{array}$ 2. $\begin{array}{r} 8.26 \\ -4.19 \\ \hline \end{array}$ 3. $\begin{array}{r} 0.516 \\ -0.384 \\ \hline \end{array}$ 4. $\begin{array}{r} 72.1 \\ -50.9 \\ \hline \end{array}$ 5. $\begin{array}{r} 0.462 \\ -0.399 \\ \hline \end{array}$ 6. $\begin{array}{r} 9.67 \\ -1.09 \\ \hline \end{array}$

7. 46.6 − 7.4 8. 5.38 − 4.1 9. 629 − 3.84 10. 80.2 − 9.3

11. 92.8 − 3.56 12. 89 − 42.8 13. 7.3 − 7.18 14. 40 − 34.8

15. 0.3 − 0.286 16. 70.6 − 2.93 17. 5.6 − 0.278 18. 0.82 − 0.007

Set 27 Multiply.

1. $\begin{array}{r} 35.8 \\ \times 2.9 \\ \hline \end{array}$ 2. $\begin{array}{r} 4.26 \\ \times 1.4 \\ \hline \end{array}$ 3. $\begin{array}{r} 57.4 \\ \times 0.35 \\ \hline \end{array}$ 4. $\begin{array}{r} 89.3 \\ \times 6.4 \\ \hline \end{array}$ 5. $\begin{array}{r} 6.52 \\ \times 18 \\ \hline \end{array}$ 6. $\begin{array}{r} 145 \\ \times 2.5 \\ \hline \end{array}$

7. $\begin{array}{r} 3.81 \\ \times 16 \\ \hline \end{array}$ 8. $\begin{array}{r} 25.3 \\ \times 0.33 \\ \hline \end{array}$ 9. $\begin{array}{r} 5.64 \\ \times 8.1 \\ \hline \end{array}$ 10. $\begin{array}{r} 78.2 \\ \times 60 \\ \hline \end{array}$ 11. $\begin{array}{r} 9.11 \\ \times 2.9 \\ \hline \end{array}$ 12. $\begin{array}{r} 6.03 \\ \times 7.5 \\ \hline \end{array}$

13. $\begin{array}{r} 9.38 \\ \times 1.21 \\ \hline \end{array}$ 14. $\begin{array}{r} 71.5 \\ \times 11.5 \\ \hline \end{array}$ 15. $\begin{array}{r} 6.35 \\ \times 34.6 \\ \hline \end{array}$ 16. $\begin{array}{r} 4.93 \\ \times 2.81 \\ \hline \end{array}$ 17. $\begin{array}{r} 38.5 \\ \times 5.63 \\ \hline \end{array}$ 18. $\begin{array}{r} 5.36 \\ \times 40.2 \\ \hline \end{array}$

Set 28 Multiply.

1. $\begin{array}{r} 0.23 \\ \times 0.3 \\ \hline \end{array}$ 2. $\begin{array}{r} 0.081 \\ \times 0.7 \\ \hline \end{array}$ 3. $\begin{array}{r} 0.074 \\ \times 0.5 \\ \hline \end{array}$ 4. $\begin{array}{r} 0.49 \\ \times 0.03 \\ \hline \end{array}$ 5. $\begin{array}{r} 0.063 \\ \times 0.08 \\ \hline \end{array}$ 6. $\begin{array}{r} 0.29 \\ \times 0.04 \\ \hline \end{array}$

7. $\begin{array}{r} 0.548 \\ \times 0.005 \\ \hline \end{array}$ 8. $\begin{array}{r} 70.3 \\ \times 0.004 \\ \hline \end{array}$ 9. $\begin{array}{r} 9.96 \\ \times 0.007 \\ \hline \end{array}$ 10. $\begin{array}{r} 78.3 \\ \times 0.38 \\ \hline \end{array}$ 11. $\begin{array}{r} 914 \\ \times 1.58 \\ \hline \end{array}$ 12. $\begin{array}{r} 59.6 \\ \times 60.8 \\ \hline \end{array}$

Set 29 Divide.

1. $3\overline{)567}$ 2. $2\overline{)586}$ 3. $3\overline{)438}$ 4. $9\overline{)288}$ 5. $7\overline{)329}$

6. $8\overline{)8.56}$ 7. $5\overline{)3.90}$ 8. $6\overline{)4.08}$ 9. $7\overline{)11.2}$ 10. $4\overline{)5.12}$

11. $12\overline{)28.8}$ 12. $15\overline{)9.45}$ 13. $11\overline{)3.85}$ 14. $26\overline{)28.6}$ 15. $35\overline{)94.5}$

16. $73\overline{)0.1314}$ 17. $55\overline{)4950}$ 18. $36\overline{)26.28}$ 19. $49\overline{)2.793}$ 20. $78\overline{)397.8}$

Set 30 Divide.

1. $0.4\overline{)1.08}$ 2. $0.03\overline{)8.19}$ 3. $0.7\overline{)58.8}$ 4. $0.2\overline{)0.836}$ 5. $0.06\overline{)2.94}$

6. $0.08\overline{)0.496}$ 7. $0.005\overline{)0.840}$ 8. $0.9\overline{)333}$ 9. $0.3\overline{)8.43}$ 10. $0.07\overline{)43.4}$

11. $0.13\overline{)0.845}$ 12. $4.7\overline{)7.05}$ 13. $2.9\overline{)17.4}$ 14. $0.83\overline{)3.818}$ 15. $0.065\overline{)0.2340}$

16. $3.4\overline{)9.894}$ 17. $0.71\overline{)11.36}$ 18. $5.2\overline{)31.72}$ 19. $0.38\overline{)771.4}$ 20. $7.6\overline{)228}$

Set 31 Divide. Round quotients to the nearest tenth.

1. $0.3\overline{)0.386}$ 2. $0.5\overline{)0.348}$ 3. $0.6\overline{)2.97}$ 4. $0.02\overline{)3.51}$ 5. $0.07\overline{)6.28}$

6. $0.08\overline{)0.347}$ 7. $0.003\overline{)0.038}$ 8. $0.05\overline{)2.97}$ 9. $0.04\overline{)0.511}$ 10. $0.9\overline{)0.629}$

11. $1.2\overline{)3.482}$ 12. $2.4\overline{)1.356}$ 13. $0.31\overline{)0.9382}$ 14. $0.43\overline{)0.3916}$ 15. $0.83\overline{)0.1374}$

16. $0.57\overline{)3.056}$ 17. $0.64\overline{)0.2973}$ 18. $3.9\overline{)5.901}$ 19. $7.8\overline{)2.643}$ 20. $4.2\overline{)0.3967}$

Set 32 Complete.

1. 3 m = ? dm 2. 5 m = ? cm 3. 7 m = ? mm
4. 200 cm = ? m 5. 4000 mm = ? m 6. 5.8 m = ? dm
7. 3.9 cm = ? mm 8. 4.6 dm = ? cm 9. 5.34 m = ? dm
10. 8.92 dm = ? mm 11. 3.96 dm = ? cm 12. 4.83 m = ? cm
13. 5.63 cm = ? mm 14. 3.82 m = ? dm 15. 2.96 cm = ? dm
16. 2.94 cm = ? mm 17. 5.42 dm = ? cm 18. 384 mm = ? cm
19. 5.39 mm = ? cm 20. 2.63 m = ? cm 21. 9.8 mm = ? dm
22. 38 km = ? m 23. 396 m = ? km 24. 2.4 km = ? m
25. 52.3 km = ? m 26. 0.74 km = ? m 27. 0.385 km = ? m
28. 8.56 m = ? cm 29. 0.3 m = ? mm 30. 75.2 m = ? km

Set 33 Complete.

1. 3 L = ? mL
2. 2.4 L = ? mL
3. 3.5 L = ? mL
4. 0.24 L = ? mL
5. 0.356 L = ? mL
6. 2000 mL = ? L
7. 1400 mL = ? L
8. 550 mL = ? L
9. 825 mL = ? L
10. 5 g = ? mg
11. 14 g = ? mg
12. 4000 g = ? mg
13. 5000 g = ? kg
14. 685 g = ? kg
15. 1445 g = ? kg
16. 58 g = ? kg
17. 2.4 kg = ? g
18. 0.42 kg = ? g

Set 34 Complete.

1. 7 ft = ? in.
2. 3 yd = ? in.
3. 9 yd = ? ft
4. 2 mi = ? ft
5. 4 mi = ? yd
6. 1 mi = ? in.
7. 156 in. = ? ft
8. 138 ft = ? yd
9. 2232 in. = ? yd
10. 40 ft = ? yd ? ft
11. 115 ft = ? yd ? ft
12. 150 in. = ? ft ? in.
13. 205 ft = ? yd ? ft
14. 2 ft 7 in. = ? in.
15. 10 ft 8 in. = ? in.
16. 23 ft 9 in. = ? in.
17. 9 yd 2 ft = ? ft
18. 17 yd 1 ft = ? ft

Set 35 Complete.

1. 36 pt = ? qt
2. 8 qt = ? gal
3. 24 c = ? pt
4. 2 gal = ? pt
5. 6 qt = ? half-gal
6. 6 gal = ? half-gal
7. 27 qt = ? gal ? qt
8. 23 pt = ? qt ? pt
9. 15 qt = ? gal ? qt
10. 2 lb = ? oz
11. 96 oz = ? lb
12. 5 T = ? lb
13. 16,000 lb = ? T
14. 16 lb = ? oz
15. 100 T = ? lb
16. 23 oz = ? lb ? oz
17. 4 lb 8 oz = ? oz
18. 1 lb 9 oz = ? oz

Set 36 Give the greatest common factor.

1. 8, 12
2. 6, 9
3. 9, 12
4. 12, 16
5. 8, 16
6. 15, 20
7. 9, 15
8. 12, 24
9. 8, 14
10. 24, 48
11. 16, 24
12. 18, 24
13. 15, 30
14. 30, 36
15. 12, 20
16. 15, 25
17. 9, 18
18. 18, 42

Set 37 Give the prime factorization.

1. 4
2. 6
3. 8
4. 9
5. 10
6. 12
7. 14
8. 15
9. 16
10. 18
11. 20
12. 21
13. 22
14. 24
15. 25
16. 26
17. 27
18. 28

Set 38 **Give the least common multiple.**

1. 2, 5
2. 3, 9
3. 4, 6
4. 4, 8
5. 6, 8
6. 9, 10
7. 4, 7
8. 8, 10
9. 4, 9
10. 2, 10
11. 7, 6
12. 6, 14
13. 18, 12
14. 16, 8
15. 15, 20
16. 6, 16
17. 14, 20
18. 16, 24

Set 39 **Solve.**

1. $n + 9 = 17$
2. $n - 6 = 15$
3. $n - 10 = 13$
4. $n + 8 = 22$
5. $n - 42 = 35$
6. $16 + n = 30$
7. $n - 30 = 16$
8. $n - 21 = 42$
9. $n - 32 = 15$
10. $n + 18 = 40$
11. $n - 14 = 23$
12. $n + 25 = 40$
13. $16 + n = 35$
14. $25 + n = 49$
15. $17 + n = 51$
16. $n - 18 = 25$

Set 40 **Solve.**

1. $3n = 12$
2. $\frac{n}{4} = 8$
3. $\frac{n}{3} = 12$
4. $5n = 20$
5. $7n = 56$
6. $\frac{n}{5} = 15$
7. $10n = 10$
8. $8n = 0$
9. $\frac{n}{4} = 72$
10. $2n = 100$
11. $\frac{n}{15} = 15$
12. $\frac{n}{9} = 9$
13. $17n = 119$
14. $4n = 76$
15. $\frac{n}{6} = 3$
16. $23n = 115$

Set 41 **Copy and complete.**

1. $\frac{1}{2} = \frac{?}{4}$
2. $\frac{1}{3} = \frac{?}{12}$
3. $\frac{2}{3} = \frac{?}{9}$
4. $\frac{5}{6} = \frac{?}{30}$
5. $\frac{3}{4} = \frac{?}{16}$
6. $\frac{3}{2} = \frac{?}{12}$
7. $\frac{5}{2} = \frac{?}{20}$
8. $\frac{4}{4} = \frac{?}{8}$
9. $\frac{7}{8} = \frac{?}{32}$
10. $\frac{1}{3} = \frac{?}{18}$
11. $\frac{1}{4} = \frac{?}{20}$
12. $\frac{3}{5} = \frac{?}{15}$
13. $\frac{1}{2} = \frac{?}{10}$
14. $\frac{3}{8} = \frac{?}{24}$
15. $\frac{3}{4} = \frac{?}{24}$

Set 42 **Change to lowest terms.**

1. $\frac{4}{8}$
2. $\frac{2}{6}$
3. $\frac{4}{6}$
4. $\frac{7}{14}$
5. $\frac{5}{10}$
6. $\frac{2}{8}$
7. $\frac{10}{15}$
8. $\frac{10}{12}$
9. $\frac{9}{15}$
10. $\frac{8}{6}$
11. $\frac{9}{12}$
12. $\frac{2}{10}$
13. $\frac{9}{6}$
14. $\frac{3}{9}$
15. $\frac{8}{12}$
16. $\frac{10}{16}$
17. $\frac{4}{20}$
18. $\frac{15}{12}$
19. $\frac{15}{30}$
20. $\frac{6}{9}$
21. $\frac{5}{20}$
22. $\frac{6}{12}$
23. $\frac{10}{4}$
24. $\frac{12}{18}$

Set 43 <, =, or >?

1. $\frac{1}{3}$ ● $\frac{1}{2}$
2. $\frac{1}{4}$ ● $\frac{1}{5}$
3. $\frac{1}{2}$ ● $\frac{6}{12}$
4. $\frac{3}{8}$ ● $\frac{1}{3}$
5. $\frac{5}{6}$ ● $\frac{15}{18}$
6. $\frac{3}{2}$ ● $\frac{2}{3}$
7. $\frac{4}{9}$ ● $\frac{1}{2}$
8. $\frac{2}{3}$ ● $\frac{5}{8}$
9. $\frac{3}{4}$ ● $\frac{5}{6}$
10. $\frac{6}{8}$ ● $\frac{3}{4}$
11. $\frac{6}{16}$ ● $\frac{3}{8}$
12. $\frac{3}{5}$ ● $\frac{6}{10}$
13. $\frac{7}{4}$ ● $\frac{5}{2}$
14. $\frac{5}{9}$ ● $\frac{9}{5}$
15. $\frac{7}{8}$ ● $\frac{5}{6}$
16. $\frac{5}{9}$ ● $\frac{2}{3}$
17. $\frac{5}{4}$ ● $\frac{5}{3}$
18. $\frac{3}{4}$ ● $\frac{4}{3}$
19. $\frac{5}{2}$ ● $\frac{15}{6}$
20. $\frac{5}{8}$ ● $\frac{5}{6}$

Set 44 **Give each sum in lowest terms.**

1. $\frac{1}{6} + \frac{1}{3}$
2. $\frac{1}{4} + \frac{5}{8}$
3. $\frac{2}{5} + \frac{3}{10}$
4. $\frac{5}{12} + \frac{1}{4}$
5. $\frac{3}{8} + \frac{1}{3}$
6. $\frac{2}{9} + \frac{1}{3}$
7. $\frac{3}{5} + \frac{1}{4}$
8. $\frac{3}{10} + \frac{1}{4}$
9. $\frac{1}{6} + \frac{5}{12}$
10. $\frac{1}{2} + \frac{1}{8}$
11. $\frac{1}{4} + \frac{2}{3}$
12. $\frac{2}{9} + \frac{1}{6}$
13. $\frac{2}{3} + \frac{1}{12}$
14. $\frac{3}{10} + \frac{1}{2}$
15. $\frac{3}{16} + \frac{1}{4}$
16. $\frac{1}{6} + \frac{7}{10}$
17. $\frac{1}{2} + \frac{5}{16}$
18. $\frac{3}{10} + \frac{1}{10}$
19. $\frac{2}{5} + \frac{1}{3}$
20. $\frac{3}{8} + \frac{1}{16}$
21. $\frac{4}{15} + \frac{3}{10}$

Set 45 **Give each difference in lowest terms.**

1. $\frac{3}{5} - \frac{1}{4}$
2. $\frac{4}{9} - \frac{1}{3}$
3. $\frac{2}{3} - \frac{1}{8}$
4. $\frac{3}{4} - \frac{5}{12}$
5. $\frac{7}{10} - \frac{2}{5}$
6. $\frac{3}{4} - \frac{5}{8}$
7. $\frac{5}{6} - \frac{1}{3}$
8. $\frac{7}{10} - \frac{1}{2}$
9. $\frac{7}{12} - \frac{1}{3}$
10. $\frac{5}{6} - \frac{2}{9}$
11. $\frac{3}{4} - \frac{2}{3}$
12. $\frac{1}{2} - \frac{1}{8}$
13. $\frac{5}{6} - \frac{7}{12}$
14. $\frac{3}{4} - \frac{7}{10}$
15. $\frac{9}{10} - \frac{7}{15}$
16. $\frac{7}{8} - \frac{5}{16}$
17. $\frac{2}{3} - \frac{3}{5}$
18. $\frac{7}{10} - \frac{1}{10}$
19. $\frac{1}{2} - \frac{3}{16}$
20. $\frac{7}{10} - \frac{1}{6}$
21. $\frac{3}{4} - \frac{7}{16}$

Set 46 **Write each mixed number as a fraction.**

1. $5\frac{1}{2}$ 2. $3\frac{1}{4}$ 3. $2\frac{2}{3}$ 4. $3\frac{3}{5}$ 5. $2\frac{5}{8}$ 6. $1\frac{5}{9}$ 7. $5\frac{3}{10}$ 8. $6\frac{1}{6}$

9. $4\frac{2}{7}$ 10. $1\frac{1}{5}$ 11. $4\frac{5}{6}$ 12. $5\frac{3}{4}$ 13. $9\frac{1}{3}$ 14. $8\frac{3}{4}$ 15. $10\frac{1}{2}$ 16. $6\frac{4}{9}$

Set 47 **Write each fraction as a mixed number.**

1. $\frac{5}{2}$ 2. $\frac{5}{3}$ 3. $\frac{23}{4}$ 4. $\frac{11}{2}$ 5. $\frac{23}{7}$ 6. $\frac{15}{4}$ 7. $\frac{15}{8}$ 8. $\frac{37}{5}$

9. $\frac{17}{4}$ 10. $\frac{19}{6}$ 11. $\frac{14}{9}$ 12. $\frac{23}{10}$ 13. $\frac{15}{2}$ 14. $\frac{19}{3}$ 15. $\frac{20}{7}$ 16. $\frac{43}{8}$

Set 48 **Give each sum in simplest form.**

1. $6\frac{1}{3} + 2\frac{1}{3}$ 2. $5\frac{1}{4} + 2\frac{1}{2}$ 3. $8\frac{1}{3} + 2\frac{1}{4}$ 4. $7\frac{3}{8} + 5\frac{1}{4}$ 5. $8\frac{1}{2} + 6\frac{1}{2}$ 6. $9\frac{3}{4} + 5\frac{1}{6}$

7. $3\frac{1}{4} + 7\frac{3}{4}$ 8. $6 + 4\frac{2}{5}$ 9. $8\frac{7}{10} + 5\frac{3}{4}$ 10. $7\frac{5}{8} + 6\frac{1}{2}$ 11. $23\frac{3}{5} + 16\frac{1}{2}$ 12. $9\frac{5}{6} + 6\frac{4}{9}$

13. $13\frac{2}{3} + 7\frac{2}{3}$ 14. $18\frac{3}{8} + 9$ 15. $18\frac{2}{5} + 10\frac{3}{4}$ 16. $20\frac{1}{6} + 8\frac{1}{3}$ 17. $23\frac{5}{9} + 15\frac{2}{3}$ 18. $16\frac{7}{8} + 3\frac{3}{8}$

Set 49 **Give each difference in simplest form.**

1. $6\frac{2}{3} - 3\frac{1}{3}$ 2. $5\frac{1}{2} - 2\frac{1}{2}$ 3. $8\frac{3}{4} - 6\frac{1}{2}$ 4. $9\frac{4}{9} - 2\frac{1}{9}$ 5. $9\frac{2}{5} - 3\frac{4}{5}$ 6. $8 - 1\frac{1}{3}$

7. $9\frac{1}{6} - 6\frac{1}{8}$ 8. $5\frac{1}{3} - 2\frac{1}{2}$ 9. $6\frac{2}{3} - 3\frac{1}{6}$ 10. $7\frac{1}{3} - 4\frac{3}{4}$ 11. $9\frac{3}{4} - 2\frac{7}{8}$ 12. $8\frac{5}{9} - 5\frac{2}{3}$

13. $14\frac{1}{3} - 5\frac{2}{3}$ 14. $17\frac{4}{9} - 8\frac{5}{6}$ 15. $15\frac{2}{5} - 5\frac{7}{10}$ 16. $19\frac{5}{12} - 4\frac{3}{4}$ 17. $21 - 9\frac{3}{5}$ 18. $18\frac{1}{4} - 6\frac{3}{5}$

Set 50 **Give each product in simplest form.**

1. $\frac{1}{3} \times \frac{1}{3}$
2. $\frac{2}{5} \times \frac{2}{5}$
3. $\frac{4}{7} \times \frac{2}{5}$
4. $\frac{2}{3} \times \frac{3}{7}$
5. $\frac{5}{9} \times \frac{3}{10}$
6. $\frac{3}{7} \times \frac{1}{8}$
7. $\frac{1}{2} \times \frac{3}{8}$
8. $\frac{1}{2} \times \frac{1}{2}$
9. $\frac{2}{3} \times \frac{3}{10}$
10. $\frac{3}{4} \times \frac{4}{3}$
11. $\frac{1}{4} \times \frac{1}{2}$
12. $\frac{7}{4} \times \frac{4}{10}$
13. $\frac{3}{4} \times \frac{8}{5}$
14. $\frac{5}{8} \times \frac{7}{5}$
15. $\frac{1}{5} \times \frac{1}{4}$
16. $\frac{4}{3} \times \frac{9}{2}$
17. $\frac{3}{4} \times \frac{8}{7}$
18. $\frac{2}{3} \times \frac{3}{2}$
19. $\frac{5}{8} \times \frac{8}{5}$
20. $\frac{3}{5} \times \frac{10}{9}$

Set 51 **Compute.**

1. $\frac{1}{2}$ of 18
2. $\frac{2}{3}$ of 27
3. $\frac{2}{3}$ of 12
4. $\frac{5}{9}$ of 36
5. $\frac{7}{4}$ of 36
6. $\frac{5}{2}$ of 10
7. $\frac{3}{8}$ of 24
8. $\frac{1}{3}$ of 21
9. $\frac{5}{7}$ of 14
10. $\frac{1}{5}$ of 20
11. $\frac{3}{5}$ of 35
12. $\frac{9}{8}$ of 24
13. $\frac{7}{8}$ of 56
14. $\frac{5}{4}$ of 40
15. $\frac{3}{4}$ of 40
16. $\frac{5}{8}$ of 16.8
17. $\frac{1}{4}$ of 12.8
18. $\frac{4}{9}$ of 19.8
19. $\frac{5}{6}$ of 73.8
20. $\frac{2}{5}$ of 12

Set 52 **Give the reciprocal of each number.**

1. $\frac{1}{2}$
2. $\frac{3}{4}$
3. $\frac{8}{5}$
4. 6
5. $\frac{7}{2}$
6. $\frac{1}{8}$
7. $\frac{3}{5}$
8. 9
9. $2\frac{1}{3}$
10. $\frac{1}{5}$
11. $4\frac{1}{2}$
12. 10
13. $2\frac{4}{5}$
14. $\frac{3}{10}$
15. $1\frac{7}{8}$
16. $\frac{7}{6}$

Set 53 **Give each quotient in simplest form.**

1. $\frac{1}{2} \div \frac{2}{3}$
2. $\frac{4}{9} \div \frac{1}{3}$
3. $\frac{3}{2} \div \frac{3}{4}$
4. $\frac{3}{4} \div \frac{1}{8}$
5. $\frac{9}{5} \div \frac{3}{8}$
6. $\frac{1}{8} \div \frac{9}{5}$
7. $\frac{1}{5} \div \frac{5}{4}$
8. $3 \div \frac{2}{3}$
9. $\frac{1}{3} \div \frac{2}{3}$
10. $\frac{5}{6} \div \frac{1}{2}$
11. $\frac{3}{10} \div \frac{1}{5}$
12. $5 \div \frac{5}{3}$
13. $6 \div \frac{3}{8}$
14. $\frac{3}{10} \div \frac{3}{4}$
15. $\frac{1}{2} \div \frac{7}{2}$
16. $\frac{5}{8} \div 2$
17. $\frac{3}{8} \div 5$
18. $\frac{1}{4} \div \frac{7}{8}$
19. $\frac{2}{3} \div \frac{1}{6}$
20. $\frac{2}{5} \div 4$

Set 54 **Simplify.**

1. $\dfrac{\frac{1}{2}}{\frac{1}{4}}$ 2. $\dfrac{\frac{7}{8}}{\frac{1}{3}}$ 3. $\dfrac{\frac{5}{2}}{\frac{3}{4}}$ 4. $\dfrac{\frac{1}{2}}{\frac{5}{6}}$ 5. $\dfrac{\frac{6}{5}}{\frac{3}{10}}$ 6. $\dfrac{\frac{3}{2}}{\frac{2}{3}}$ 7. $\dfrac{\frac{3}{5}}{\frac{1}{3}}$ 8. $\dfrac{\frac{3}{8}}{\frac{1}{4}}$

9. $\dfrac{\frac{4}{5}}{\frac{2}{3}}$ 10. $\dfrac{\frac{1}{4}}{\frac{5}{8}}$ 11. $\dfrac{\frac{5}{6}}{\frac{2}{3}}$ 12. $\dfrac{\frac{5}{8}}{\frac{1}{2}}$ 13. $\dfrac{\frac{2}{3}}{\frac{3}{4}}$ 14. $\dfrac{\frac{5}{12}}{\frac{1}{4}}$ 15. $\dfrac{\frac{3}{8}}{\frac{3}{4}}$ 16. $\dfrac{\frac{1}{2}}{\frac{1}{3}}$

Set 55 **Change to a fraction or mixed number in simplest form.**

1. 0.25 2. 0.3 3. 0.5 4. 0.625 5. 0.8 6. 1.1 7. 1.5

8. 1.75 9. 2.4 10. 2.6 11. 0.04 12. 1.08 13. 2.05 14. 0.76

Set 56 **Change to a decimal.**

1. $\frac{1}{5}$ 2. $\frac{7}{20}$ 3. $\frac{3}{8}$ 4. $\frac{3}{5}$ 5. $\frac{7}{8}$ 6. $1\frac{1}{4}$ 7. $1\frac{1}{2}$

8. $2\frac{3}{10}$ 9. $2\frac{4}{5}$ 10. $3\frac{1}{25}$ 11. $\frac{3}{25}$ 12. $\frac{1}{50}$ 13. $\frac{1}{20}$ 14. $\frac{3}{4}$

Set 57 **Change to a mixed decimal.**

1. $\frac{1}{3}$ 2. $\frac{1}{6}$ 3. $\frac{5}{7}$ 4. $\frac{2}{3}$ 5. $\frac{1}{9}$ 6. $\frac{10}{3}$ 7. $\frac{5}{6}$ 8. $\frac{5}{9}$

9. $\frac{11}{6}$ 10. $\frac{7}{3}$ 11. $\frac{4}{9}$ 12. $\frac{2}{9}$ 13. $\frac{5}{3}$ 14. $\frac{4}{3}$ 15. $\frac{10}{9}$ 16. $\frac{7}{6}$

Set 58 **Give each product.**

1. $4\frac{1}{2} \times 2$ 2. $2\frac{1}{3} \times 7$ 3. $7\frac{1}{4} \times 6$ 4. $6\frac{1}{3} \times 5$ 5. $3\frac{1}{4} \times 4$

6. $2\frac{2}{3} \times 2\frac{1}{2}$ 7. $3\frac{1}{4} \times 3\frac{1}{3}$ 8. $3\frac{3}{4} \times 4\frac{1}{4}$ 9. $1\frac{3}{8} \times 1\frac{1}{4}$ 10. $2\frac{1}{5} \times 1\frac{1}{4}$

11. $4\frac{2}{3} \times 3\frac{1}{2}$ 12. $1\frac{1}{5} \times 2\frac{3}{4}$ 13. $1\frac{1}{2} \times 1\frac{1}{2}$ 14. $2\frac{3}{8} \times 3\frac{1}{4}$ 15. $4\frac{2}{3} \times 1\frac{3}{4}$

16. $1\frac{3}{4} \times 1\frac{1}{6}$ 17. $3\frac{1}{3} \times 2\frac{1}{5}$ 18. $2\frac{1}{9} \times 2\frac{1}{4}$ 19. $3\frac{2}{3} \times 2\frac{1}{2}$ 20. $2\frac{1}{4} \times 2\frac{3}{4}$

Set 59 **Give each quotient.**

1. $1\frac{1}{4} \div 1\frac{1}{4}$
2. $3\frac{1}{2} \div 1\frac{3}{4}$
3. $5 \div 1\frac{1}{2}$
4. $7\frac{1}{3} \div 2\frac{1}{4}$
5. $7\frac{2}{3} \div 4$
6. $2\frac{2}{9} \div 1$
7. $9\frac{1}{2} \div 3\frac{1}{2}$
8. $3\frac{3}{4} \div 3$
9. $0 \div 5\frac{1}{3}$
10. $3 \div 6\frac{2}{3}$
11. $6\frac{2}{3} \div 8\frac{3}{4}$
12. $2\frac{7}{8} \div 7\frac{2}{3}$
13. $8\frac{1}{2} \div 2\frac{1}{8}$
14. $8\frac{1}{4} \div 4\frac{2}{3}$
15. $9\frac{1}{3} \div 2\frac{3}{5}$
16. $4\frac{5}{8} \div 6\frac{3}{4}$
17. $4\frac{1}{2} \div 3\frac{1}{6}$
18. $5\frac{1}{4} \div 5\frac{4}{5}$
19. $6\frac{1}{3} \div 7\frac{1}{5}$
20. $6\frac{3}{5} \div 8\frac{2}{3}$

Set 60 **Give the perimeter (circumference). Use 3.14 as an approximation for π.**

1.
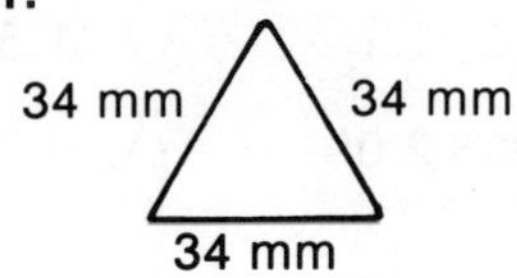

2.
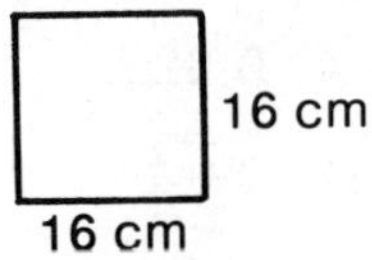

3.
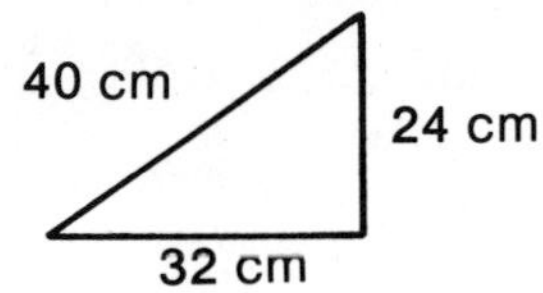

4.
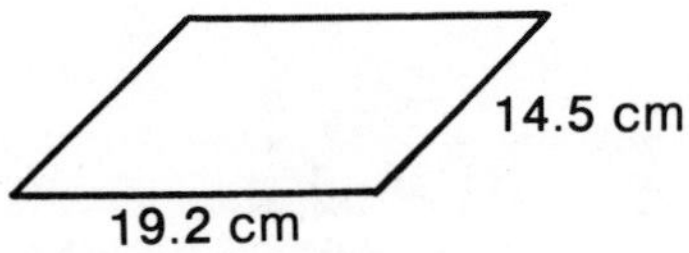

5.
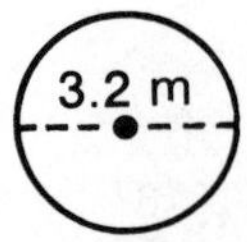

6.
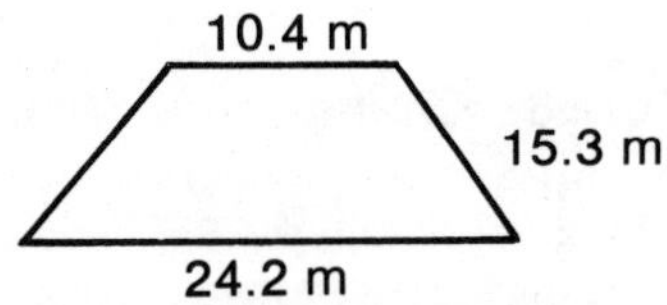

7.
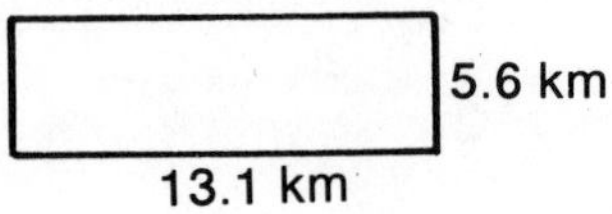

8.
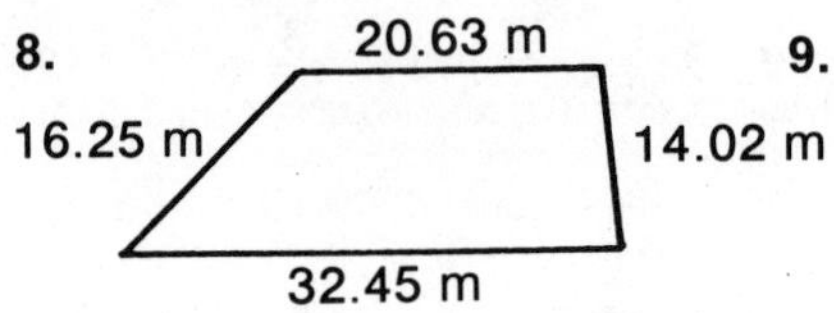

9.
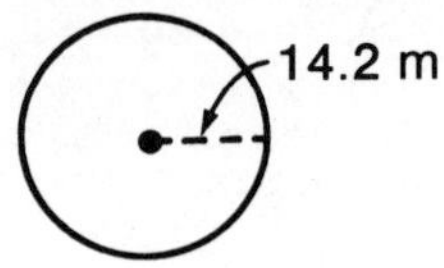

Set 61 **Give the area. Use 3.14 as an approximation for π.**

1.
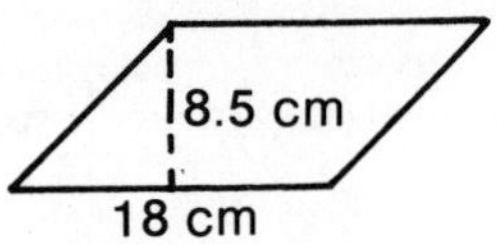

2.
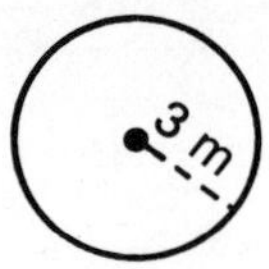

3.
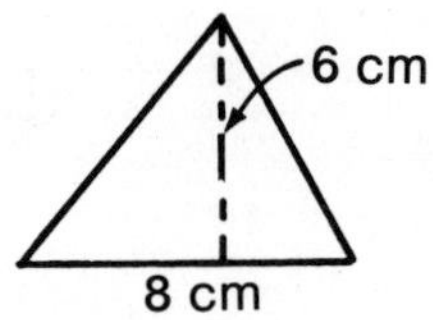

4.
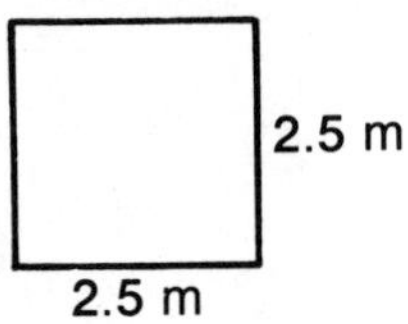

5.
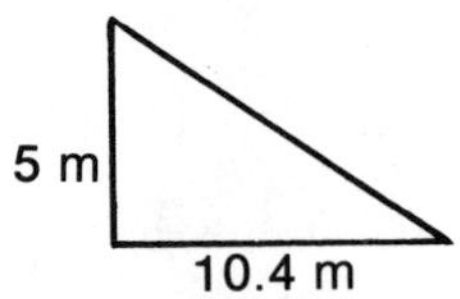

6.
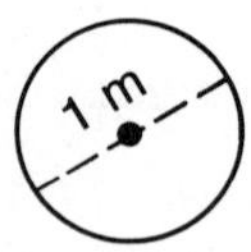

7.
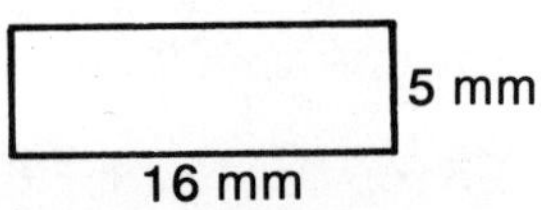

8.

9.
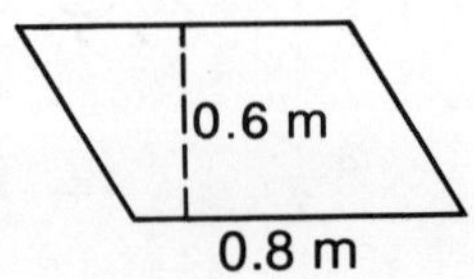

Set 62 Give the surface area.

1. 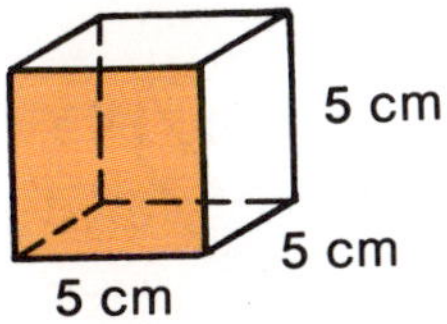

2. 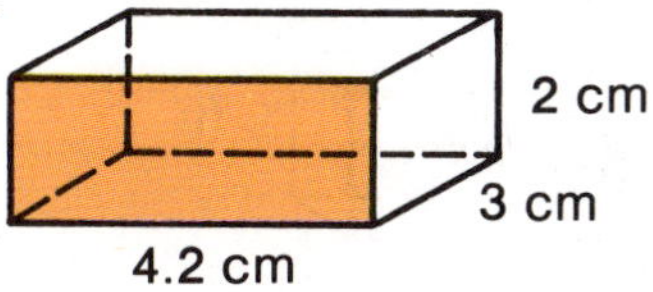

3. 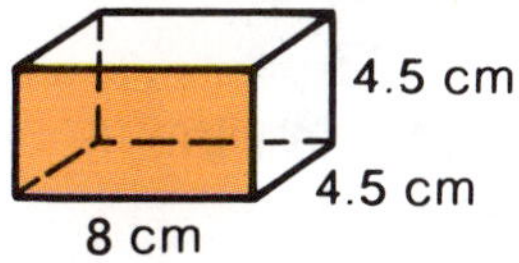

4. 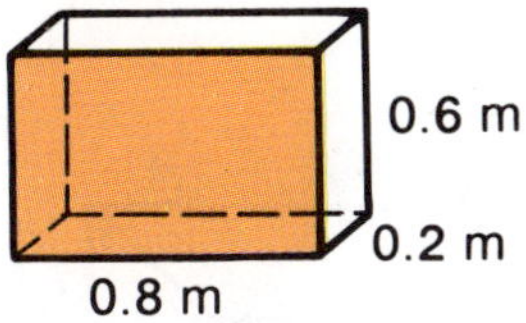

5. 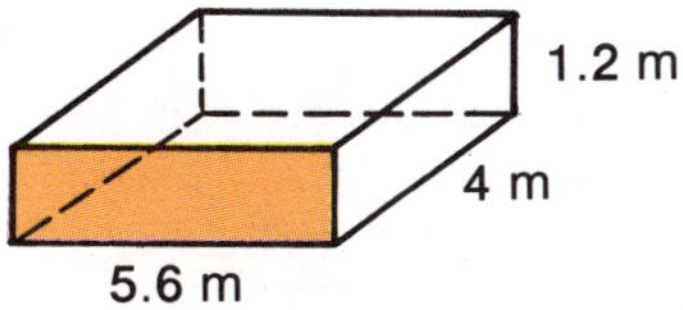

6. 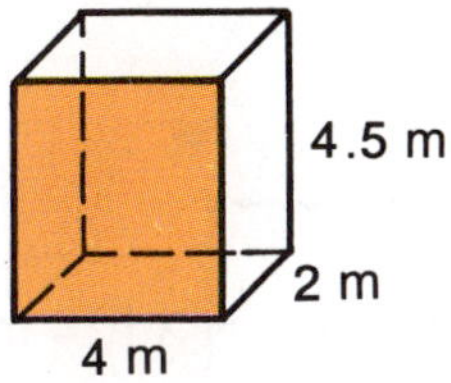

Set 63 Give the volume.

1. 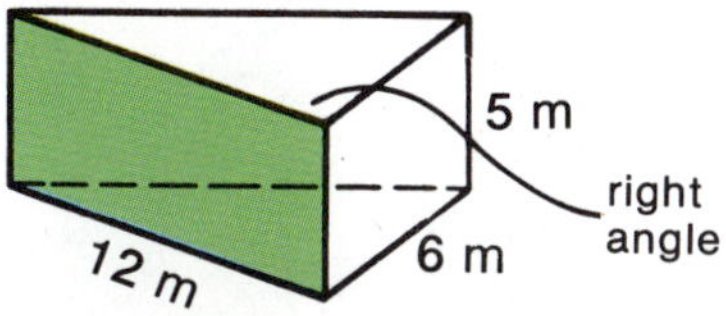

2. 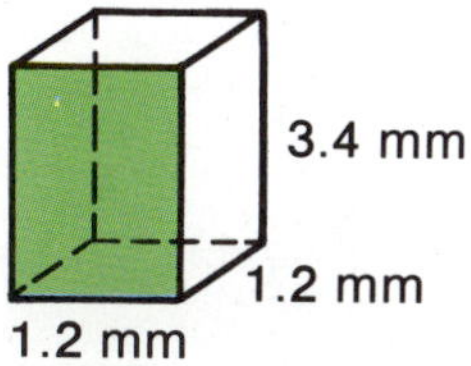

3. 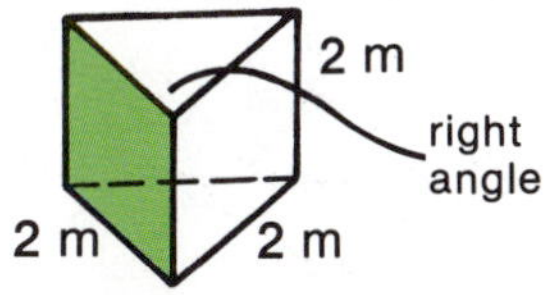

4. 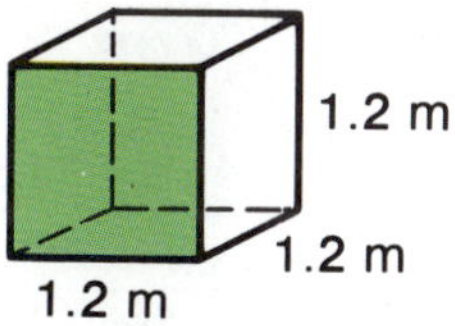

5. 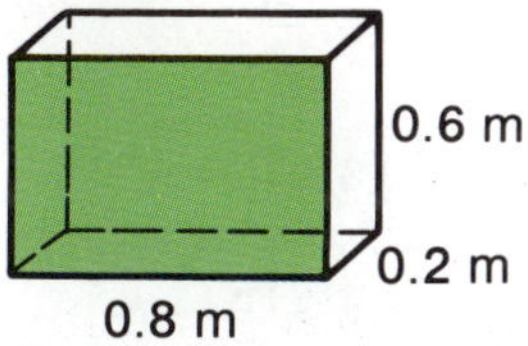

6. 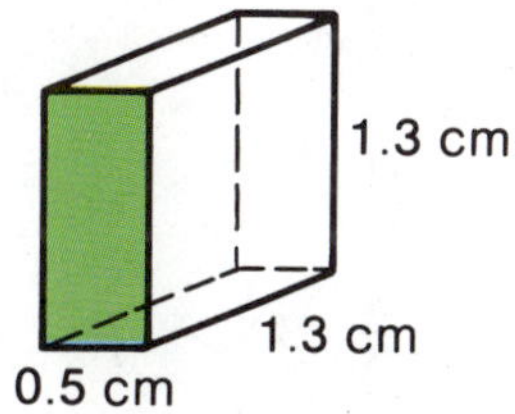

Set 64 Give the area. Use 3.14 as an approximation for π.

1. 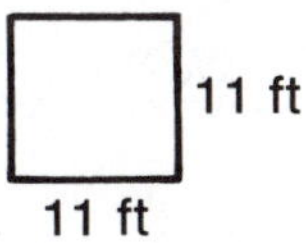

2. 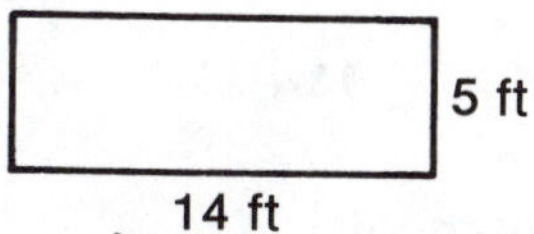

3. 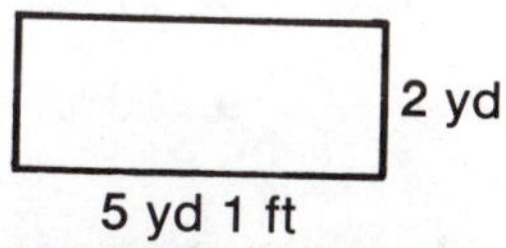

4. 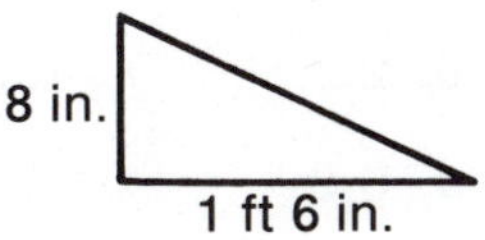

5. 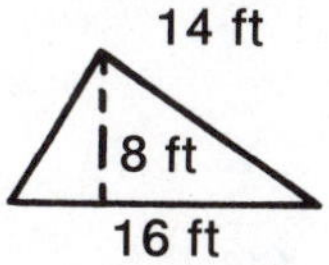

6. 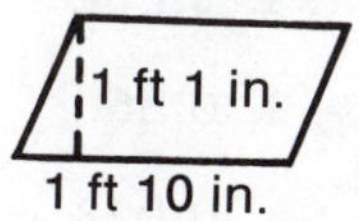

7. 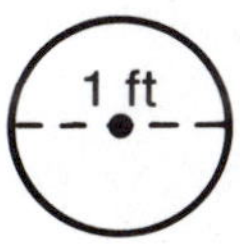

8. 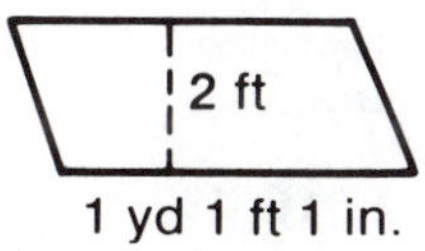

9. 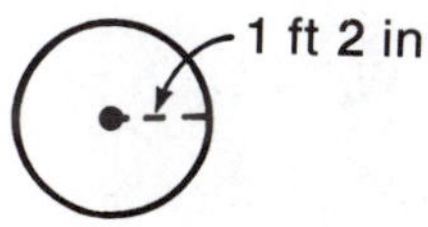

Set 65 **Give the volume.**

1.

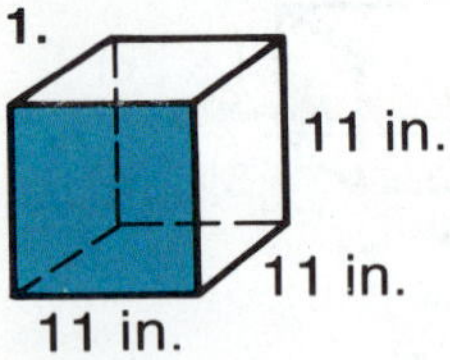

2.

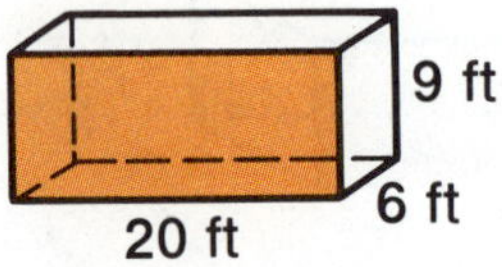

3.

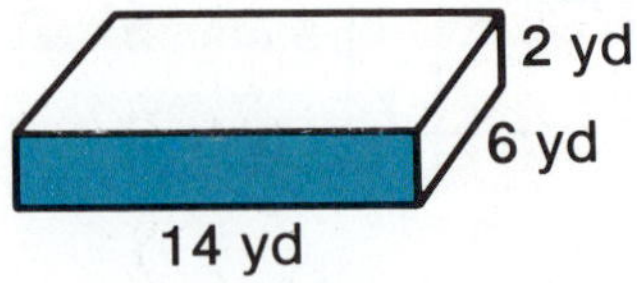

4.

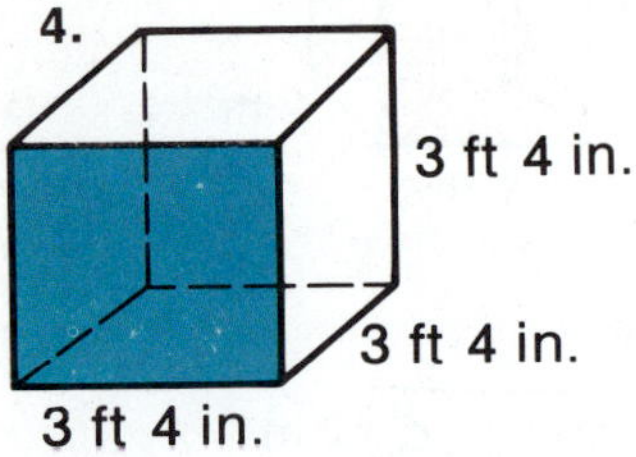

5.

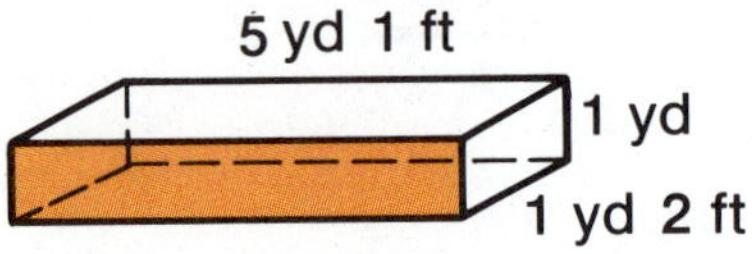

6.

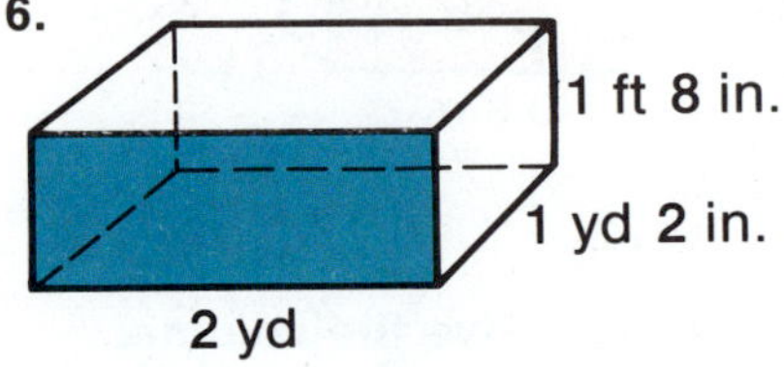

Set 66 **Solve.**

1. $\frac{2}{3} = \frac{x}{9}$ 2. $\frac{5}{9} = \frac{20}{x}$ 3. $\frac{5}{8} = \frac{x}{96}$ 4. $\frac{24}{x} = \frac{3}{5}$ 5. $\frac{9}{8} = \frac{x}{6}$

6. $\frac{3}{8} = \frac{5}{x}$ 7. $\frac{7}{6} = \frac{8}{x}$ 8. $\frac{8}{x} = \frac{5}{6}$ 9. $\frac{9}{x} = \frac{3}{7}$ 10. $\frac{8}{3} = \frac{x}{9}$

11. $\frac{6}{1} = \frac{8}{x}$ 12. $\frac{3}{4} = \frac{x}{5}$ 13. $\frac{8}{5} = \frac{7}{x}$ 14. $\frac{x}{11} = \frac{7}{8}$ 15. $\frac{13}{x} = \frac{4}{5}$

16. $\frac{12}{x} = \frac{4}{3}$ 17. $\frac{x}{14} = \frac{11}{2}$ 18. $\frac{15}{x} = \frac{3}{2}$ 19. $\frac{17}{x} = \frac{5}{7}$ 20. $\frac{16}{x} = \frac{9}{5}$

Set 67 **Write each percent as a fraction in lowest terms.**

1. 21% 2. 37% 3. 55% 4. 16% 5. 40% 6. 3% 7. 34%

8. 8% 9. 10% 10. 25% 11. 6% 12. 78% 13. 68% 14. 45%

Set 68 **Write each decimal as a percent.**

1. 0.2 2. 0.35 3. 0.415 4. 1.8 5. 0.07 6. 2.3 7. 1.253

8. 0.71 9. 0.028 10. 2.06 11. 0.5 12. 10.1 13. 0.9 14. 1.15

Set 69 **Write each percent as a decimal.**

1. 24% 2. 4% 3. 60% 4. 16.4% 5. 82.5% 6. 3.8% 7. 101%

8. 230% 9. 9% 10. 18% 11. 80% 12. 0.5% 13. 1.6% 14. 45%

Set 70 **Write each fraction as a percent.**

1. $\frac{7}{100}$ 2. $\frac{13}{100}$ 3. $\frac{7}{50}$ 4. $\frac{3}{25}$ 5. $\frac{1}{20}$ 6. $\frac{9}{10}$ 7. $\frac{8}{5}$ 8. $\frac{7}{4}$

9. $\frac{5}{8}$ 10. $\frac{3}{50}$ 11. $\frac{11}{5}$ 12. $\frac{7}{20}$ 13. $\frac{1}{25}$ 14. $\frac{1}{2}$ 15. $\frac{3}{16}$ 16. $\frac{7}{2}$

Set 71 **Write each fraction as a percent.**

1. $\frac{2}{3}$ 2. $\frac{1}{6}$ 3. $\frac{3}{8}$ 4. $\frac{1}{12}$ 5. $\frac{7}{9}$ 6. $\frac{9}{7}$ 7. $\frac{11}{6}$ 8. $\frac{13}{8}$

9. $\frac{9}{16}$ 10. $\frac{4}{15}$ 11. $\frac{10}{9}$ 12. $\frac{8}{3}$ 13. $\frac{5}{6}$ 14. $\frac{7}{12}$ 15. $\frac{1}{7}$ 16. $\frac{11}{15}$

Set 72 **Solve.**

1. 25% of 80 = n
2. 50% of 72 = n
3. 20% of 60 = n
4. 40% of 10 = n
5. 100% of 63 = n
6. 150% of 8 = n
7. 125% of 20 = n
8. $33\frac{1}{3}$% of 15 = n
9. $12\frac{1}{2}$% of 40 = n
10. 25% of 28 = n
11. 75% of 78 = n
12. 60% of 90 = n
13. 200% of 12 = n
14. 500% of 30 = n
15. 50% of 30 = n
16. 5% of 30 = n
17. 30% of 60 = n
18. 60% of 30 = n

Set 73 **Solve.**

1. 6% of 100 = n
2. 50% of 39 = n
3. 20% of 45 = n
4. 12% of 25 = n
5. 175% of 12 = n
6. 8% of 50 = n
7. 125% of 200 = n
8. $37\frac{1}{2}$% of 24 = n
9. 25% of 48 = n
10. $87\frac{1}{2}$% of 40 = n
11. 10% of 60 = n
12. 150% of 52 = n
13. $12\frac{1}{2}$% of 40 = n
14. 200% of 19 = n
15. $16\frac{2}{3}$% of 42 = n
16. 40% of 8 = n
17. $83\frac{1}{3}$% of 24 = n
18. $33\frac{1}{3}$% of 37 = n

Set 74 **Solve.**

1. 4% of n = 2
2. 75% of n = 12
3. 9% of n = 9
4. 50% of n = 13
5. 100% of n = 29
6. $66\frac{2}{3}$% of n = 18
7. $33\frac{1}{3}$% of n = 23
8. 25% of n = 31
9. 10% of n = 8
10. 42% of n = 21
11. $12\frac{1}{2}$% of n = 19
12. $83\frac{1}{3}$% of n = 25
13. $62\frac{1}{2}$% of n = 25
14. 80% of n = 74
15. 200% of n = 39
16. $16\frac{2}{3}$% of n = 43
17. 20% of n = 18
18. $37\frac{1}{2}$% of n = 12

Set 75 **Compute the unit price (cents per gram). Round to the nearest hundredth of a cent.**

1. 39 g for 78¢
2. 100 g for 60¢
3. 60 g for 54¢
4. 40 g for 92¢
5. 120 g for 84¢
6. 165 g for 43¢
7. 65 g for 48¢
8. 58.5 g for 52¢
9. 95.5 g for 75¢
10. 62.5 g for 56¢
11. 68.75 g for 64¢
12. 200 g for 72¢

Set 76 **Compute the prices.**

1. 2 for 89¢
   a. 1 costs ?
   b. 3 cost ?
2. 3 for 71¢
   a. 1 costs ?
   b. 2 cost ?
3. 4 for 57¢
   a. 1 costs ?
   b. 3 cost ?
4. 3 for 79¢
   a. 1 costs ?
   b. 4 cost ?
5. 5 for 83¢
   a. 1 costs ?
   b. 4 cost ?
6. 4 for 95¢
   a. 1 costs ?
   b. 5 cost ?
7. 2 for 93¢
   a. 1 costs ?
   b. 5 cost ?
8. 3 for $1.13
   a. 1 costs ?
   b. 5 cost ?
9. 4 for $1.93
   a. 1 costs ?
   b. 3 cost ?
10. 2 for $1.19
   a. 1 costs ?
   b. 7 cost ?
11. 3 for $2.15
   a. 1 costs ?
   b. 8 cost ?
12. 5 for $2.74
   a. 1 costs ?
   b. 7 cost ?

Set 77 **Copy and complete.**

	1.	2.	3.	4.	5.	6.	7.	8.	9.	10.
Regular Price	$80	$60	$50	$18	$43	$62	$73	$81	$56	$42
Percent of Discount	10%	15%	12%	18%	25%	32%	30%	50%	$33\frac{1}{3}\%$	$66\frac{2}{3}\%$
Sale Price										

	11.	12.	13.	14.	15.	16.	17.	18.	19.	20.
Regular Price	$18.50	$16.75	$23.40	$26.80	$52.40	$36.75	$81.50	$72.30	$65.45	$72.48
Percent of Discount	10%	20%	15%	30%	50%	$37\frac{1}{2}\%$	$33\frac{1}{3}\%$	$66\frac{2}{3}\%$	$62\frac{1}{2}\%$	$12\frac{1}{2}\%$
Sale Price										

Set 78 Copy and complete.

	Principal	*Rate*	*Time*	*Interest*
1.	\$100	16%	1 year	
2.	\$300	9%	2 years	
3.	\$150	12%	$1\frac{1}{2}$ years	
4.	\$275	18%	6 months	
5.	\$450	10%	8 months	
6.	\$600	15%	9 months	
7.	\$1000	$12\frac{1}{2}\%$	3 months	
8.	\$850	10%	2 months	
9.	\$1125	10%	4 months	
10.	\$3600	$9\frac{1}{2}\%$	18 months	
11.	\$2100	$10\frac{1}{4}\%$	$2\frac{1}{2}$ years	
12.	\$8000	$12\frac{3}{4}\%$	$1\frac{3}{4}$ years	

Set 79 Complete.

Event:
Toss this number cube.

1. $P(1)$ = ?

2. P(even number) = ?

3. $P(2)$ = ?

4. P(not 2) = ?

5. P(odd number) = ?

6. P(prime number) = ?

7. P(number less than 2) = ?

8. P(number greater than 3) = ?

9. P(factor of 12) = ?

10. P(multiple of 2) = ?

Set 80 Complete.

First Event:
Spin this spinner.

Second Event:
Then (without looking) pick a marble.

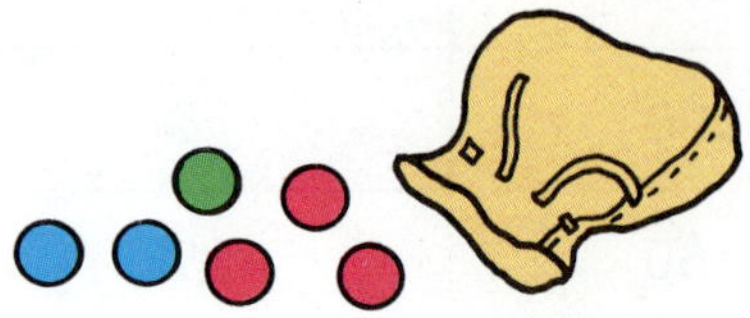

1. $P(1, \text{red}) = \underline{?}$
2. $P(2, \text{green}) = \underline{?}$
3. $P(\text{odd}, \text{red}) = \underline{?}$
4. $P(\text{odd}, \text{blue}) = \underline{?}$
5. $P(\text{even}, \text{blue}) = \underline{?}$
6. $P(\text{even}, \text{red}) = \underline{?}$
7. $P(3, \text{blue}) = \underline{?}$
8. $P(\text{odd}, \text{green}) = \underline{?}$
9. $P(\text{number} > 3, \text{red}) = \underline{?}$
10. $P(\text{number} < 3, \text{green}) = \underline{?}$

Set 81 Find the range, mode, mean, and median.

1. 3, 5, 8, 8, 10, 11, 14
2. 5, 6, 9, 9, 11, 12
3. 32, 38, 46, 49, 50
4. 26, 32, 34, 41, 42, 42, 46
5. 68, 69, 70, 71, 72
6. 39, 43, 56, 57, 65
7. 45, 45, 45, 45, 45, 45
8. 83, 86, 86, 89, 90, 91
9. 126, 124, 132, 128, 130
10. 153, 162, 147, 165, 147

Set 82 $<$ or $>$?

1. $^{+}5$ ● $^{+}4$
2. $^{-}5$ ● $^{+}4$
3. $^{-}5$ ● $^{-}4$
4. $^{+}5$ ● $^{-}4$
5. $^{+}9$ ● $^{-}8$
6. $^{+}6$ ● $^{+}5$
7. $^{-}9$ ● $^{-}8$
8. $^{-}5$ ● $^{+}4$
9. $^{-}14$ ● $^{+}16$
10. $^{+}9$ ● $^{-}12$
11. $^{+}1$ ● 0
12. 0 ● $^{-}1$
13. $^{+}12$ ● $^{-}13$
14. $^{+}12$ ● $^{+}13$
15. $^{-}12$ ● $^{+}13$
16. $^{-}12$ ● $^{-}13$
17. $^{+}18$ ● $^{+}19$
18. $^{+}32$ ● $^{+}33$
19. $^{-}16$ ● $^{-}15$
20. $^{+}12$ ● $^{-}12$
21. $^{-}36$ ● $^{+}35$
22. $^{-}13$ ● $^{+}13$
23. $^{-}23$ ● $^{-}24$
24. $^{-}18$ ● $^{+}19$

Set 83 Give each sum.

1. $^{+}5 + ^{+}9$
2. $^{+}9 + ^{-}5$
3. $^{-}8 + 0$
4. $^{-}5 + ^{-}6$
5. $^{+}9 + ^{+}3$
6. $^{-}9 + ^{-}5$
7. $^{-}6 + ^{-}6$
8. $^{+}9 + 0$
9. $^{-}15 + ^{+}11$
10. $^{-}11 + ^{-}3$
11. $^{+}8 + ^{-}9$
12. $^{-}9 + ^{+}13$
13. $^{+}11 + ^{-}11$
14. $^{-}1 + ^{-}10$
15. $^{+}7 + ^{+}9$
16. $^{-}14 + ^{+}14$
17. $^{+}13 + ^{+}5$
18. $^{-}19 + 0$
19. $^{-}13 + ^{+}13$
20. $^{+}13 + ^{-}20$
21. $^{-}16 + ^{+}16$
22. $0 + 0$
23. $^{+}12 + ^{-}5$
24. $^{+}11 + ^{+}9$

Set 84 Give each difference.

1. ${}^{+}6 - {}^{+}4$
2. ${}^{+}6 - {}^{-}4$
3. ${}^{+}7 - 0$
4. ${}^{-}5 - {}^{+}7$
5. $0 - {}^{-}6$
6. ${}^{-}5 - {}^{-}3$
7. ${}^{-}3 - {}^{+}9$
8. ${}^{+}12 - {}^{+}5$
9. ${}^{-}2 - {}^{-}6$
10. ${}^{-}9 - {}^{+}8$
11. ${}^{+}8 - {}^{+}3$
12. ${}^{-}8 - {}^{-}2$
13. ${}^{+}3 - {}^{-}9$
14. $0 - 0$
15. ${}^{+}12 - {}^{-}3$
16. ${}^{-}2 - {}^{-}9$
17. ${}^{+}3 - {}^{+}8$
18. ${}^{-}4 - {}^{+}11$
19. ${}^{-}9 - {}^{-}4$
20. ${}^{+}4 - {}^{-}6$
21. ${}^{-}4 - {}^{+}6$
22. ${}^{+}13 - {}^{-}6$
23. ${}^{+}6 - {}^{+}14$
24. ${}^{-}1 - {}^{-}5$

Set 85 Give each product.

1. ${}^{+}5 \times {}^{+}4$
2. ${}^{+}7 \times {}^{-}3$
3. ${}^{-}8 \times {}^{-}8$
4. ${}^{-}6 \times {}^{+}6$
5. ${}^{-}6 \times {}^{+}2$
6. $0 \times {}^{-}4$
7. ${}^{+}5 \times {}^{+}7$
8. ${}^{-}4 \times {}^{+}3$
9. ${}^{-}3 \times {}^{-}9$
10. ${}^{+}9 \times {}^{-}9$
11. ${}^{-}5 \times {}^{-}8$
12. ${}^{-}9 \times {}^{-}5$
13. ${}^{-}7 \times {}^{+}9$
14. ${}^{+}6 \times {}^{+}4$
15. ${}^{+}5 \times {}^{-}9$
16. ${}^{-}8 \times {}^{-}9$
17. ${}^{+}9 \times {}^{+}8$
18. ${}^{+}7 \times {}^{-}7$
19. ${}^{-}4 \times {}^{-}8$
20. ${}^{+}8 \times {}^{-}9$
21. ${}^{-}9 \times {}^{-}6$
22. 0×0
23. ${}^{+}6 \times {}^{-}8$
24. ${}^{+}6 \times {}^{+}9$

Set 86 Give each quotient.

1. ${}^{+}24 \div {}^{-}6$
2. ${}^{+}56 \div {}^{-}7$
3. ${}^{-}32 \div {}^{+}8$
4. ${}^{-}27 \div {}^{-}3$
5. ${}^{-}35 \div {}^{+}7$
6. ${}^{-}36 \div {}^{-}6$
7. $0 \div {}^{-}7$
8. ${}^{+}32 \div {}^{+}8$
9. ${}^{+}36 \div {}^{+}4$
10. ${}^{-}8 \div {}^{-}8$
11. ${}^{+}48 \div {}^{-}6$
12. ${}^{-}45 \div {}^{-}9$
13. ${}^{-}36 \div {}^{+}9$
14. ${}^{+}48 \div {}^{+}6$
15. ${}^{-}54 \div {}^{+}6$
16. ${}^{-}21 \div {}^{+}3$
17. ${}^{+}49 \div {}^{-}7$
18. ${}^{-}28 \div {}^{-}4$
19. ${}^{+}63 \div {}^{-}9$
20. $0 \div {}^{+}9$
21. ${}^{+}81 \div {}^{+}9$
22. ${}^{+}72 \div {}^{-}9$
23. ${}^{-}48 \div {}^{-}8$
24. ${}^{+}64 \div {}^{-}8$

Extra Problem Solving

Set 1

1. Mr. Carson and his daughter Barbara went on a white-water-rapids raft trip. The cost was \$68 for each of them. What was the total cost?

2. They had to drive 245 miles to the camp where the raft trip began. After driving 150 miles in 3 hours, they stopped for lunch and gas. How much farther did they have to drive?

3. They spent \$18.75 for gas and \$9.53 for lunch. How much did they spend for gas and lunch?

4. They drove 3 hours before lunch, took 1 hour to eat lunch, and drove 2 hours after lunch. They arrived at the camp at 3:30 P.M. What time had they started their trip?

5. They arrived at the camp at 3:30 P.M. The raft trip was to begin at 8:15 the next morning. How long did they have to wait?

Set 2

1. Mr. Carson bought 2 postcards for 15¢ each and a book about rafts for \$3.85. What was the total cost?

2. Barbara bought a small roll of film with 12 exposures and a large roll of film with 20 exposures. Then she took 5 pictures. How many exposures did she have left?

3. The large roll of film cost \$3.20. The small roll of film cost \$1.80. How much more did the large roll cost?

4. The 72 people going on the raft trip went to the river in 2 buses. The larger bus took 45 people. How many people did the smaller bus take?

5. Each raft carried 12 people. How many other people were on the raft with Mr. Carson and Barbara?

Set 3

1. There were 72 people going on the raft trip. Each raft carried 12 people. How many rafts were used?

2. The total weight of 12 people and their equipment in Mr. Carson and Barbara's raft was 2040 pounds. The equipment weighed 240 pounds. What was the average weight per person?

3. The raft trip began at 8:15 A.M. They stopped for lunch at 12 noon. How long were they on the raft before lunch?

4. At lunch each of the 72 people ate 2 hot dogs. How many hot dogs were eaten?

5. Barbara bought 6 postcards for 15¢ each and a model raft for $2.95. What was the total cost?

Set 4

1. They were on the raft 3 hours 45 minutes in the morning and 3 hours 20 minutes in the afternoon. How long were they on the raft altogether?

2. Barbara took 32 pictures. The cost of the film was $5 and the cost of developing the pictures would be $11. What is the average cost per picture?

3. They left the camp at 9:30 A.M. the next day and arrived home 10 hours later. What time did they arrive home?

4. They drove the 245 miles back home, with many stops, in 10 hours. What was their average speed?

5. The whole trip lasted 56 hours and cost $352.80. What was the average cost per hour?

Set 5

1. At the end of the wrestling season Ben weighed 43.5 kg. At the beginning of the next wrestling season he was 5.7 kg heavier. How much did he weigh then?

2. Steve weighed 57.7 kg. In order to wrestle in a lower-weight class, he lost 3.46 kg. How much did he weigh then?

3. When Betty began training for track, she weighed 53 kg. 10.6 kg of her body weight was fat. How much of her body weight was not fat?

4. While training for a cross-country run, Kent ran 7.5 km. Then, to work on his speed, he ran 0.4 km in 80 seconds 6 times. How many kilometers did he run?

Set 6

1. Carl ran 100 meters in 12.15 seconds. After practicing his starts for a week, he reduced his time by 0.2 second. What was his new time?

2. In one running season Donna reduced her time in the 400-meter run from 62.3 seconds to 57.5 seconds. How much did she reduce her time?

3. David ran 4 times around a 400-meter track in 4.82 minutes. What was his average time to run around the track?

4. Marilyn's at-rest heart rate is 64 beats per minute. When she is running hard, her peak heart rate is 2.7 times as fast. What is her peak heart rate?

5. Through training, Bill reduced his average heart rate from 75 beats per minute to 58 beats per minute. How many fewer times does his heart beat in a day now?

Set 7

1. In a 10,000-meter race there were 895 entries. In the same race one year later there were 1,521 entries. What was the increase in entries?

2. Maria ran 10,000 meters in 57.2 minutes. How many seconds more than 57 minutes is that?

3. Marcia is a figure skater. She practices 2.5 hours every day. How many hours of practice will she have in a year?

4. In gymnastics 4 judges rate a performer. Then the highest and the lowest scores are discarded, and the remaining 2 scores are averaged. In the floor exercise Janet's scores were 7.3, 8.1, 7.6, and 7.9. What was her final score?

5. Art's scores on the parallel bars were 8.3, 8.3, 8.7, and 8.4. What was his final score? (See exercise 4 for the rules of scoring.)

Set 8

1. In one swimming season Terry lowered his 100-meter freestyle time from 68.24 seconds to 65.37 seconds. How much did he lower his time?

2. When Kyle began power lifting, he could bench-press 37.5 kg. After a year he could lift 1.92 times as much. How much could he bench-press then?

3. In the dead-lift, Kyle can lift 182 kg. This is 2.08 times as much as he could lift when he began the sport. How much could he dead-lift then?

4. In a test of physical conditioning, June ran 1.9 km in 12 minutes. After a year of training, she ran 1.42 times as far in the same time. How many kilometers did she run then?

5. At the beginning of a training program, James could inhale 0.84 L of air. After a year of jogging, he had increased that to 1.27 L. What was the increase in inhaled air?

Set 9

Mary Jo is a lapidary. That is, she collects and polishes gems and makes jewelry from them.

1. Mary Jo saw an ad in a lapidary magazine for blue-gold turquoise at $35 a kilogram. She ordered 0.75 kg. How much did she have to pay?

2. Another ad offered rose quartz at $.85 for 1 kg and $3.50 for 5 kg. How much is saved per kilogram by buying the larger quantity instead of the smaller quantity?

3. Red jasper from Africa cost $.70 a kilogram. How much would 2.5 kg cost?

4. Two grades of opal were advertised: Grade B for $288 per 25 g and Grade A for $450 per 25 g. How much more will 75 g of Grade A opal cost than 75 g of Grade B opal?

5. Mary Jo ordered 3 kg of golden agate at $.80 per kilogram and 10 grams of Brazilian opal for $3.40 per gram. She sent a check for $42.10, which included the cost of postage. How much was the postage?

Set 10

Fine gems are weighed in *carats*. One carat (ct) is 200 milligrams.

1. The largest diamond ever found, the Cullinan diamond, weighed 3106 carats. How many grams did the Cullinan weigh?

2. The Cullinan diamond was cut into 9 large stones and 96 smaller stones, with about 6 carats of small chips left. What was the average weight in grams of the cut stones?

3. The Hope diamond, another famous diamond, is blue and weighs 45.52 ct. How many grams is that?

4. An average-sized diamond in an engagement ring weighs about 67 mg. How many carats is that?

5. The Iranian XX diamond is a peach-colored diamond weighing 10.87 grams. How many carats is that?

Set 11

1. Kubla Khan, a pacer, won \$1959 in 1978 and \$3041 in 1979. How much more did Kubla Khan win in 1979 than in 1978?

2. Spinster Bull, a trotter, won \$41,090 in 1978 and \$4,590 in 1979. How much less did Spinster Bull win in 1979 than in 1978?

3. As a 7-year-old, Symbol Art had a record of 5 firsts, 5 seconds, and 8 thirds in 22 races. In what fraction of the races was Symbol Art third or better?

4. As a 6-year-old, Symbol Art won $\frac{4}{11}$ of the races he ran. As a 7-year-old, Symbol Art won $\frac{5}{22}$ of the races he ran. In which year did he win the greater fraction of races? How much greater?

5. As a 5-year-old, Hi Ike was first in $\frac{1}{3}$ of the races he ran and second in $\frac{4}{15}$ of the races he ran. In what fraction of the races was Hi Ike either first or second?

Set 12

1. In a 1-mile pace, The Slammer paced the first quarter in 29 seconds, the second quarter in $31\frac{4}{5}$ seconds, the third quarter in $29\frac{1}{5}$ seconds, and the fourth quarter in $30\frac{1}{5}$ seconds. What was his time for the mile?

2. In a 1-mile trot, Clay Port's times for the four quarters were $29\frac{1}{5}$, $30\frac{3}{5}$, $30\frac{2}{5}$, and $31\frac{1}{5}$. What was his time for the mile?

3. At one time the 1-mile record for a trotter was $1{:}54\frac{4}{5}$ (1 minute and $54\frac{4}{5}$ seconds), held by Nevele Pride. The record for a pacer was 1:52, held by Steady Star. How much faster was the pace record?

4. The best time as a 4-year-old by High Bret, a pacer, was $2{:}04\frac{2}{5}$. His best time as a 5-year-old was $2{:}02\frac{3}{5}$. How much faster was his 5-year-old best time?

Set 13

The number of calories needed daily by a person depends on the kinds of activity engaged in and the age, height, weight, and sex of the person.

1. A 75-kg male who does office work needs only $\frac{7}{9}$ of the 3600 calories needed by a 75-kg male carpenter. How many calories are needed by the office worker?

2. A 65-year-old male needs 100 calories more than $\frac{2}{3}$ of the 3600 calories needed by a 16-year-old male. How many calories does the 65-year-old male need?

3. A male 160 cm tall and weighing 59 kg needs 2700 calories daily. A 190-cm male weighing 82 kg needs $\frac{1}{3}$ more calories. How many calories are needed by the larger male?

4. A 61-kg male needs 100 calories more than 1.2 times the 2500 calories needed by a female of the same size. How many calories are needed by the male?

5. A female 11 to 14 years old needs 2400 calories a day. A male 11 to 14 years old needs 2800 calories a day. How many times as many calories does the male need than the female?

Set 14

Protein and minerals such as calcium and iron are also needed daily.

1. A child 4 to 6 years old needs 30 g of protein daily. A child 7 to 10 years old needs $\frac{1}{5}$ more grams. How many grams of protein does the older child need?

2. A female 15 to 18 years old needs 48 g of protein daily. A female 11 to 14 years old needs $\frac{1}{12}$ less. How many grams of protein does the younger female need?

3. A female 19 to 22 years old needs 800 mg of calcium daily. A female 11 to 14 years old needs 1.5 times that amount. How many milligrams of calcium does the younger female need?

4. A male 11 to 14 years old needs 18 mg of iron daily. A male 19 to 22 years old needs $\frac{4}{9}$ less than that amount. How many milligrams of iron does the older male need?

5. A female more than 50 years old needs 10 mg of iron daily. A female 11 to 14 years old needs $\frac{4}{5}$ more than that amount. How many milligrams of iron does the younger female need?

Set 15

Sesame rolls	$.96 per dozen
Steak rolls	$1.20 per dozen
Onion rolls	$1.60 per dozen

1. What is the cost of $\frac{1}{2}$ dozen steak rolls?

2. What is the cost of $2\frac{1}{2}$ dozen onion rolls?

3. What is the cost of 1 sesame roll?

4. What is the cost of 17 steak rolls?

5. What is the cost of $1\frac{3}{4}$ dozen onion rolls?

6. What is the cost of 7 sesame rolls?

7. Arthur spent $1.50 for steak rolls. How many dozen steak rolls did he buy?

Set 16

Egg rolls	$1.50 per dozen
Bagels	$1 per $\frac{1}{2}$ dozen

1. What is the cost of 2 dozen bagels?

2. What is the cost of $1\frac{2}{3}$ dozen egg rolls?

3. What is the cost of $3\frac{1}{2}$ dozen bagels?

4. What is the cost of $\frac{5}{6}$ dozen egg rolls?

5. What is the cost of $\frac{3}{4}$ dozen bagels?

6. What is the cost of $2\frac{1}{3}$ dozen egg rolls?

7. Darlene bought $2\frac{1}{2}$ dozen pan biscuits for $1.75. What is the cost of 1 dozen pan biscuits?

Set 17

The pressure at the surface of the ocean (from the atmosphere above) is about 14.7 lb per in.^{2} Water pressure increases about 0.44 lb per in.^{2} for each foot of depth below the surface of the ocean.

1. The deepest part of the oceans is the Challenger Deep in the Pacific Ocean. It is 36,198 ft deep. What is the pressure at the bottom of the Challenger Deep?

2. The deepest point in the Atlantic Ocean is the Puerto Rico Trench. The depth is 28,374 ft. What is the pressure at that depth?

3. The greatest depth of the Indian Ocean is 25,344 ft. What is the pressure at that depth?

4. The deepest point in the Arctic Ocean is 17,880 ft deep. What is the pressure at that depth?

5. Suppose that an experimental underwater craft is designed to withstand pressures up to 725 lb per in.^{2} What is the greatest depth to which it can go?

Set 18

1. There are 335.5 million cubic miles of water on the earth. Only $\frac{3}{100}$ of this is fresh water. How many cubic miles of fresh water are there?

2. Three fourths of the fresh water is frozen in polar ice caps and in glaciers. How many cubic miles of fresh water are frozen and unavailable for use? (See exercise 1.)

3. The total surface area of the earth is about 139,660,400 mi^2. About $\frac{7}{10}$ of the surface is covered by water. How many square miles of land are there?

4. The diameter of the earth at the equator is 7926.41 mi. Find the circumference of the earth at the equator. ($C = \pi d$, $\pi \approx 3.14$)

5. The Pacific Ocean is about 11,000 mi wide at the equator. The Pacific Ocean stretches what fraction of the way around the earth? Give your answer as a decimal correct to the nearest hundredth. (See exercise 4.)

Set 19

1. During the first 8 months of 1979 Tom Watson, a golfer, won $447,636. If he continued to win at the same rate for the rest of the year, how much would he win in the year?

2. Tom Watson's 8-month winnings were $203,881 more than the 8-month winnings of the golfer in second place, Larry Nelson. How much did Nelson win? (See exercise 1.)

3. The prizes paid to women golfers are not as high as those paid to the men. Nancy Lopez, the leading woman golfer, had an 8-month winning total of about $\frac{2}{5}$ of Watson's winnings. About how much did Lopez win? (See exercise 1.)

4. Bjorn Borg, a tennis player, won $442,867 in the first 8 months of 1979. This was only $34,880 more than John McEnroe won. Jimmy Conners was only $23,582 behind McEnroe. How much did Conners win?

5. After 8 months of 1979, Martina Navratilova was the leading money-winner in women's tennis. She won $405,335. If she continued to win at the same rate, how much would she win in the year?

Set 20

1. Darrell Waltrip was the leading NASCAR driver in the first 8 months of 1979. He was $64,505 ahead of the second-place driver, Richard Petty, who won $312,040. How much did Waltrip win?

2. If Waltrip won at the same rate all year, how much would he have won in the first 6 months? (See exercise 1.)

3. Bowling does not pay as well as golf, tennis, or auto racing. The top bowler, Mark Roth, had 8-month winnings that were $176,249 less than the third-place money-winner in women's tennis, Tracy Austin. Austin won $258,323. How much did Roth win?

4. Golfers, tennis players, racing drivers, and bowlers win large amounts of money, but the top thoroughbred racehorses win even more. In the first 8 months of 1979 Affirmed won $200,800 less than $1 million. How much did Affirmed win?

Set 21

1. According to a recent report, the average human intake of water in a lifetime is 60,600 liters. If the average life span is 69.2 years, what is the average daily intake of water?

2. Besides getting water by drinking, you get water by eating foods that contain water. For example, a tomato is 95% water. How much water is in a tomato that weighs 325 g?

3. The human body is made up of 65% water by weight. How much water is in the body of a 75-kg person? How much water is in your body?

4. The total daily use of water in an average home is 260 liters. How much water is used in a year (365 days)?

5. Each minute under a shower uses 19 liters of water. How much water is used in a 4 5-minute shower?

Set 22

1. A dishwasher uses about 38 liters of water per load. Washing dishes by hand takes about 4.5 liters. Suppose that you run 1 load in a dishwasher per day or wash by hand 3 times per day—how much water is saved each week by hand-washing?

2. Industries use much more fresh water than homes do. For example, the ratio of water used to paper produced in a paper mill is 250:1. How much water is used to produce a metric ton (1000 kg) of paper?

3. Steel-making uses 8% more water than paper-making to produce a metric ton of material. How much water is used to produce 1 metric ton of steel? (See exercise 2.)

4. Green plants pass water into the atmosphere. A birch tree gives off 260 liters on a sunny summer day. It gives off 15% less water on a cloudy day. How much water is given off on a cloudy day?

5. Corn plants give off about 37,000 liters of water per hectare per day. (1 hectare = 10,000 m^2.) How much water is given off in a week by a 21.3-hectare cornfield?

Glossary

acute angle An angle that measures between 0° and 90°.

addend A number used in an addition problem.

adding 0 property The sum of any number and 0 is the number.

$$29 + 0 = 29$$

angle A figure formed by two rays with the same endpoint.

arc Part of a circle.

area The number of unit squares that it takes to cover a region.

associative property of addition Changing the grouping of the addends does not change the sum.

$$(9 + 4) + 6 = 9 + (4 + 6)$$

associative property of multiplication Changing the grouping of the factors does not change the product.

$$(7 \times 25) \times 4 = 7 \times (25 \times 4)$$

average The sum of all the numbers divided by the number of numbers.

axes Two perpendicular lines used as a reference for graphing ordered pairs.

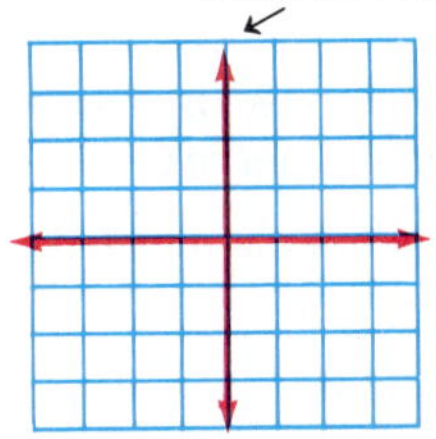

The horizontal line is the first coordinate axis and the vertical line is the second coordinate axis.

base (of an exponent) The number that is raised to a power.

base → 2^3 ← power

bisect To cut into halves.

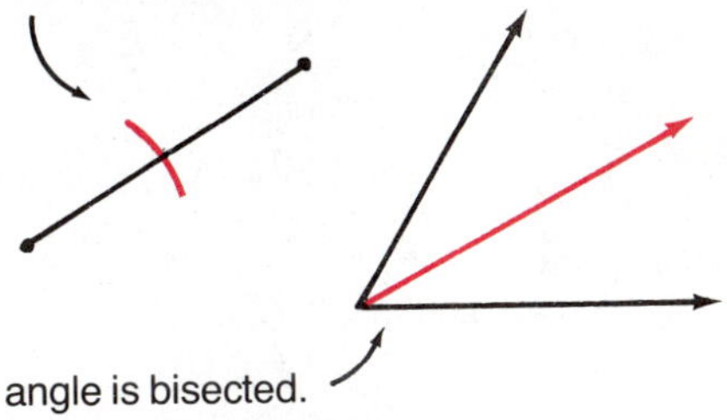

budget A plan for using your money.

canceling Dividing both the numerator and the denominator by a common factor before multiplying fractions.

$$\frac{\overset{4}{\cancel{8}}}{15} \times \frac{1}{\underset{3}{\cancel{6}}}$$

Celsius temperature scale (°C) The metric temperature scale in which 0°C is the freezing point of water and 100°C is the boiling point of water.

centimeter A metric unit of length.
1 centimeter = 0.01 meter.

circle A curved plane figure with all points a given distance from the center.

circumference The distance around a circle.

commission The part of the total sales that goes to the salesperson.

common denominator A common denominator for $\frac{1}{2}$ and $\frac{1}{3}$ is 6, because $\frac{1}{2} = \frac{3}{6}$ and $\frac{1}{3} = \frac{2}{6}$. A common denominator is a common multiple of the denominators of two fractions.

common factor 2 is a common factor of 4 and 6, because 2 is a factor of 4 and a factor of 6.

common multiple 30 is a common multiple of 5 and 6, because it is a multiple of both 5 and 6.

commutative property of addition Changing the order of the addends does not change the sum.

$$23 + 89 = 89 + 23$$

cummutative property of multiplication Changing the order of the factors does not change the product.

$$19 \times 54 = 54 \times 19$$

compass An instrument used to draw circles.

complex fraction A fraction in which either the numerator or the denominator or both are fractions.

$$\frac{\frac{6}{7}}{\frac{2}{3}} \qquad \frac{8}{\frac{3}{4}} \qquad \frac{\frac{5}{2}}{6}$$

composite number A whole number other than 0 that has more than two factors.

compounding interest Paying interest on the interest.

congruent figures Figures that have the same size and shape.

coordinates An ordered pair of numbers that are used to graph a point on a grid.

corresponding parts In congruent figures, the sides and angles that match are called corresponding parts.

cube A rectangular prism whose six faces are squares.

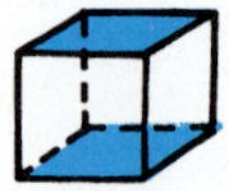

customary system The system of measurement that uses foot, quart, pound, and degrees Fahrenheit temperature.

cylinder A solid figure formed by two congruent circles in parallel planes and a "curved" rectangular region.

data Numerical pieces of information.

decagon A polygon with ten sides.

decimal or decimal fraction A number written in our place-value system with a decimal point before the tenths place.

decimeter A metric unit of length.
1 decimeter = 0.1 meter.

degree A unit for measuring angles. This is a 1° (1-degree) angle.

dekameter A metric unit of length.
1 dekameter = 10 meters.

denominator In the fraction $\frac{2}{3}$ the denominator is 3.

diameter The distance across a circle through its center.

difference The answer to a subtraction problem.

digits The basic symbols used to write numerals in a place-value system. In our base-ten system the digits are 0, 1, 2, 3, 4, 5, 6, 7, 8, and 9.

discount An amount a price is decreased.

distributive property of multiplication A product can be written as the sum of two products.

$$3 \times (10 + 2) = (3 \times 10) + (3 \times 2)$$

dividend The number that is divided.

$3\overline{)18}$ (quotient 6) $\quad 18 \div 3 = 6$

dividend (18)

divisible When one number divides another number and leaves no remainder. 84 is divisible by 2, since $84 \div 2$ leaves no remainder.

divisor The number that one divides by.

$4\overline{)36}$ (quotient 9) $\quad 36 \div 4 = 9$

divisor (4)

down payment The first amount paid when buying on an installment plan.

equally likely outcomes Outcomes that have the same chance of occurring.

equation A sentence with an equal sign, such as $3 \times 9 = 27$.

equilateral triangle A triangle with 3 lines of symmetry and 3 congruent sides.

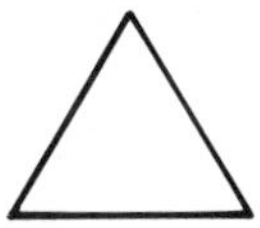

equivalent fractions Fractions that name the same number. $\frac{1}{2}$, $\frac{2}{4}$, and $\frac{3}{6}$ are equivalent fractions.

error in measurement The difference between the true measurement and the recorded measurement.

estimate To use rounded numbers to check whether an answer is correct. To estimate 47 + 32, you would add 50 + 30. The sum should be about 80.

even number Zero and the multiples of 2.

expanded numeral A numeral used to show the place value of each digit in a standard numeral.

$$248 = 200 + 40 + 8$$

exponent An exponent tells how many times a number is used as a factor.

exponent

$$2^3 = \underbrace{2 \times 2 \times 2}_{\text{3 factors}}$$

factors Numbers used in a multiplication problem.

$$\begin{array}{r l} 8 & \leftarrow \text{factor} \\ \times 6 & \leftarrow \text{factor} \\ \hline 48 & \leftarrow \text{product} \end{array}$$

Fahrenheit temperature scale (°F) The customary temperatute scale in which 32°F is the freezing point of water and 212°F is the boiling point of water.

finance charge Buying an item on an installment plan costs more than paying cash. The difference is called the finance charge.

formula A general way of expressing a relationship using symbols.

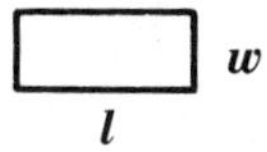

$$A = l \times w$$

fraction A numeral for part of a group or for part of a region. $\frac{1}{2}$, $\frac{4}{6}$, and $\frac{6}{5}$ are fractions.

frequency distribution A table showing the number of times different events occur.

function A set of number pairs of which no two first numbers are the same.

Function rule

$n \rightarrow n + 3$

(0, 3)

(1, 4)

(2, 5)

gram A metric unit of weight (mass).
1 gram = 0.001 kilogram.

graph A picture used to show numerical information. A graph can be a bar graph, a picture graph, a circle graph, or a line graph.

greatest common factor (GCF) The greatest number that is a factor of each of two or more numbers.

8: 1, 2, 4, 8

12: 1, 2, 3, 4, 6, 12

4 is the GCF of 8 and 12.

greatest possible error Half of the unit used in making the measurement.

hectare A metric unit for measuring land area. A square that is 100 meters on a side has an area of 1 hectare.

hectometer A metric unit of length.
1 hectometer = 100 meters.

heptagon A polygon with seven sides.

hexagon A polygon with six sides.

installment buying A way of buying expensive items. You pay part of the cost (the down payment) when you get the item and then agree to pay a certain amount each month for a certain number of months.

integers The numbers ..., ⁻5, ⁻4, ⁻3, ⁻2, ⁻1, 0, ⁺1, ⁺2, ⁺3, ⁺4, ⁺5, ...

interest The amount a borrower pays for using the money.

intersecting lines Lines that meet at only one point.

intersecting segments (or rays) Segments (or rays) that are parts of intersecting lines.

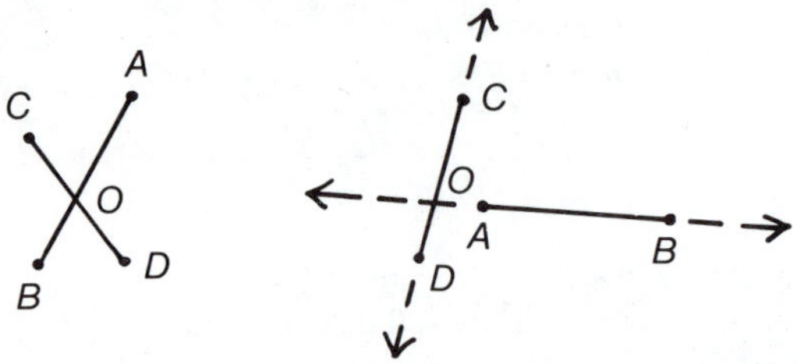

$\overline{AB}$ intersects $\overline{CD}$ at point O.

isosceles trapezoid A trapezoid with 1 line of symmetry through opposite sides and 2 congruent sides.

isosceles triangle A triangle with 1 line of symmetry and 2 congruent sides.

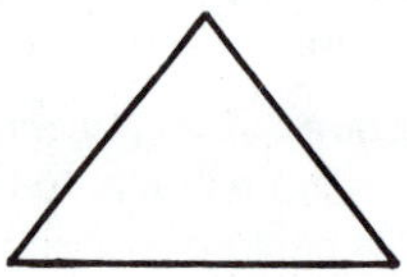

kilogram A metric unit of weight.
1 kilogram = 1000 grams.

kilometer A metric unit of length.
1 kilometer = 1000 meters.

kite A quadrilateral with a line of symmetry through opposite vertices and 2 pairs of congruent sides.

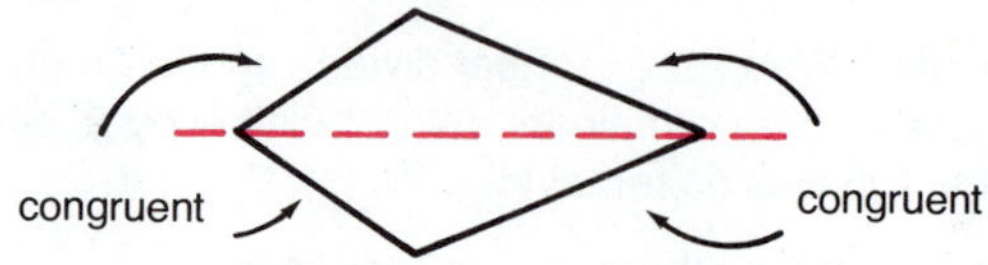

least common denominator The least common multiple of the denominators of two or more fractions.

least common multiple The least (smallest) common multiple of two or more numbers.
The least common multiple of 6 and 15 is 30.

line of symmetry If a figure can be folded along a line so the two halves match, the fold line is a line of symmetry.

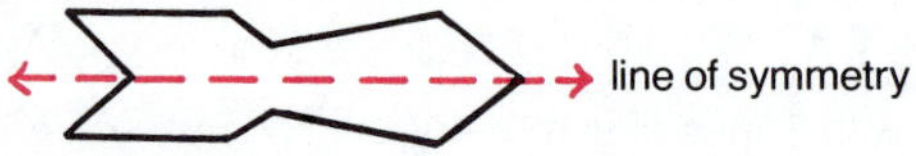

liter A metric unit of volume.

lowest terms A fraction is in lowest terms if the greatest common factor of the numerator and the denominator is 1.

mean The average of all the numbers.

median If there is an odd number of numbers, the median is the number in the middle. If there is an even number of numbers, the median is the average of the two numbers.

median of a triangle A segment joining a vertex of a triangle to the midpoint of the opposite side.

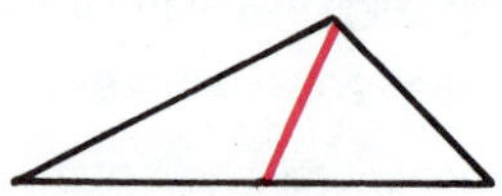

meter A metric unit of length.
1 meter = 100 centimeters.

metric system An international system of measurement that uses meter, liter, gram, and degrees Celsius temperature.

midpoint The point that bisects a segment.

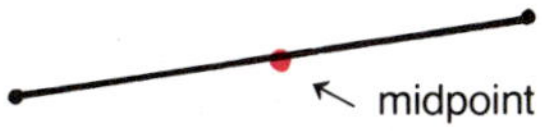

milligram A metric unit of weight (mass). 1 milligram = 0.001 gram.

milliliter A metric unit of volume 1 milliliter = 0.001 liter.

millimeter A metric unit of length. 1 millimeter = 0.001 meter.

mixed decimal A number that has a decimal part and a fraction part. $0.25\frac{1}{2}$ is a mixed decimal.

mixed number A number that has a whole-number part and a fraction part. $2\frac{3}{4}$ is a mixed number.

mode The number that occurs most often.

multiple A product. 4, 8, 12, 16, 20, and so on, are multiples of 4.

multiplying by 1 property The product of any number and 1 is the number.

$$8 \times 1 = 8$$

multiplying by 0 property The product of any number and 0 is 0.

$$24 \times 0 = 0$$

nonagon A polygon with nine sides.

number line A line with its points labeled with numbers.

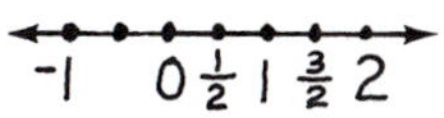

numeral A name or symbol for a number. Numerals for ten: 10, 2 × 5, 12 – 2, X

numerator In the fraction $\frac{2}{3}$, the numerator is 2.

obtuse angle An angle that measures between 90° and 180°.

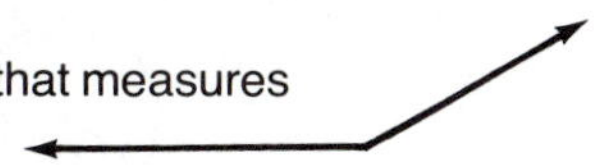

octagon A polygon with eight sides.

odd number A whole number that is not divisible by 2. The numbers 1, 3, 5, 7, 9, 11, and so on, are odd.

odds The ratio of the number of ways that an outcome can occur to the number of ways that the outcome cannot occur.

opposites Two numbers are opposites if their sum is 0.

$^-3 + {}^+3 = 0$
↖ ↗
opposites

ordered pair A pair of numbers that give the coordinates of a point on a grid.

origin The point where axes intersect.

outcome A possible result.

parallel lines Lines in a plane that do not intersect.

parallelogram A quadrilateral with a point of symmetry and opposite sides parallel.

pentagon A polygon with five sides.

percent (%) *Percent* means "per hundred." 5% (5 percent) equals $\frac{5}{100}$.

perimeter The distance around a figure. The sum of the lengths of the sides.

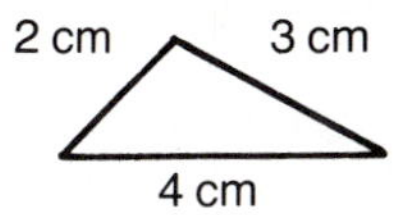

The perimeter is 9 cm.

perpendicular bisector A line that bisects and is perpendicular to a segment.

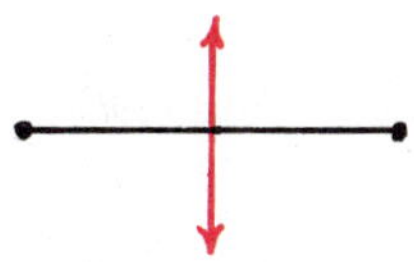

perpendicular lines Two lines that intersect to form right angles.

perpendicular segments (or rays) Segments (or rays) that are parts of perpendicular lines.

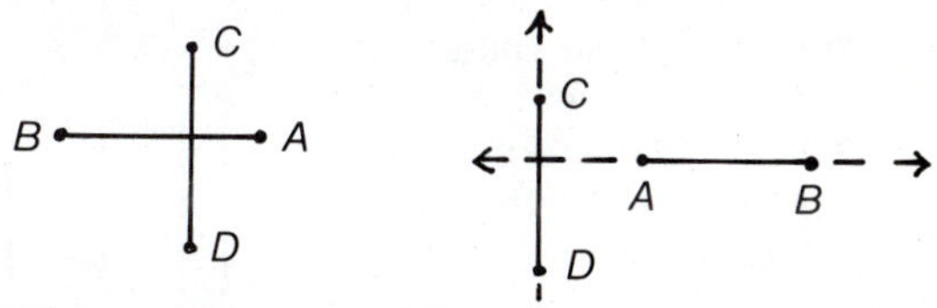

$\overline{AB}$ is perpendicular to $\overline{CD}$.

pi The number that is the ratio of the circumference of a circle to its diameter. It is represented by the Greek letter π and is approximately equal to 3.14.

place value A system for writing numbers in which the value of a digit is determined by its position.

plane A flat surface that extends endlessly in all directions.

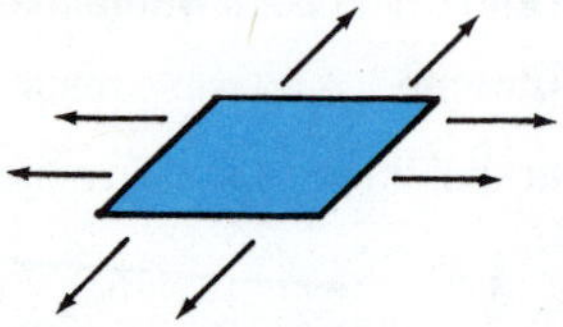

point of symmetry If a figure fits itself after a half-turn about a point, the point is a point of symmetry of the figure.

polygon A plane figure made up of segments.

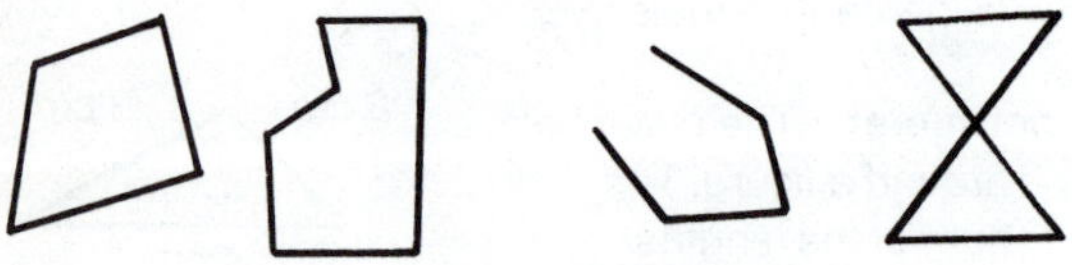

polygons not polygons

power An exponent. The number that tells how many times a number is used as a factor.

$2^3 \leftarrow$ power

prime factorization Expression of a composite number as a product of prime numbers. The prime factorization of 18 is $2 \times 3 \times 3$.

prime number A whole number that has exactly two factors. 2, 3, 5, 7, 11, 13, and so on, are prime numbers.

principal An amount of money borrowed.

prism A solid figure with two congruent parallel faces that are polygons and with the remaining faces parallelograms.

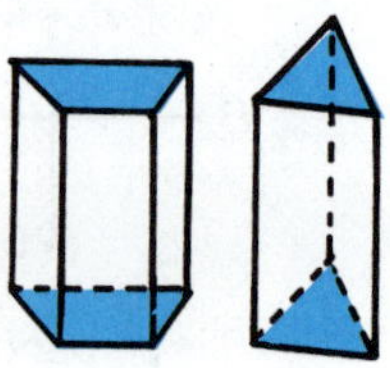

probability The ratio of the number of favorable outcomes to the total number of outcomes.

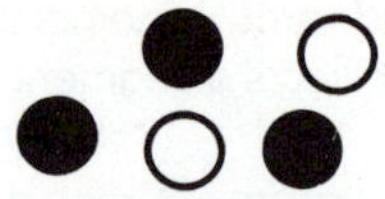

probability of picking black $= \frac{3}{5}$.

product The answer to a multiplication problem.

proportion An equation stating that two ratios are equal.

$$\frac{5}{8} = \frac{30}{48}$$

protractor An instrument used to measure angles.

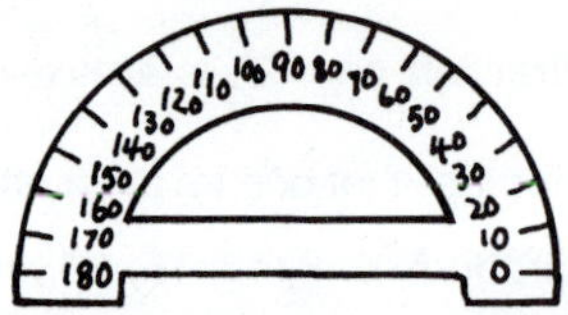

pyramid A solid figure with a face (known as the base) that is any polygon and with all other faces, which are triangles, sharing a common vertex.

quadrilateral A polygon with four sides.

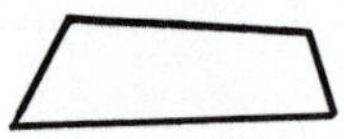

quotient The answer to a division problem.

radius The distance from the center of a circle to the circle. The radius is equal to one half the diameter.

range The difference between the least and the greatest numbers.

rate A comparison by division of two quantities.

$$\frac{\text{87 kilometers}}{\text{2 hours}}$$

ratio A comparison of two quantities by division. Below, the ratio of squares to circles is 4 to 2, 4:2, or $\frac{4}{2}$.

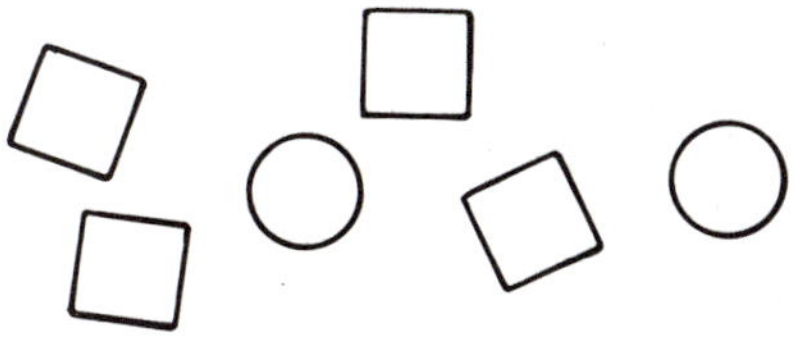

ray A part of a line that has one endpoint. This is ray *AB*.

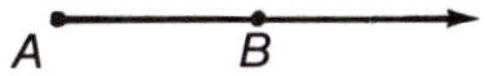

reciprocal Two numbers are reciprocals when their product is 1.

$$\frac{3}{4} \times \frac{4}{3} = 1$$

↖ ↗
reciprocals

rectangle A parallelogram with a point of symmetry, a line of symmetry through opposite sides, and four right angles.

rectangular prism A prism whose six faces are rectangles.

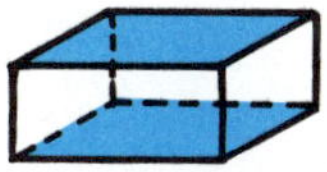

regular polygon A polygon that has all of its sides congruent and all of its angles congruent.

remainder In a division problem, the number that is "left over." When it is added to the product of the divisor and quotient, the sum is the dividend.

```
   3
8)29
 -24
   5 ← remainder
```

repeating decimal A decimal in which a digit or a group of digits repeats forever.

0.3333... 1.47474747...

rhombus A parallelogram with a point of symmetry, a line of symmetry through opposite vertices, and four congruent sides.

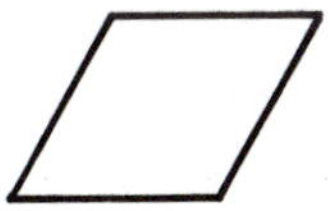

right angle An angle whose measure is 90°.

round a number To replace a number by another one that is eaiser to use. You round a number to the nearest ten by choosing the nearest multiple of 10. (5 is rounded up.)

13 → 10 27 → 30 45 → 50

You round a number to the nearest hundred by choosing the nearest multiple of 100.

487 → 500 1238 → 1200 550 → 600

sales tax A tax paid when you make a purchase.

sample A small group, chosen from a larger group, that is examined carefully in order to make predictions about the larger group.

scale drawing A drawing of an object such that the ratio of a unit of length on the drawing to a unit of length on the object is fixed.

scalene triangle A triangle with no lines of symmetry and no congruent sides.

scientific notation A notation for writing a number as the product of a number between 1 and 10 and a power of ten.

$$186.3 = 1.863 \times 10^2$$

segment A part of a line that has two endpoints.

side of an angle One of the rays that make up an angle.

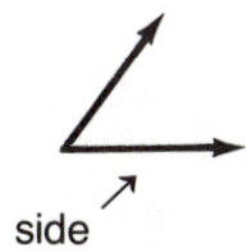

side of a plane figure One of the segments that make up a figure.

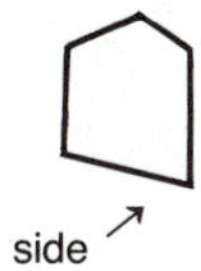

solve an equation To find the numbers that make an equation true.

square A rectangle with a point of symmetry, four lines of symmetry, and four congruent sides.

standard numeral A numeral written in its simplest place-value form. 178

statistics A branch of mathematics that studies numerical facts as a basis for drawing general conclusions and making predictions.

substitute Replace a variable with a numeral.

$$7a + 3$$

$$7 \cdot 6 + 3$$

sum The answer to an addition problem.

surface area The sum of the areas of all the surfaces of a solid figure.

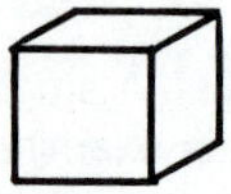

terms (of a fraction) The numerator and denominator of a fraction.

$\frac{3}{4}$ terms

trapezoid A quadrilateral with two parallel sides

triangle A polygon with three sides.

triangular prism A prism with two triangular faces in parallel planes.

unit A fixed quantity used as a standard for measuring length, area, volume, weight, etc.

unit price The cost per unit (weight, volume, etc.) of an item.

variable A symbol, usually a letter, that holds the place for a number.

$$8x + 19 = 23$$

vertex The point at the "corner" of an angle, plane figure, or solid figure.

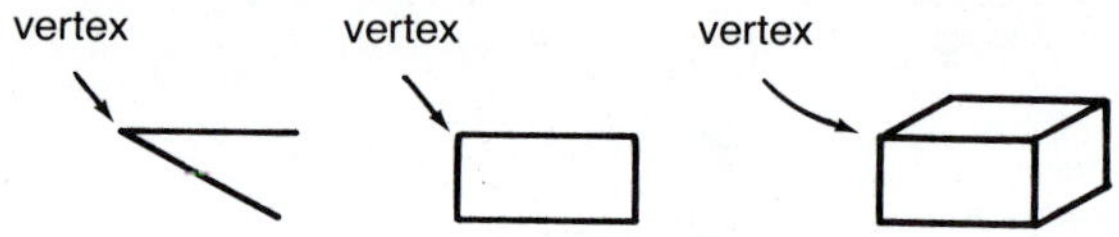

volume The measure of the space inside a solid figure.

whole number Any of the numbers 0, 1, 2, 3, 4, and so on.

Symbols

$<$	is less than	$\overline{AB}$	segment AB
$>$	is greater than	$\overrightarrow{AB}$	ray AB
$^\circ$	degree	$\angle$	angle
$'$	foot	$\parallel$	is parallel to
$''$	inch	$\perp$	is perpendicular to
%	percent	$\triangle$	triangle
$0.\overline{34}$	0.34343434 . . .	$\cong$	is congruent to
$\approx$	is approximately equal to	$\leftrightarrow$	corresponds to
π	pi	$P(2)$	the probability of the outcome 2
$\neq$	is not equal to	$^+6$	positive 6
$a:b$	the ratio of $a:b$	$^-6$	negative 6

Formulas

$P = 4s$	Perimeter of a square
$P = 2l + 2w$	Perimeter of a rectangle
$C = \pi d$	Circumference of a circle
$A = lw$	Area of a rectangle
$A = s^2$	Area of a square
$A = bh$	Area of a parallelogram
$A = \frac{1}{2}bh$	Area of a triangle
$A = \pi r^2$	Area of a circle
$V = lwh$	Volume of a rectangular prism
$V = Bh$	Volume of a prism (where B is the area of a base)
$V = \pi r^2 h$	Volume of a cylinder
$I = prt$	Interest

Index

3 4 5 6 7 8 9 0